Aristotle was both a metaphysician and the inventor of formal logic, including the logic of possibility and necessity. *Aristotle's Modal Logic* presents a new interpretation of Aristotle's logic by arguing that a proper understanding of the system depends on an appreciation of its connection to his metaphysics.

Richard Patterson develops three striking theses in this book. First, there is a fundamental connection between Aristotle's logic of possibility and necessity and his metaphysics, a connection extending far beyond the widely recognized tie to scientific demonstration and relating to the more basic distinction between the essential and accidental properties of a subject. Second, although Aristotle's development of modal logic depends in very significant ways on his metaphysics, this does not entail any sacrifice in logical rigor. Third, once one has grasped the nature of that connection, one can better understand certain genuine difficulties in the system of logic and also appreciate its strengths in terms of the purposes for which it was created.

# Aristotle's modal logic

# Aristotle's modal logic

## Essence and entailment in the *Organon*

RICHARD PATTERSON

EMORY UNIVERSITY

CAMBRIDGE
UNIVERSITY PRESS

Published by the Press Syndicate of the University of Cambridge
The Pitt Building, Trumpington Street, Cambridge CB2 1RP
40 West 20th Street, New York, NY 10011-4211, USA
10 Stamford Road, Oakleigh, Melbourne 3166, Australia



First published 1995

Printed in the United States of America

Library of Congress Cataloging in Publication applied for.

A catalog record for this book is available from the British Library.

ISBN 0-521-45168-x hardback

# Contents

*Contents*

# Acknowledgements

I owe special thanks to Peter Geach, whose seminar on the *Prior Analytics* at the University of Pennsylvania in 1973 first aroused my interest in Aristotle's modal logic. He would certainly not agree with all of Aristotle's ideas on the subject, or with all of my ideas about Aristotle, but he is nonetheless responsible for much of anything that may be found useful in this book.

For ideal working conditions and generous financial support I am grateful to the National Humanities Center, the Institute for Advanced Study, and the University Research Council of Emory University.

Many readers have helped me make improvements in various versions of the manuscript over the last ten years. Besides the two anonymous and extremely helpful readers for the Press, these include Michael Ferejohn, John Corcoran, Robin Smith, Howard Stein, William Rumsey, Brian Chellas, Charles Kahn, Morton White, Henry Mendel, Betsey Devine, Allan Silverman, Henry Mendel, Jim Goetsch, and Laura Wedner.

# Chapter 1

# Introduction

## 1.1. BACKGROUND TO THE PRINCIPAL ISSUES

The chapters of the *Prior Analytics* devoted to modal arguments are notoriously difficult, controversial, and, according to numerous weighty authorities, deeply confused. Accordingly, one major aim of this study will be to examine in detail the internal workings of Aristotle's modal logic – his logic not just of statements simply asserting the application of a predicate to a subject but also of those asserting a necessary or possible or contingent relation between subject and predicate – in order to understand and assess its strengths and its weaknesses. A second aim will be to establish a fundamental connection between Aristotle's metaphysical essentialism (along with his theory of scientific demonstration) on the one hand and his modal logic on the other. These two goals are closely connected, or so it will be argued here, in that the logical system itself must be understood from the start in the light of basic points of syntax and semantics deriving from Aristotle's views on what there is and on the various ways in which we can speak and reason about what there is.

There has always been healthy interest in Aristotle's metaphysical essentialism – interest heightened recently by work on essentialism as such, and especially by work deriving, like Aristotelian essentialism, from intuitions about the natures or essences of things.[1] Such developments have contributed at least indirectly to the study of Aristotle by provoking careful thought about how essentialism might be formulated and how different objects (individual living things, the "natural kinds" of chemistry or physics or biology, sets, numbers) might involve very different sorts of essential properties, discoverable only through a variety of approaches. It has not, however, led to a broad interest in the details of Aristotle's modal syllogistic. This apparently can be attributed, in some quarters, to lack of interest in this more formal side of things, in others to an assumption that

Aristotle's modal logic can be perfectly well formulated using now-familiar modal systems based on non-categorical logic, and in still others to a supposition that Aristotle's own system is either too weak or too confused to be worth disinterring at this late date.

Some of the slack has been taken up by scholars more directly interested in modal logic. Here, too, contemporary work – in particular the recent emergence and wide appeal of "possible-worlds" modal semantics,[2] along with the extensive development of modal logic from a purely formal point of view – has led at least a few commentators to apply these modern means of formalization to Aristotle's modal syllogistic.[3] But generally speaking, these commentators have not taken a comparably detailed interest in Aristotle's metaphysics. This may be due, again, to simple lack of interest, or perhaps to the idea that in Aristotle's work there is no significant dependence of logic on metaphysics, or perhaps to a suspicion that consorting with metaphysics can only lead to the corruption of logic. Thus, even Günther Patzig, who has given us important work on both the metaphysics and logic of Aristotle, is noticeably grudging in his admission of any conscious, fundamental dependence of the latter on the former. He views with a jaundiced eye the tendencies of several earlier German commentators (e.g., Prantle, Waitz, Maier, Trendelenburg) to see Aristotle's logic as a kind of "philosophical logic" or a "conceptual metaphysics" or the like, and he concludes that

> the validity of the propositions in Aristotle's syllogistic can, neither in fact nor in Aristotle's opinion, be thought dependent on the *truth* of certain ontological propositions. It is consistent with this view both that Aristotle's *presentation* of his syllogistic is unconsciously influenced in many ways by his ontological predilections, and also that the marrow of Aristotle's ontology contains views which mirror his logical tenets. If a causal connection between Aristotle's logic and his ontology must be found, it seems to me more correct to base his ontology on his logic than the other way about.[4]

By contrast, I shall argue that even the most basic formal aspects of the modal system of the *Prior Analytics* cannot be accurately understood – except by luck, as in the case of Aristotle's fellow who chanced upon buried treasure while digging in the garden – without serious consideration of his essentialist metaphysics, along with his related views on scientific demonstration. More specifically, Aristotle believed in a distinction between the essential and accidental properties of a thing. He held also that there were only a few ways in which a property could be related predicatively to a subject [i.e., as its genus, differentia, species, *idion* (proprium), or accident] and that all these relations were either necessary or accidental. Both points were related, in turn, to his view that scientific demonstrations proceeded from *per se* predications in their premises to a *per se* conclusion.

All of those tenets motivated Aristotle's modal logic and shaped its foundations. At a basic level, because on Aristotle's view modal propositions differed from non-modal ones in asserting one or another special *connection* between predicate and subject, Aristotle's modal syntax incorporated modal *copulae* or linking expressions ('necessarily applies to all of', 'possibly applies to all of'), rather than today's more familiar sentence or predicate operators, to express the various possible connections between predicate and subject. Extra-logical considerations also determined the sorts of propositions – plain (assertoric), necessary, one-way possible, two-way possible (problematic, contingent) – whose logical relations were to be investigated, for although he was interested in determining what followed from what in a general sense, Aristotle investigated systematically only syllogisms containing various possible combinations of plain, necessary, and contingent categorical premises.[5] Why just those, and not also syllogisms with one-way possibility premises – the kind of possibility so central to contemporary modal logic? Evidently because the former were the sorts of propositions he thought could exhaustively express the necessary and accidental connections of subject to predicate constituting everything that might be the case. Within that framework, and given his views on science, he needed to investigate syllogisms involving necessary or two-way possible premises and conclusions, for those (speaking very roughly for the moment) were the sorts of propositions he thought could be used in constructing scientific demonstrations. Again, Aristotle failed to take up syllogisms with premises involving one-way possibility: Unlike two-way possibility, it reflects neither any of the primary ways a predicate can relate to a subject nor any kind of scientific proposition.[6]

Other, more local connections between Aristotle's metaphysics and logic will emerge as we proceed.[7] However, we can say that the influence of his metaphysics on his logic is pervasive, in that it decisively influences the basic structure of his modal propositions and the kinds of propositions whose logical relations are to be studied. And because the question of the internal structure of premises and conclusions is crucial for any study of his logic, whether from a logical or more philosophical point of view, it is necessary to consult those metaphysical views in order to establish the very starting points of Aristotle's modal syllogistic.

On the other hand, once the starting points have been fixed, the investigation becomes more purely logical. Indeed, Aristotle pursues the properly logical question of what follows from what with characteristic alacrity and perseverance. So it should not be imagined that we shall find Aristotle constantly doing logic by way of metaphysics; on the contrary, most of the *Prior Analytics* is concerned with strictly logical questions. Thus the extra-logical background will be consulted extensively in the laying of the foundations, but much less frequently, and for more narrowly prescribed reasons, thereafter. Exactly how this is so is a long story; in the following pages I shall try to convey briefly the essentials of the tale through a preliminary discussion of three traditional approaches to modality.

In the *Prior Analytics*, Aristotle recognizes four modally distinct types of propositions: plain, or assertoric (e.g., '*A* applies to every *B*'); necessity ('*A* necessarily applies to every *B*'); possibility ('*A* possibly applies to every *B*'); and two-way possibility – sometimes called "contingent" or "problematic" or "two-sided" propositions ('*A* possibly applies and possibly does not apply to every *B*').[8] Within each type there obtains a fourfold distinction according to quantity (universal or particular) and quality (affirmative or negative), so as to give universal and particular affirmatives, and universal and particular negatives, of each modality. [These four types will be represented here, as in "traditional" syllogistic, by the letters *A, I, E,* and *O* – or, within a given proposition, by their lowercase counterparts, respectively (as in '*A a B*', '*A i B*', etc.). Lowercase subscript letters will indicate modality: $A_n$ for a universal affirmative necessity proposition, $A_{pp}$ for a universal affirmative two-way possibility proposition, and so on. Plain *A* without a subscript will then stand for an assertoric universal affirmative, e.g., '*B* applies to all *C*'.] Thus, the basic propositions of each modality – putting aside some important complications to be indicated as we proceed – will be written as follows:

| | | | |
|---|---|---|---|
| Assertoric | *A*: | *A a B* | (*A* applies to every *B*) |
| | *E*: | *A e B* | (*A* applies to no *B*) |
| | *I*: | *A i B* | (*A* applies to some *B*) |
| | *O*: | *A o B* | (*A* does not apply to some *B;* i.e., there is some *B* to which *A* does not apply) |
| Necessity | $A_n$: | *A N a B* | (*A* necessarily applies to every *B*) |
| | $E_n$: | *A N e B* | (*A* necessarily fails to apply to every *B;* i.e., of every *B* it is true that *A* necessarily fails to apply to it)[9] |

| | | |
|---|---|---|
| | $I_n$: *A N i B* | (*A* necessarily applies to some *B*) |
| | $O_n$: *A N o B* | (*A* necessarily fails to apply to some *B;* i.e., there is some *B* to which *A* necessarily fails to apply) |
| Two-way possibility | $A_{pp}$: *A PP a B* | (*A* two-way possibly applies to every *B;* i.e., *A* possibly applies and possibly does not apply to every *B*) |
| | $E_{pp}$: *A PP e B* | (*A* two-way possibly fails to apply to every *B*) |
| | $I_{pp}$: *A PP i B* | (*A* two-way possibly applies to some *B*) |
| | $O_{pp}$: *A PP o B* | (*A* two-way possibly fails to apply to some *B*) |

One-way possibility propositions will parallel those given here. As various complications arise, we shall find the varieties of modal formulae multiplying, sometimes thick and fast. For convenient reference, all the formulations used in this study, along with the traditional nicknames (''Barbara,'' ''Celarent,'' etc.) of Aristotle's syllogisms, are collected in the Appendix.

Roughly put, then, Aristotle's general aim in *Prior Analytics* (*Pr. An.*) A.1–22 was to specify which pairs of propositions logically implied which conclusions, where the two premises might both be plain (as in *Pr. An.* A.4–7), both necessary (A.8), one plain and one necessary (A.9–11), both two-way possible (A.14, 17, 20), and so on, inexorably, through the various sorts of premise pairs involving plain, necessary, or contingent propositions.

Aristotle's plain syllogistic (*Pr. An.* A.4–6), having been worked out with great clarity and, in the metalogical remarks of chapter 7, much elegance as well, went on to become, until recently, *the* logic of the West and much of the East.[10] Meanwhile, his modal syllogistic suffered the opposite fate: Theophrastus and Eudemus immediately challenged Aristotle on basic points. In later centuries, those chapters of the *Prior Analytics* were not routinely studied even by the learned, and in some quarters that pernicious subject was banned altogether.[11] In our own time, at least one distinguished logician has concluded that "Aristotle's modal syllogistic is almost incomprehensible because of its many faults and inconsistencies."[12]

Still, some order was introduced into modern commentary on the subject by Albrecht Becker, who, writing in 1933, saw most of those apparent faults and inconsistencies as the results of an unwitting vacillation on Aristotle's part between two sorts of modalities, or two ways of understanding modal propositions.[13] If one says, for example, that all lions are necessarily animals, one might mean either (1) it is a *necessary truth* that all lions be animals or (2) it is true, *of* each and every lion, that being an animal necessarily applies to *it.* Both these statements are true (let us suppose, for the moment). On the other hand, given that everything lying down in a given place is in fact a lion, one could say that it is true, *of each and everything lying down there*, that being a lion necessarily applies to it. But it is not a *necessary truth* that all things lying down in said place be lions: It is entirely possible that the lion and the lamb lie down there together. So in this case, one reading of our modal statement ("everything lying down . . . is necessarily a lion") comes out true, and the other false.

From as least as far back as Abelard the contrast between these two ways of interpreting modal statements has been framed in terms of *de dicto* vs. *de re* modality.[14] On the former, modalities are regarded as modes of truth of entire statements, so that necessity, for example (or, being necessarily true), is a property not of things or of their properties but of linguistic statements or of the propositions they express (*dicta*).[15] On the latter, necessity is supposed to apply to the *things* about which some *dictum* is asserted (as in "It is true, of each thing now reading this manuscript, that it is necessarily rational"), and this will explain any necessary truth there may be. More precisely, the *res* in question is the subject(s) signified by the subject term of a given statement; the statement attributes, say, a necessary property to that *res*, or asserts that some property necessarily belongs to it.[16] Among various other ways of describing this distinction, one of the most useful for our purposes will be that the modality of *de dicto* modal statements depends on assigning a property to a subject only as that subject is considered under one description or another. So, adapting Quine's example slightly, it is necessarily true that a certain bicycling mathematician, *qua* mathematician, is rational, but equally – and with equal necessity – true that *qua* bicyclist he is an exerciser. Here necessary truth derives from a direct connection between the descriptions involved (or between the concepts or universals or natures signified by those descriptions). By contrast, a *de re* ascription assigns essential properties to a subject independently of whatever description one may happen to use in picking it out. Thus the cycling mathematician, being human, is essentially (and necessarily) rational, whether described by anyone as a mathematician or not: It is simply true *of this subject* – this cyclist, or this person wearing striped pantaloons – that he is necessarily rational. But the same person is only accidentally or contingently a cyclist or a wearer of striped pantaloons, and this will be true of that person no matter how he is picked out or described. So on a *de re* reading 'This cyclist is necessarily rational' is true, and 'This bicyclist is necessarily an exerciser' is false.

The distinction makes a great deal of difference as to what follows from what. For example, from the premises

> Necessarily: Every human is rational

and

> Everything standing in the conference room on Monday morning is human

it does not follow that

> Necessarily: Everything standing in the conference room on Monday morning is rational

It may be a (*de dicto*) necessary truth that every human is rational, and true simply as a matter of fact that everything standing in the conference room on Monday morning is a human, so that both premises are true. But these would not entail that it is necessarily, as opposed to contingently, true that everything standing in the conference room on Monday morning be rational.

By contrast, from the premises

> *Rational* necessarily applies to every *human*

and

> Everything *standing in the conference room on Monday morning* is *human*

it does follow that

> *Rational* necessarily applies to everything *standing in the conference room on Monday morning*

The conclusion does not say it is *necessarily true* that all such standing things are rational; it now says only that it is true, of each thing that happens to be standing in the room on Monday morning, that that thing is necessarily rational. And this will be true in any situation in which all humans are necessarily rational and it happens that everything standing in the room is a human. But these were precisely the premises laid down. (Actually, the interpretation of this particular syllogism is hotly contested; see Chapter 4, Section 4.1.[17])

Commentators on Aristotle have long been aware that even after putting aside a few Aristotelian slips, no single formulation, whether *de dicto* or *de re*, can give all the logical results Aristotle propounds in *Pr. An.* A. Some sections, such as the one on conversion of necessity propositions,[18] seem to require a *de dicto* reading; others, such as chapter 9 on "complete" or "perfect"[19] syllogisms with one plain and one necessity proposition, including the example just surveyed, seem to require a *de re* reading. In some cases a syllogism that is valid only when read *de re* is shown valid by a proof that is itself valid only on a *de dicto* reading.[20]

Consequently, one often reads of a fundamental inconsistency, or of vacillation on Aristotle's part, between *de dicto* and *de re* modalities.[21] Indeed, the single largest issue dividing modern commentators has been whether one must rest content with recording the fact that Aristotle alternates between *de dicto* and *de re* readings of necessity – and with the project of recording where the one reading must be invoked, and where the other – or whether there is a different way of regarding the entire system such that a single, unambiguous reading will suffice to give (more or less all of) Aristotle's results.

I have already suggested that resolution of the issue depends on establishing the relation between Aristotle's modal syllogistic and the essentialism of the *Organon.* More specifically, I would like to propose, as a first step toward the interpretation of Aristotle's modal logic and its place in his philosophy as a whole – and at the same time toward understanding why Aristotle *appears* to vacillate in the way just mentioned – a revision of the terms in which the topic is today ordinarily framed. Notice first that *de re* propositions are nowadays usually treated, by those commentators who remain at least in part within a categorical framework, as involving modalized predicates, as in '*Being necessarily an animal* belongs to all *human*'. In fact, one frequently encounters a hyphenated modal predicate, as in '*necessary-human* applies . . . '.[22] The disquieting fact about any approach based on a dichotomy of modalized *dictum* vs. modalized predicate is that Aristotle himself speaks in a third way, on which modality attaches neither to predicate nor to *dictum*, but rather to *the manner of the predicate's applying* to the subject. It is the copula or linking expression between the terms to which Aristotle, in the *Prior Analytics*, ordinarily attaches his modal operators, as in 'Animal applies to all Human' (plain), 'Animal *necessarily* applies to all Human', 'Animal *possibly* applies to all Human', and so on. Commentators, too, frequently speak in this way, at least when expressing themselves in a natural language rather than in the more technical terms of a proposed interpretation or formalization. Abelard himself, for example, along with several other major medieval figures (William of Sherwood, Albert the Great, Thomas Aquinas), took this "modalized copula" interpretation as fundamental. So I am not, thus far, proposing anything at all new.[23] Nonetheless, this reading, insofar as it receives any particular attention, is nowadays regularly identified, either explicitly or implicitly, with a *de dicto* or (modal predicate) *de re* reading.

Neither identification is by any means arbitrary. On the one hand, it is natural enough to suppose that the *plain* copula indicates a *combination* of subject(s) and predicate and that the assertion of such a combination is simply the content or sense of the *dictum* taken as a whole. Thus the sense of the *dictum* would be *that* one thing *is predicated of* some subject. So it would be easy to view the modalization of the copula as, in effect, a modalization of the content of the original sentence as a whole: Subject and predicate are not simply conjoined, but *necessarily conjoined.* And because what one intends to express is the necessity of the content of the original assertoric proposition as a whole, the modal operator might very sensibly stand at the front of the original sentence, with appropriate notation to indicate that its scope is the entire sentence, as in 'nec: *A* all *B*' or '$\Box$(*A* all *B*)'.[24] (Here the grammatically internal modal operator of '*A* necessarily applies to all *B*' is similar in scope to an internal negation – as in 'Socrates is not a Satyr' – wherein the "not" serves to negate an entire proposition, or the content of the *dictum* taken as a whole, by grammatically negating the copula. And, of course, in modern propositional and predicate logic, negations then find expression in an external sign of negation whose scope is the whole of the proposition to which it is prefixed.) Thus does the modal *copula* come to be expressed as a sentential operator indicating the modality of a given *dictum.* This would not be objectionable except that the label "*de dicto* necessity" is sometimes used rather vaguely, without due notice of the fact that it can cover a variety of underlying conceptions, including the now familiar approach on which the ground-level explanation of necessary truth is a matter of the truth at all times, or in all possible worlds, or the like, of the relevant assertoric proposition, as well as any approach based simply on a primitive notion of necessary truth, or the more properly copulative approach on which the primary explanation of necessary truth is a matter of the essential connection between predicate and subject (as, for example, when they are related as genus to species). Some of these ways of looking at necessary truth are more appropriate to Aristotle than others.

On the other hand, the modal copula is often taken up into the predicate, as opposed to the subject, of the initial proposition. This is entirely harmonious with the ancient and modern idea of the "sign of predication" being included in the predicate [cf. *De Interpretatione* (*De Int.*) 3, 16b6–25), and also with the practice of including everything but the ontological subject (the kitchen sink, say) in the predicate, so that the subject term serves simply to designate those items to which the predicate applies. It is then a short step, especially within an essentialist context, to the familiar idea of "necessary properties" being predicated of subjects, where modality now becomes a part of the predicate term proper.[25] Of course, there is at the very least a syntactic distinction between a modal predicate term ('necessary-Animal') used with a plain copula and a plain predicate term ('Animal') used with a modal copula ('necessarily applies'). But ordinarily neither this distinction nor its possible implications are thought worth pursuing in the literature on Aristotle's modal logic, so that his modal copula winds up in this case as part of a modal predicate.

This is not to say that either of these ways of reading a modal copula is in itself an error. The point is rather that it has become almost standard to approach Aristotle's modal logic in terms of a supposedly exclusive modal *dictum*–modal predicate dichotomy. And this does seem to me an error. In any event, one essential tenet of the interpretation of Aristotle's modal logic offered here is that for a variety of important reasons the modalized copula reading must not be assimilated to either of those approaches. For one thing, the obvious syntactic differences among modal copulae, *dicta*, and predicates are of great importance for revealing how Aristotle represented to himself the structure of his many arguments for the validity of conversion principles and syllogisms. And the aim here is not just to obtain end results that tally with Aristotle's, but to be able to think through Aristote's discussions and arguments from the inside. It will be argued that certain syntactic properties of representations in terms of modal *dicta* or modal predicates preclude that possibility.[26]

At the same time, the importance of the underlying semantics for Aristotle's modal logic hardly needs emphasizing, and it will be a major aim of the reading developed here to show precisely how the background distinction between essence and accident, and the theory of the "four predicables," inform his invention of modal logic in *Pr. An.* A.3 and 8–22. So the "modal copula" approach defended here should be seen as involving both syntactic and semantic components, where the former should reflect the latter.

## 1.2. MAIN THEMES OF THIS WORK

As remarked earlier, the idea of a modal copula is far from new. More important, but generally unnoticed, is the fact that the modal copula reading of Aristotelian necessity (and other modalities) is itself already ambiguous between two interpretations. One sort of *de copula* (or *cop*, for short) reading asserts a definitional relation either of entailment or exclusion between its subject and predicate *terms*, where (Aristotelian) definitions are accounts of the natures or essences signified by such terms rather than of the meanings of linguistic subject and predicate. On the other *cop* reading, a necessity proposition asserts a necessary relation between its own predicate term and the items referred to by its subject term, where those two terms themselves may or may not bear anything more than an accidental relation to one another. The latter type of *cop* necessity would include 'Cat necessarily applies to all Things on the Mat', which simply asserts, of whatever things may be on the mat, that they are necessarily feline; no necessary connection is asserted between their being on the mat and their being cats. The former sort of *cop* proposition includes 'Animal necessarily applies to all Human', where (1) the predicate *P* applies necessarily to whatever falls under the subject term *S* (as with the cat-on-the-mat case just considered), and (2) being *P* is entailed by what-it-is-to-be-(an)-*S*.

The origin of the distinction lies simply in the fact that some properties of a thing apply only accidentally, and others essentially, to it. Thus one might pick out certain objects (Socrates, Coriscus) by reference to one of their accidental properties, then predicate of them some one of their essential properties (as in 'Animal necessarily applies to every White Thing on the Mat'). In the other sort of case, one uses as subject term some essential property of the subject, then predicates of that subject another of its essential properties, as in 'Animal necessarily applies to every Human'. In both cases it is true that the predicate applies necessarily to the designata of the subject term. But in the former case there need be no essential connection between the predicate and subject terms themselves ('Animal', 'White Thing on the Mat'), whereas in the latter there is a connection either of entailment or exclusion between the terms ('Animal', 'Human'). Again, the possibility of two readings of modal propositions –

or of two sorts of truth conditions of such propositions – arises directly from Aristotle's distinction between accidental and essential properties of a thing.

I shall refer to the former sort of statement as "weak *cop*" necessity (written '$A\ N_w$ all $B$') and to the latter as "strong *cop*" necessity (written '$A\ N_s$ all $B$'). '$A\ N\ a\ B$', with no subscript, is neutral among all the possible readings of Aristotle's statement. Thus it results, among other things, that there will be eight basic necessity statements, where before there were four (e.g., instead of just '$A\ N\ a\ B$' we shall have '$A\ N_w\ a\ B$' and '$A\ N_s\ a\ B$'; these are included in the complete list of modal propositions in the Appendix).

The following chapters will defend this weak/strong necessity distinction – and its analogues for other modalities – by showing that both *cop* readings are thoroughly Aristotelian in a way in which modal *dictum* and modal predicate readings are not. By the same token, we shall see that the two *cop* readings give rise to logical results very similar to those for *de dicto* and modal predicate interpretations, which helps explain the persistence of attempts to read Aristotle in one of these two ways, or in terms of a vacillation between them. It will also be argued, however, that there is a fundamental connection (besides the obvious syntactic one) between the two *cop* readings, a connection based on Aristotle's essentialism and one much closer than any evident connection between *de dicto* and *de re* propositions – which will help explain how Aristotle could have unwittingly incorporated two readings of modal propositions.[27]

With regard to that basic division among commentators ("vacillates between *de dicto* and *de re*" versus "uses consistently only one reading of modality"), this means that the view to be developed here has something important in common with both camps. Like the first, it finds a basic ambiguity in Aristotle's modal propositions; like the second, it finds a strong underlying unity in the system. It differs from most representatives of both in basing itself from the start on connections between Aristotle's logic and his metaphysics and also in denying that the inclusion of two readings of modal propositions is in itself a defect in the system (an "inconsistency" or a "vacillation," and a blot on the good name of the Master). This is not to deny either that Aristotle ought to have recognized the ambiguity and worked out its implications or that the presence of semantically ambiguous propositional forms is a defect in the system as it stands. The point I wish to emphasize is rather that his system *should* include both sorts of readings if it is to express what *he* wants to express, given his metaphysical essentialism and the dialectical, philosophical, and scientific purposes for which he devised the system. As we just saw, there are two basic types of situation in which, say, a universal affirmative necessity proposition ('*A* necessarily applies to all *B*') will be true, two situations in which the predicate does apply necessarily to the subject: one in which subject and predicate themselves stand in some essential relation to one another, and one in which they do not. The crucial point is that Aristotle formulates and reasons about both sorts of cases in *Prior Analytics* A not because he negligently failed to hold on to any one reading (as if he ought to have devised a system that, like modern propositional modal logic, and, apparently, like Theophrastus' system, used only one sort of necessity proposition) or because his necessity propositions, like their counterparts in modern English, simply are in fact open to two kinds of readings, but rather because his essentialism already implies two important types of truth conditions for propositions of necessity. And these require in turn two distinct, if closely related, ways of asserting a necessary connection between subject and predicate. Thus the inclusion of necessity propositions of two different types, or the reading of such propositions in two ways, is not in itself a mistake from which interpreters should try their best to rescue Aristotle, but is entirely correct and even necessary if all the essentialist facts of life are to be expressed and reasoned about.

But this means that even those who do correctly find a basic ambiguity in Aristotle's modal propositions may yet be faulted on other important grounds. First, they have tended too quickly to identify those readings with traditional *de dicto* and *de re* conceptions of modality. Second, they have often concerned themselves too narrowly with the project of identifying where one reading or the other is required to make things work out as Aristotle wants, rather than demanding to know *why* two readings – and why *these two in particular* – show up in an Aristotelian modal logic, and how they are at bottom related to one another.

In sum, the interpretation to be developed here holds that (1) Aristotle's modal propositions utilized modal copulae rather than modal predicates or modally qualified *dicta*; (2) the *cop* reading represents an alternative distinct not only syntactically but also semantically from both of these now more familiar conceptions; (3) neither the modal predicate reading nor modal *dictum* reading represents a genuinely Aristotelian understanding of propositions of necessity, nor is the more general contrast between predication of *dicta* and of *things* appropriate to the Aristotelian modalities; (4) the *cop* reading lends itself naturally to two genuinely Aristotelian readings in a way that helps reveal their underlying unity even as it explains the appearance of vacillation between *de re* and *de dicto* modality; (5) the *cop* reading, in both its Aristotelian versions, arises from facts about Aristotle's essentialism – above all, the basic contrast between essential and accidental properties – and is closely tied to leading ideas of the *Categories* (in particular, about the ten kinds of things there are and the main types of relations among them), the *Topics* (the ten categories of predication and the "four predicables"), and the *Posterior Analytics* (the theory of scientific demonstration and of *per se* predication).

The task in the following chapters is to work out and defend these claims in detail, specifically in relation to Aristotle's treatments of conversion (both "term" and "qualitative" conversions[28]), of modal syllogisms of all sorts (including the celebrated two Barbaras[29]), of "ampliation,"[30] of scientific demonstration (in its relation both to necessity and to two-way possibility propositions), of the temporality of modal propositions, of the "completeness" or "perfection" of modal syllogisms, and a host of more local curiosities. In the end it will be possible to lay out more formally a consistent modal system incorporating both weak and strong *cop* necessity and their counterparts for the other modalities. But my goal is not so much to produce a formal model of the system as to determine why Aristotle devised a modal logic in the first place; how, in full detail, this logic is built up and how it works; how and why his treatment in the *Prior Analytics* fell short of realizing some of his own larger aims; and how the principles and insights there introduced might yet provide an adequate basis for the essentialist logic of the *Categories*, the *Topics*, the *Sophistical Refutations*, and the *Posterior Analytics*.

# *Chapter 2*

# The basic modal proposition

## 2.1. ARISTOTLE'S GENERAL INTRODUCTION TO THE MODALITIES

In the terminology of '*A* belonging to or applying to *B*' ('*A huparchei* + dat. *B*'), dominant in *Pr. An.* A.4–22, the three basic readings described earlier go as follows: On a *de dicto* version, Aristotle employs only one copulative expression, *huparchei* ('belongs to', 'applies to'), but three sentential operators for possibility, necessity, and two-way possibility, each attaching to a plain proposition to form a new modal *dictum* asserting that the original statement is necessarily true, possibly true (in the sense of not necessarily false), or contingently true (i.e., neither necessarily true nor necessarily false). A modalized predicate reading also calls for the plain copulative expression *huparchei,* but now with three term-forming operators on terms: *n,* let us say, which attaches to a given term *A* to form the term 'necessarily *A*' or 'necessary-*A*' (*nA*), and the operators *p* and *pp* for 'possibly *A*' (*pA*) and 'two-way possibly *A*' (*ppA*). Finally, the copulative reading involves no sentential operators and no term-forming operators on terms, but rather four expressions linking Aristotle's general terms: *huparchei,* 'belongs to, applies to'; *ex anangkēs huparchei,* 'necessarily belongs to' (symbolized as '*A N* all *B*'); *endechetai* (or *dunatai*) *huparchein,* 'possibly applies to' ('*A P* some *B*') or 'two-way possibly applies to' ('*A PP* all *B*').[1] (Negation and quantification pose other questions and may well be, in Aristotle's view, copula operators. The matter is discussed at the end of this section.)

Turning from the secondary literature, which on the whole adopts one or the other of the first two readings, to the relevant Aristotelian texts, one must be surprised at how directly and forcefully the latter support a modal copula reading. This is especially true in those lines of chapter 8 that effect a transition from plain to modal syllogisms and that indicate how Aristotle wishes the reader or auditor to think of his basic modal propositions:

> To apply, to apply of necessity, and to possibly apply are different (*heteron estin huparchein te kai ex anangkēs huparchein kai endechetai huparchein*). (29b29–30)

The wording indicates that the predicate of a proposition may *apply in different ways* to the items referred to by the subject term: *A* might simply *apply* to all *B*'s or *necessarily apply* or *possibly apply* to all *B*'s. As Aristotle continues,

> for many things apply, but without applying of necessity; and yet others neither apply of necessity nor apply at all, but still possibly apply (*polla gar huparchei men, ou mentoi ex anangkēs, ta d' out ex anangkēs outh' huparchei holōs, endechetai d'huparchein*). (29b30–32)

Here again Aristotle's modal terms modify adverbially the *copula:* The predicate relates in modally distinct ways to the subject.

As for the *de dicto* interpretation, the passage carries no suggestion that Aristotle is thinking about different modes or manners in which sentences may be true or false. At most, one might argue (on behalf of *de dicto*) that *huparchein, ex anangkēs huparchein,* and so forth, should be read as 'obtains', 'necessarily obtains', and the like, rather than as '(necessarily) applies', where these are predicates of entire *dicta* rather than copulae. This is not in itself impossible. But if these terms are not taken as copulative expressions, then they might well be read metalogically (rather than *de dicto*) neutrally as regards the issue at hand. I favor the copulative reading not simply because that is the predominant use of these terms throughout A.3–22 (for there are, after all, exceptions to this; recall note 1) but also because in the next chapter (A.9) Aristotle will use *ex anangkēs huparchein* in a clearly copulative sense. See, for example, 30a17–18: "for example, if *A* is taken as applying of necessity to *B*" (*hoion ei to men A tōi B ex anangkēs eilēptai huparchon*) (cf.a20–22). He then moves to a metalogical use ("but if the *A–B* ⟨premise⟩ is not necessary . . . but the *B–C* ⟨premise⟩ is necessary, the conclusion will not be necessary," *ei de to men AB mē estin anagkaion, to de BC anagkaion, ouk estai to sumperasma anagkaion,* 30a23–25), and then immediately back to a clearly copulative expression ("For if it is, it will follow that *A* applies of necessity to some *B*," *sumbēsetai to A tini tōi B huparchein ex anangkēs,* 30a25–26). Here the metalogical use is simply shorthand for a fuller copulative expression.

Moreover, to return now to chapter 8, Aristotle's next remark will unequivocally employ a copulative reading:

> Regarding necessity ⟨propositions⟩, things are almost the same as for those of ⟨plain⟩ applying; for with the same arrangement of terms in propositions of applying and in ones saying that something necessarily applies or does not apply, there will and will not be a syllogism; they will differ only in adding to the terms 'necessarily applies' or '⟨necessarily⟩ does not apply' (*plēn dioisei tōi proskeisthai tois horois to ex anangkēs huparchein ē mē huparchein*). 29b36–30a2

Whereas in plain propositions 'applies' or 'does not apply' would be added to the terms (cf. 24b16: 'is' or 'is not' may be added to the terms of a plain proposition), in necessity propositions 'necessarily applies' or 'necessarily does not apply' is added to the terms: One forms modally distinct types of propositions by simply adding one copulative expression to the terms rather than another. At the initial, syntactic level, this once again counts equally against the modal *dictum* and modal predicate readings. In a more positive vein, it fulfills the important task of extending the basic syntax of the system from assertoric syllogistic propositions to modal ones.

This is enough to establish the general picture: In the assertoric case, one has two terms and a linking expression; the modal cases then modify the linking expression as needed. However, this leaves some important questions unanswered. First, what is the status of negation, or, rather, of the negative element of Aristotle's *E* and *O* propositions? [Remember that to deny *A* of *B*, as in '*A e B*', is not always to negate (give the contradictory of) a statement affirming *A* of *B*: '*A a B*' and '*A e B*' are not contradictories, but contraries.] Aristotle's answer is tolerably clear: He seems to regard the negative element of his negative statements, at least for purposes of setting up the logical system of *Pr. An.* A, as a copula operator. Thus he has a positive and a negative copula. In the assertoric case, we have seen this indicated at 24b16–18 (following Ross's text): "I call a term that into which a premise is divided, i.e., that which is predicated and that of which it is predicated, 'is' or 'is not' being added." The idea of a negative copula is relatively easy to accept, not only because it is so common in textbook presentations of categorical logic but also because it makes good sense: A simple statement is one either affirming or denying something of something. As Aristotle puts it at *Pr. An.* A.24a16–17, a premise is a "statement affirming or denying something of something" (*logos kataphatikos ē apophatikos tinos kata tinos*). (Cf. *De Int.* 17a20–

21, and 17a25: "An affirmation is a statement affirming something of something, a negation is a statement denying something of something.") He will argue at *De Int.* 12, 21a38–21b33, that it is by the addition of 'is' and 'is not' to such terms as 'man' and 'white' that an affirmation and a negation are produced.

Of course, Greek as a natural language does not require that simple affirmations include the copula as a separate element. Still, the presence of a copula, whether in a natural or artificial language, will only make explicit that something is being affirmed or denied of something, and one can always recast an affirmation that does not contain a copula as one that does (see *De Int.* 21b23). In *Pr. An.* A.1–22, Aristotle consistently uses terms that can trade places within a proposition, serving indifferently as subject term or predicate term. Given this, he uses the copula (usually 'applies to' or 'does not apply to,' rather than 'is' or 'is not') as an additional element to signify one or another relation between his terms. Thus he does not, in the *Prior Analytics,* as opposed to *De Interpretatione,* use two sorts of terms, *onomata* and *rhēmata,* where only the latter signifies time and is always a "sign of what holds of a subject" (*De Int.,* 16b6–10). But I take it that this important difference between *De Interpretatione* and the *Prior Analytics* does not affect the point that a separate copula, where it does appear, functions to signify the affirmation or denial of some predicate of a subject.

Similarly, the negative *modal* cases are entirely parallel to their assertoric counterparts. (And we should observe that a modal copula will be made explicit unless the context makes this unnecessary: There is, for example, no verbal suffix to serve as a "sign of what holds necessarily of a subject.") Although at the very opening of *Pr. An.* 8 (quoted just above, 29b36ff.) Aristotle mentions only the positive modal copulae, he adds in the same sentence that propositions of necessity differ from assertoric ones by "adding to the terms 'necessarily applies' or '⟨necessarily⟩ does not apply'." (The parallel to 24b16–18 is, I think, evidence for Ross's reading of the earlier passage. I put aside here a full discussion of the various readings that have been proposed.) In chapter 9, where he considers the mixed Barbara *N,A/N,* Aristotle says three times in the space of four lines that a certain term may "of necessity . . . apply or not apply" to its subject (30a18–21), as if the necessity attaches to the positive or negative copula to produce what I am calling a modal copula. One could add 33a25–27, where "two-way possibly not applying" is treated as parallel to "two-way possibly applying," and 36a17–22, where "two-way possibly not applying" is parallel to its negative assertoric counterpart (see esp. 36a20–21: The conclusion will pertain not to "not applying, but to possibly not applying,'' *ou tou mē huparchein alla tou endechesthai mē huparchein*). There is even a passage in *De Interpretatione* that appears to treat the assertoric copula as a kind of subject, of which 'possibly' is then predicated: ''For as in the previous examples 'to be' and 'not to be' are additions, while the actual things that are subjects are white and man, so here 'to be' serves as subject, while 'to be possible' and 'to be admissible' are additions. . . .''[2]

Other passages could be added in support of the claim that Aristotle treats positive and negative copulae, plain and modal, on a par; but more important than that is a certain qualification called for in light of Aristotle's definition of ''applies to all ⟨of⟩.'' That can be best addressed, however, after we have considered in its own right the difficult issue of quantification. The syllogistic propositions of *Pr. An.* A 1–22 are all, or almost all, at least implicitly quantified as *A, E, I,* or *O* statements. [Some of Aristotle's ''indesignate'' propositions (*adioriston,* having 'applies' or 'does not apply', without 'universally' or 'particularly', 24a19–20) are implicitly particular. They may also include singular statements, but this is uncertain, as is the role of quantification, if any, in such statements. See note 14, this chapter. This does not matter for the present point, however.] They are not of the ''Socrates is wise'' or ''Theaetetus flies'' variety, which can be analyzed exhaustively into two terms and, sometimes, a copula.

Although there is scant evidence to go on, there is some indication that Aristotle considered quantification as part of the copula. At least, this seems to be suggested by his identifying 'predicated of all ⟨of⟩' (*kata tou pantos*) and 'predicated of none ⟨of⟩' (*kata tou mēdenos*) as fundamental building blocks of his syllogistic. In the opening paragraph of *Pr. An.* A.1 he highlights ''being predicated of all or of none'' (*to kata pantos ē mēdenos katēgoreisthai,* 24a14–15) as one of six basic items to be explicated. He takes up this item at the end of the chapter, saying that ''to be predicated of all'' is the same as ''for one thing to be in another as a whole'' (24b26–28), then offering this definition: ''We say 'predicated of all ⟨of⟩' (*legomen de to kata pantos katēgoreisthai*) when none of the subject can be taken of which the other ⟨term⟩ will not be said. And similarly for ⟨is predicated⟩ of none ⟨of⟩'' (*to kata mēdenos,* 24b28–30). (The '⟨of⟩' is not essential, but could be justified by the fact that *kata tou pantos* is shorthand for *kata tou pantos tou . . . ,* both of which appear at 25b37–40: *ei gar to A kata pantos tou B kai to B kata pantos tou C, anangkē to A kata pantos tou C katēgoreisthai. Proteron gar eirētai pōs to kata pantos legomen.* But in fact his standard linking expressions in *Pr. An.* A.1–22 will be *huparchei mēdeni tōi* and *huparchei panti tōi, huparchei tini tōi* and *huparchei mē tini tōi,* rather than *kata pantos katēgoreisthai* and the like.) So here, in the case of the universal negative, we see the negative element being merged with quantification, and these two then joined with 'applies' to produce a new linking expression, 'applies to none ⟨of⟩'. Indeed, because the premises and conclusions of *Pr.An.* A.4–22 are regularly quantified, Aristotle's standard assertoric copula will in fact be 'applies to all/none ⟨of⟩', and so forth, rather than simply 'applies to'.

Aristotle does not fully discuss the topic of quantification in its own right. There are other scattered remarks that would bear on the subject (besides the definition of *kata tou pantos* in *Pr. An.* A.1, we should at least mention the important remark of *De Int.* 10, 20a13, that "all" does not in its own right signify a universal, but only that a term is taken as a whole, or universally), but nothing to provide a clear answer to the question, never actually formulated by Aristotle, of whether or not the quantifiers of the *Prior Analytics* should be regarded as part of the relation between terms of a proposition. We can ask, however, whether or not the affirmative answer we have seen suggested by various passages, and by Aristotle's practice, makes sense. I think it does make good sense, and just as much sense as his concept of one thing being, as a whole, in another. In the latter case we have a simple assertion that *A* and *B* (two groups or wholes) are related in a certain way: For example, '*A a B*' says that the relation 'being included, as a whole, within' relates *B* to *A*. This is the sort of conception that can lead, on the one hand, to the now familiar use of Venn diagrams or Euler circles to represent assertoric *A, E, I,* and *O* propositions in terms of geometric relations between two circles, and, on the other hand, to set-theoretic models of categorical syllogistic, and of some portions of modal syllogistic (one such model is discussed later, in Chapter 6). Similarly, the expressly equivalent (24b26–28) "*A* is predicated of all of *B*" can be understood as "the relation 'predicated of all of' relates *A* to *B*."

The other basic categorical propositions would be as follows: *E,* "*applies to none of* relates *A* to *B*"; *I,* "*applies to some of* relates *A* to *B*"; *O,* "*does not apply to all of* relates *A* to *B*" (following the *mē panti huparchein* at 24a19). Aristotle wants to treat all four in parallel fashion, and so says at A.4, 26b25–33, that all four basic types of conclusion, *A, E, I,* and *O*, can be proved in the first figure: . . . *kai to panti kai to mēdeni kai to tini kai to mē tini huparchein. . . .* (He evidently realizes that the last of these, "does not belong to some of," is ambiguous between an *O* and an *E* statement. It is presumably for this reason that at the first mention of *O* propositions he says "*mē tini ē mē panti huparchein*" to show that his particular negative is to be taken as "*A* does not apply to all *B*,"

24a19.) *E, I,* and *O* would then be defined, following the definition of *kata pantos,* as "there is no *B* one can take such that *A* applies to it," "there is some *B* such that *A* applies to it," and "there is some *B* such that *A* does not apply to it," respectively.

In sum, we see Aristotle thinking of the terms *A* and *B* as standing for groups of things that can, in the assertoric case, be related in one of four ways indicated by an expression that combines a positive or negative element with a sign of universality or particularity. Although we are not used to including quantification in the copula – that is, as part of a two-place relation relating one term to another – this is a perfectly coherent notion, and one readily represented by simple geometric relationships between circles (or line segments). Indeed, Aristotle's description of plain Barbara at A.4, 25b32–40, first sets forth the situation as "the extreme ⟨term⟩ being in the middle as a whole and the middle being or not being in the first as a whole." He then redescribes the situation in terms of his definition of *kata pantos* (25b37–40, quoted earlier: "If *A* is ⟨predicated⟩ of all of *B*, and *B* of all of *C*," etc.).

All we need to do now is pick up the modal operators, which will give us 'necessarily applies to none of', 'necessarily applies to all of', and so on, as linking expressions. Here the basic relation *applies to all/none of*, and so forth, is modified by 'necessarily'. Again, where a proposition is not quantified, or where Aristotle does not have quantification in view, he will speak of 'is' or 'applies' as the basic copula, and use 'necessarily applies' and the like to express modality. But either way – that is, whether a given modal statement is quantified or not – it can be broken down into two terms, each serving indifferently as subject or predicate term, and a modal copula.

At the same time, it should be acknowledged that Aristotle's quantificational definitions are reminiscent of the conception of universal quantification utilized in modern systems of predicate logic: His thought that '*A a B*' is true if and only if there is no individual *B* one can take of which *A* is not predicated, might just as easily find expression in '$(x)(Bx \rightarrow Ax)$' or '$\neg\ (\exists x)(Bx$ and $\neg\ Ax)$' – where an explicit existence assumption could be added, as in the categorical '*A* applies to all of *B*'. Moreover, we have just remarked that the assertoric '*A* applies to none ⟨of⟩ *B*' could be defined as "there is no *B* one can take such that *A* is predicated of it"; and this does not involve either a negative or a quantificational copula.[3] In modern predicate logic it would simply be expressed as '$(x)(Bx \rightarrow \neg\ Ax)$'. So it would be mistaken to claim that Aristotle's definitions of 'applies to all' and 'applies to none' essentially involve negative or quantificational copulae, or even that they essentially involve a categorical propositional form at all. One can say that (1) Aristottle did develop a formal logic that was categorical, and (2) he incorporated "the negation of something of something," and universal and particular quantification, by building these items into a complex expression relating predicate to subject. Modality was then signified by further appropriate modification of this relational expression or copula.[4]

This result has an interesting, and confirming, implication for how we read quantified modal categorical propositions. One might well be tempted to read '*A N e B*' as '*necessarily applies to* relates *A* to none of the *B*'s.' Read this way, it in effect makes the scope of 'necessarily' be 'applies to', with 'none' (or 'all' or 'some') as an addition indicating to how much of *B* the predicate *A* is related by the modal relation 'necessarily applies to'. And read this way, it would be equivalent to '*possibly fails to apply to* relates *A* to each and every *B*.' But this is obviously quite different from saying what we want our universal negative to say, namely, that for each *B, A* necessarily does not apply to it. At this point we might try '*A N*-not *a B*'. This does now say what we want to say, but it gives a proposition having the form of a universal affirmative rather than a universal negative. Or we might retain the *E*-proposition form and try to remove possible ambiguity by writing '*A p e B*', reading this as 'there is no *B* to which *A* possibly applies,' or '*possibly applies* links *A* to no *B*.' This, too, says what we want to say, but now it has the form of a possibility, rather than a necessity, proposition.

None of these awkward consequences will arise if we read '*A N e B*' with the kind of copula suggested earlier, as '*necessarily applies to none of* relates *A* to *B*', where 'necessarily' modifies 'applies to none of', and where this is to be spelled out as "one can take no *B* to which *A* does not necessarily fail to apply." Thus it is equivalent to '*A* does not possibly apply to any *B*'. Analogously, '*A p e B*' will read '*possibly applies to none of* relates *A* to *B*,' with 'possibly' modifying 'applies to none of', which is equivalent to '*A* does not necessarily apply to any *B*'. This will give the desired contradictory to '*A N i B*', or '*necessarily applies to some* ⟨*of*⟩ relates *A* to *B*'. '*A p o B*' will be '*possibly does not apply to all of* relates *A* to *B*', and this will give the proper contradictory to '*A N a B*'.

There will be an additional interesting consequence for two-way possibility propositions. '*A pp e B*' is, as Aristotle says, equivalent to '*A pp a B*'; it should be read, in line with our previous results, as '*possibly applies to all of and possibly applies to none of* relates *A* to *B*'. Notice that here, rather than a conjunctive proposition ('*A* possibly applies to all of *B* and *A* possibly applies to none of *B*') that does not, without significant ado of a sort not found in Aristotle, fit Aristotle's categorical syntax, we have instead a conjunctive (copulative) relation. And although this does not fit the same simple formula we found in the cases of one-way possibility and necessity (i.e., assertoric copula plus single modal operator), it does at least function as part of a properly categorical proposition.

Finally, I think we can already see a potential, and potentially critical, distinction emerging between a copula relating two natures or essences or attributes *A* and *B* and one relating some predicable *A* to each thing that belongs among the *B*'s. Such statements as '*B* is included within *A*' are ambiguous in this respect: Is *B* itself essentially included in *A* (as when *A* is the genus Animal and *B* is the species Horse), or might it just happen that all the *B*'s are *A*'s (as when *B* = In the Agora and it happens that everything in the Agora is an Animal). One wants to be able to make either sort of statement, as one's purposes require. But Aristotle never raises this matter directly in *Pr. An.* A.1–22. At the assertoric level this presents no problems, as one would normally read both sorts of statements in the same way, as "There is no *B* to which *A* does not apply." But where one wishes to express some necessary or contingent relation between two natures themselves (or between possession of one and possession of the other), the distinction becomes crucial. As we have already seen, the necessity statements of the *Prior Analytics* can be true under two sorts of conditions, one that does involve a *per se* connection between its terms, and one that does not. The latter sort of statement could be handled along the lines of the assertoric case: '*A N a B*' says that there is no *B* to which *A* does not necessarily apply. But for the former, we will have to explicitly specify that *A* and *B* are themselves appropriately related. This happens in *Post. An.* A.4, but not in the *Prior Analytics,* where Aristotle draws on examples of both sorts and employs conversion principles and arguments that require now one sort of reading, now the other. As we proceed, we shall find a number of quite specific reasons why Aristotle did not, in the *Prior Analytics,* realize the need to distinguish two readings of his modal propositions.

## 2.2. SOME INITIAL PROBLEMS ABOUT CONVERSION

In addition to its syntactic difficulties, the modal predicate reading also fails to preserve the validity of Aristotle's conversion principles. We shall see in detail, starting with Chapter 3, that Aristotle's treatment of modal syllogisms follows closely that of plain syllogisms in *Pr. An.* A.4–7: For each major group of syllogisms (with a few exceptions discussed in Chapter 7), he identifies a set of "complete" or "perfect" (*teleios*) ones (those whose validity is "obvious," *phaneros*) and then shows the rest to be valid by "reducing" (*anagein*) them to one or another perfect syllogism. Aristotle's conversion principles are by far his most important tool in carrying out the reduction of imperfect syllogisms. Among plain propositions, *I* and *E* do entail their converses ['*A* applies to some (no) *B*' entails '*B* applies to some (no) *A*']; the *A* proposition converts not "simply" (*haplōs*) but "particularly" (*kata meros*) to an *I* proposition ('*A a B*' entails '*B i A*', but not '*B a A*')[5]. Aristotle reduces all his valid second- and third-figure moods to four manifestly valid first-figure moods, in most cases by use of these conversions. Thus he validates the third-figure mood Datisi

| | | |
|---|---|---|
| (1) | *A a B* | (*A* applies to every *B*, or all *B*'s are *A*'s) |
| (2) | *C i B* | (*C* applies to some *B*, or some *B* is a *C*) |
| | *A i C* | (*A* applies to some *C*, or some *C* is an *A*) |

by converting (2) to '*B i C*' and then combining that with (1) to obtain the perfect first-figure mood Darii:

(1) *A a B*
(3) *B i C*
*A i C*

This yields a proof of validity for plain Datisi: If the syllogism Darii and the conversion of plain *I* are valid – and both clearly are – then Datisi is also valid.

Chapter A.2 had established the plain (assertoric) conversion principles; A.3, on modal principles, is a direct continuation of A.2, maintaining that the necessity and one-way possibility propositions convert in the same way as their plain counterparts (i.e., *E* to an *E*, *I* to an *I*, *A* to an *I*). (Two-way possibility involves special problems that we shall treat only briefly here, but shall discuss at length in Chapters 5 and 6.) Thus the third-figure mood $A_nI_nI_n$

| | | |
|---|---|---|
| (1) | *A N a C* | (*A* necessarily applies to all *C*) |
| (2) | *B N i C* | (*B* necessarily applies to some *C*) |
| | *A N i B* | (*A* necessarily applies to some *B*) |

can be validated by converting (2) to

(3) *C N i B* (*C* necessarily applies to some *B*)

which then combines with (1) to give the syllogism

| | | |
|---|---|---|
| (1) | *A N a C* | (*A* necessarily applies to all *C*) |
| (3) | *C N i B* | (*C* necessarily applies to some *B*) |
| | *A N i B* | (*A* necessarily applies to some *B*) |

But this is just the complete or perfect first-figure mood (Darii) $A_nI_nI_n$. So given the obvious validity of this mood, plus the validity of the conversion of the $I_n$ proposition, the third-figure mood Datisi $A_nI_nI_n$ is also valid.

By such means does Aristotle, in the space of a few lines (ch. 8, 29b29–30a14), reduce all but two second- and third-figure necessity syllogisms (Baroco and Bocardo require different methods) to four complete first-figure moods.[6] The problem is that the modal conversion principles used in these proofs are, in terms of the traditional distinction, valid if read *de dicto* but invalid if read with a modal predicate. Read *de dicto*, $A_n$, $I_n$, and $E_n$ do convert, and $O_n$ fails to convert, just as Aristotle says. [All three conversions would follow from the familiar principles of modal propositional logic that if *p* (strictly) entails *q*, then nec: *p* entails nec: *q*, and if *p* entails *q*, then poss: *p* entails poss: *q*. It is important to recognize, however, that this is not the sort of proof Aristotle gives here and that it is doubtful he ever recognized the quoted principles of modern modal logic. This vexed issue receives a full hearing in Chapter 6.]

By contrast, there are problems for conversion of $A_n$ and $I_n$ on any *de re* reading, including the popular modal predicate version. For example, it is true, given that rationality is essential to human beings, and that all philosophers are human beings, that Necessary-rational applies to all Philosopher. But it is false that Necessary-philosopher applies to some Rational. $E_n$ also fails to convert on a modal predicate reading. Let $A$ = 'Animal' and $B$ = 'White Thing lying on the Mat', and suppose that all such white things are in fact sheets of paper.

With regard to two-way possibility propositions, however, the story is quite different, for when in chapter 17 Aristotle provides his *only* detailed discussion of *any* modal conversion, he quite clearly does not have a *de dicto* interpretation in mind and does conclude (correctly) that that conversion is invalid.[7] Near the beginning of chapter 17, Aristotle inserts (36b25–37a31) a lengthy discussion of why $E_{pp}$ ('*A* is two-way possibly inapplicable to all *B*') does not convert. That it should not convert is rather startling, since the *A*, *I*, and *E* propositions had consistently converted throughout the previous discussion of plain, necessity, and one-way possibility statements, while the *O* statement alone had consistently failed to entail its converse. But now, in the case of two-way possibility, it appears that this usual situation does not obtain: Although $A_{pp}$ and $I_{pp}$ do convert, $E_{pp}$ does not.[8] Aristotle takes great care to establish this unexpected result, giving three arguments to show either that the $E_{pp}$ conversion is invalid or that some specific attempt to validate it must fail. His second argument (37a4–9) consists in giving a counterexample to the conversion in question: Whereas White may two-way possibly fail to apply to all Human, the converse, 'Human two-way possibly fails to apply to all White', does *not* hold, since there are some white things (cloaks or snow) of which Human is necessarily, hence *not* two-way possibly, false. Aristotle's use of the example is perfectly correct, at least insofar as it involves either an appropriately interpreted modal copula (i.e., as weak *cop*) or a modal predicate rather than a *de dicto* reading.

What is equally significant, however, and what Aristotle does not seem to realize, is that the same sort of example, adapted in an obvious way, would show the non-convertibility of $E_p$ ('*P* one-way possibly fails to apply to all *S*'), as well as $E_{pp}$. White, for example, one-way possibly fails to apply to all Human, but that does not show Human to be one-way possibly false of all White, since Human may in fact necessarily apply to some things that are white (e.g., Coriscus, Socrates). And again, the same terms show that we might have true a *necessity* version of 'Human applies to all/some White' without the necessity version of 'White some Human' being true. (This would come about in a situation in which all/some white things were human beings.) And this shows that on one quite natural (i.e., *de re*) reading, neither $A_n$ nor $I_n$ converts. In fact, Aristotle's counterexample for the case of two-way possibility could just as well show, with easy, minor adaptations, the non-convertibility of all the necessity and one-way possibility conversions he has previously asserted, and on which he relies heavily in validating syllogisms outside the first figure.

As if this were not sufficiently disturbing, Aristotle's own very brief discussion of those necessity and one-way possibility conversions in chapter 3 simply "proves" the one by appeal to the other. At 25a27–34 he validates the conversions of $E_n$, then $A_n$ and $I_n$, by *reductio* arguments that appeal to the conversions of $I_p$, then $E_p$:

> It will be the same [as in the case of plain *A, I, E,* and *O* conversions] with the necessity premises. For the universal negative converts to a universal, and each affirmative to a particular. For if it is necessary that *A* belong to no *B,* it is necessary also that *B* belong to no *A*. For if ⟨it is possible that *B* belong⟩ to some *A,* then it would also be possible that *A* belong to some *B* [i.e., $I_p$ converts]. And if, on the other hand, *A* belongs of necessity to all or some *B,* then it is necessary that *B* belong to some *A*. If this is not necessary then neither would *A* belong of necessity to any *B* [i.e., $E_p$ converts].

Here the proof of the conversion of $E_n$ appeals to conversion of $I_p$; the proof for conversion of $A_n$ and $I_n$ appeals to conversion of $E_p$. A few lines later (25a40–25b3, as the text stands) Aristotle will validate the conversion of $I_p$ on the basis of the conversion of $E_n$, which later principle, he says, "has already been proved" (25b2–3). And a little further on, he will argue for the conversion of $E_p$ by appeal to that of $I_n$ (25b11–13). One could debate the authenticity of 25a29–34, or 25b2–3 and 25b11–13, and hence the circularity in these proofs,[9] but that would not in any case resolve the difficulties posed by the counterexamples surveyed earlier.

If one were to remove the circularity by excising 25b2–3 and 25b11–13, one might then furnish Aristotle with an argument for the conversion of $I_p$ by supposing that he has in mind an *ekthesis* argument similar to the one he used to prove the conversion of assertoric '*A e B*'. There he argued (25a14–17) that if *A e B*, then *B e A*. For suppose that *B* applies to some *A*, say *c*. Then since *c* is by hypothesis both a *B* and an *A*, it will follow that *A i B*. But this contradicts *A e B*. (Here I take *c* to be an appropriate individual rather than some subset of the *A*'s; otherwise the proof would appeal to a syllogism that has not yet been validated or even introduced. On this type of *ekthesis*, see Section 3.7.) Now Aristotle may well have thought in a similar way that if *A N e B*, then *B N e A*. For suppose that *B P i A*, say *c*. Then there is nothing impossible about *B* actually applying to *c*. But *if B* actually applies to *c*, then by hypothesis *B* actually applies to some *A*, for *c* just is some *A*. And if *B* applies to something to which *A* applies, then *A* applies to something to which *B* applies. So there should be nothing impossible about *A* applying to some *B* (i.e., *A P i B*). But this contradicts *A N e B*. (An equally direct proof could be given for conversion of $E_p$: If nothing impossible follows from supposing that *A* is disjoint from *B*, then obviously nothing impossible follows from supposing that *B* is disjoint from *A*. Thus if *A P e B*, then *B P e A*.)

With its appeal to what is or may be true about some individual, the reasoning might at first glance appear to prove the *de re* conversion of $I_p$. This appearance might be reinforced by imagining a concrete instance: Let *A* = White, *B* = Animal, and *c* = Socrates. But that it cannot possibly do, for the *de re* conversion is invalid: Let *B* = White and *A* = Animal, where some animal is possibly white and all white things are stones or cloaks. In fact, the argument, in using the conversion of plain *I* to move from 'there is nothing impossible about *A i B*' to 'there is nothing impossible about *B i A*', would *in effect* use the *de dicto* principle that if $p \vdash q$ and poss: *p*, then poss: *q*. (Read 'poss:' as 'there is nothing impossible about:'. Aristotle *may* use this sort of inference more explicitly in chapter A.15. See Chapter 6 herein.) Or, in terms of the concrete instance, the argument moves from the possibility of both White and Animal being true of Socrates to that of Both Animal and White being true of Socrates – which hinges simply on the compatibility of the terms *A* and *B* themselves.

There is no adequate basis, however, for adopting this conjecture (and excising 25b2–3 and 25b11–13) in order to remove the circularity of 25b1–3. Still, it helps make clear the important point that just as with propositions of necessity, there are two natural ways to read Aristotelian one-way possibility propositions, only one of which gives valid conversions.

'*A P* all *B*', for example, might be read in either of two ways:

(1) Being *A* is not incompatible with being *B*
(2) Being *A* is not incompatible with being anything in the essence of any *B*'s

These readings diverge because the *B*'s may be only accidentally *B* but essentially *D;* in such a case, *A* could be compatible with *B* itself, but incompatible with the essence of the *B*'s (i.e., *D* itself). For example, condition (1) is met if *A* = Human and *B* = White, *no matter what things are white.* But this leaves open the question whether or not condition (2) is met. And if all white things are stones or cloaks, then (1) is met, but (2) is not. As the phrase 'any *B*'s' in (2) indicates, one cannot simply consider the nature or attribute *B* and its relation to *A,* but must first look to the actual *B*'s to discover their essence (which may or may not include *B*), and then consider the relation of that essence to *A* itself.

As for the alleged converse ('*B P i A*'), it will follow from (1) that

Being *B* is not incompatible with being *A*

hence that *B P i A*. So if $A_p$ is taken simply to assert (1), it does convert to a particular.

By contrast, to test the conversion on reading (2) [or, *a fortiori,* on a reading that consisted of the conjunction of (1) and (2)], one would have to consider, in verifying '*B P i A*', what things are in fact *A*'s, then establish what those *A*'s "really are" (what they essentially are), and finally establish that the essence of at least one of the *A*'s was compatible with *B* itself. It may be true that *A* (White, say) possibly applies to all *B* (Human), but nonetheless impossible that *any* of the things that are actually *A* (perhaps these are all stones or cloaks) be or become a *B*. Thus, whereas (1) entails its converse, neither (1) nor (2) entails the following:

*B* is compatible with everything in the essence of some *A*

So let us say, subject to further discussion, that of two reasonable ways of looking at statements of the form '*A P a B*', one converts, whereas the other does not. Notice, however, that the distinction between them is to be drawn not in terms of the *de dicto/de re* distinction in any of its common presentations, but rather in the language of Aristotelian essentialism and theory of definition. In a nutshell, we have two different sorts of cases: one in which compatibility (or incompatibility) obtains between the natures signified by one's predicate and subject terms, and another in which the original predicate is compatible with the essence of all/some of those things to which the subject applies. One-way possibility statements are convertible when viewed from the former point of view simply because the relation of incompatibility (or compatibility) between two terms or properties is symmetrical. But two terms may be compatible in and of themselves even if one of them is incompatible with the essence of all/some of the things applied to by the other. This sort of thing is simply a fact of life with metaphysical essentialism.

A similar situation obtains with regard to necessity propositions, whose conversion was supposed to be proved by use of conversion of $E_p$ and $I_p$. In the vast majority of Aristotle's own examples of $A_n$ and $I_n$ propositions, conversion does preserve truth. These will be the sorts of cases in which *A* is necessarily applicable to all/some *B* because *A* is part of what-it-is-to-be-a-*B*, as with Animal and Human. In every such case, at least one underlying subject will be essentially *both A* and *B* – *A* being that subject's genus, say, and *B* its species – so that it follows that *B* will necessarily belong to some *A*: As all/some humans are essentially animals, so some animal is essentially human. So if restricted to this sort of case, the truth of $I_n$ does entail the truth of its converse, and this can be verified by thinking through the truth conditions, in terms of genus-species relationships, of the propositions in question. (We shall complicate this reasoning slightly later in this chapter by the addition of differentiae and *idia*.) By contrast, where Human belongs essentially to some White, the latter term need not represent an essential property of humans, but might only pick out, by use of an accidental property of some white things, objects (e.g., Socrates, Coriscus) that are essentially human. In this instance, the converse (White *N* some Human) does *not* follow and is in fact false, because White is related only accidentally, not necessarily, to any human beings. Indeed, one could argue more generally that there is no way to infer from 'Human *N* some White' that White is related necessarily to *anything* at all – nor, *a fortiori*, to any humans.

More will be said in due time about the interpretation and validity of various conversions. Let us say at this point that the issue of vacillation, or of ambiguity in the treatment of modality, arises already *within* the specific realm of modal conversion. The "scorecard" approach does not give a total victory even here for a *de dicto* reading. But more important, a simple comparison of results claimed by Aristotle in one place or another with results obtained on this or that reading of a modal proposition will never tell us how two different readings of a conversion or a syllogistic premise might from Aristotle's point of view be related to one another – nor, therefore, how Aristotle might have come to conflate or to alternate between the two. In the following sections we shall work our way clear of the modal *dictum*–modal predicate dichotomy to a different pair of readings, both more clearly Aristotelian in content, both involving a modal copula structure, and both grounded in Aristotle's essentialist metaphysics.

## 2.3. *COP* AND ITS COMPETITORS: PROBLEMS FOR MODAL PREDICATES

Although on both the weak *cop* and modal predicate readings all of Aristotle's conversion principles are invalid, there is yet a prior problem for the modal predicate version: Interchanging the terms as given does not even yield the desired converse in a purely formal sense. For example, the converse of '*nA* all *B*' ('necessary-*A* applies to all *B*') would be '*B* all *nA*' ('*B* applies to all necessary-*A*'). But what Aristotle would consider the converse in the modal predicate format would be '*nB a A*' ('necessary-*B* applies to all *A*'). This would remain as a problem for attributing to Aristotle a modal predicate conception even if we put aside the fact that on this interpretation his conversion principles are invalid. By contrast, a weak *cop* reading at least gives the formally required converse: Interchanging the terms of '$A$ $N_w$ all $B$' does give '$B$ $N_w$ all $A$'.

Prior even to this, one finds oddities in the very formulation of modal propositions. $O_n$ and $E_n$ will read something like

| | |
|---|---|
| *pA o B* | (possible-*A* fails to apply to some *B*) |
| *pA e A* | (possible-*A* fails to apply to any *B*) |

And the negative one-way possibility counterparts ($O_p$, $E_p$) will become

| | |
|---|---|
| *nA o A* | (necessary-*A* fails to apply to some *B*) |
| *nA e B* | (necessary-*A* fails to apply to any *B*) |

The point here is not to deny that these are *logically* incorrect (for they are logically equivalent to Aristotle's formulations), but simply to point out that this is not the way Aristotle thought of these propositions. He routinely formulated his necessity and one-way possibility propositions "in their own terms" as it were: necessity ones using 'necessarily applies/does not apply to all or to none', possibility ones using 'possibly applies/fails to apply'. And even if he is aware that certain pairs of necessity and possibility propositions are interdefinable, the point remains that there is no ready way to formulate all of these propositions, in a way that mirrors Aristotle's formulations, using modal predicates. By contrast, the modal copula versions do precisely reflect Aristotle's formulations.

Third, the modal predicate reading will produce an enormous number of ill-formed syllogisms. In the first-figure mood Barbara, for example, with plain major premise and two-way possibility minor premise, one would have

$$(1)\quad \begin{array}{c} A \text{ all } B \\ \underline{ppB \text{ all } C} \\ pA \text{ all } C \end{array}$$

Aristotle considers this syllogism at A.15, 34a34ff. But with modalized predicates it becomes a five-termed monstrosity. So, too, with necessity major and two-way possibility minor:

$$(2)\quad \begin{array}{c} nA \text{ all } B \\ \underline{ppB \text{ all } C} \\ pA \text{ all } C \end{array}$$

The result with two two-way premises, or with two-way major and assertoric minor, or with two-way major and necessary minor, is only slightly less disturbing:

$$(3)\quad \begin{array}{c} ppA \text{ all } B \\ \underline{ppB \text{ all } C} \\ ppA \text{ all } C \end{array} \qquad (4)\quad \begin{array}{c} ppA \text{ all } B \\ \underline{B \text{ all } C} \\ ppA \text{ all } C \end{array} \qquad (5)\quad \begin{array}{c} ppA \text{ all } B \\ \underline{nB \text{ all } C} \\ ppA \text{ all } C \end{array}$$

Syllogism (3) could be read, with "ampliation" of the two-way premise (to give '*ppA a ppB*'), as a well-formed – and, by the way, valid – syllogism. Syllogism (4) is well formed as it stands. Syllogism (5) still has four terms. These reduce to three if one weakens the minor premise to '*B a C*' (on grounds that it follows from '*nB a C*'). However, Aristotle says nothing in his treatment of these syllogisms about performing such an operation.

Meanwhile, those five-termed mutants will require more extensive surgery. In (1), we shall first have to ampliate the *assertoric* premise (to get '*nA a ppB*') – a procedure nowhere recognized by Aristotle – then derive the conclusion '*ppA a C*', and finally weaken this to obtain the conclusion '*pA a C*'. But all of this yields a proof quite unlike what Aristotle gives in chapter A.15 or anywhere else. With (2) we must first ampliate the *necessity* premise – again something Aristotle never considers – then weaken that premise to give '*pA* all *ppB*', and finally derive '*pA* all *C*'. There are other routes to this conclusion, but none of them any closer to Aristotle's.

This is a problem for those who wish to see Aristotle as thinking in terms of modal predicates, or even as alternating between that and modal *dicta,* for it is customary to read the modal *syllogisms,* if not the modal conversions, in modal predicate terms, simply because that way they come out valid almost exactly the way Aristotle says, whereas only a small minority are valid when read *de dicto.* But now it is evident that even if that is so, it overlooks the prior problem that on a modal predicate reading a great many of these inferences, valid as they may be in the end, are not even properly formed Aristotelian syllogisms.

By contrast, on a weak *cop* reading, not only do these syllogisms turn out valid (again almost exactly) as Aristotle says, but both the syllogisms themselves and the conversions used to validate imperfect moods are properly formed from the start.

Finally, I would suggest that the reason Aristotle sets things up in terms of modal *copulae* rather than modal predicates is that he simply does not, from the metaphysical point of view, believe in four different properties associated with, for example, the quality White ['being White', 'being (one-way) possibly White', etc.]. Rather, there is one quality White, which is related in different ways to different subjects. It applies one- and two-way possibly and most of the time actually to some cloaks, but it applies necessarily to swans and snow (at least, for purposes of certain illustrative examples; see, e.g., *Pr. An.* 36b10–12). White is, according to the *Categories,* a quality, and there need be no *qualitative* difference between the whiteness of a cloak and that of snow. The difference would be that in one case White belongs (or does not belong) accidentally, and in the other essentially, to the underlying subject. Nor are there three distinct qualities [Being (plainly) White, Being possibly White, Being two-way possibly White] that attach simultaneously to some cloak, but rather one quality that belongs, and both one- and two-way possibly belongs, to a certain cloak.

Thus Aristotle's explicit adoption of the *cop* syntax makes good sense: It is workable and natural, in a way that modal predicates are not, (1) as an expression of individual facts that consist in some property's applying in a certain way to a subject, (2) for investigating various questions involving conversion, and (3) for the construction of modal syllogisms.

## 2.4. FURTHER PROBLEMS FOR *DE DICTO* AND FOR A MODAL *DICTUM*/MODAL PREDICATE ALTERNATION

As with the modal predicate reading, close inspection reveals structural problems with any *de dicto* reading of Aristotle's modal propositions. Although full discussion of two-way possibility is reserved for Chapters 5 and 6, we may at least consider here one feature of the "qualitative conversion" [as opposed to the (term) conversion we have been discussing] of contingent propositions. As Aristotle rightly says, the positive and negative universal two-way propositions, and the positive and negative particular ones, are equivalent: If *A* two-way possibly applies to some/every *B*, then *A* two-way possibly fails to apply to some/every *B*. But to preserve the usual *de dicto* structure would mean writing the universal affirmative as

(1) poss. and poss. not (*A a B*)

and the universal negative as

(2) poss. and poss. not (*A e B*)

But these are not equivalent. Nor do they say what Aristotle wants to say. The universal affirmative (1) is equivalent to

(3) poss. (*A a B*) and poss. not (*A a B*)

or

(4) poss. (*A a B*) and poss. (*A o B*)

Meanwhile, the universal negative (2) is equivalent to

(5) poss. (*A e B*) and poss. not (*A e B*)

or to

(6) poss. (*A e B*) and poss. (*A i B*)

But (4) is obviously not equivalent to (6). To express what Aristotle wants to say – and at the same time to preserve the equivalence of $A_{pp}$ and $E_{pp}$ – the *de dicto* approach would have to read $A_{pp}$ as the conjunction of two *de dicto* propositions:

(7) poss. (*A a B*) and poss. (*A e B*)

$E_{pp}$ will obviously be

(8) poss. (*A e B*) and poss. (*A a B*)

which is equivalent to (7).

But this raises two problems. First, there are no conjunctions in Aristotle's logic, so it leaves one at a loss as to how this reading is to be integrated into the object language of Aristotle's various discussions and proofs of two-way syllogisms and qualitative conversions. This is not a problem for the *cop* interpretation of Aristotle's intent. His object language will simply contain *A endechetai huparchei panti tōi B,* which is in fact identical with his usual object-language expression for *one*-way possibility. This can be formulated categorically as '*possibly applies to all of and possibly applies to none of* relates *A* to *B*' (see Section 2.1 herein) and, in light of Aristotle's definition of "applies to all of" and his definition of an accident at *Topics* 102b4–7 (see Section 2.6 herein), defined as 'you can take no *B* of which it is not true that *A* possibly applies to it and possibly fails to apply to it'.

Second, although it preserves qualitative conversion, this *de dicto* formulation raises a question of how *term* conversion is to be conceived. With a single *de dicto* operator prefixed to '*A e B*', say, one can at least produce the converse by the usual means of reversing the terms. But with the conjunctions (7) and (8), this does not work: One must perform two reversals rather than, as Aristotle had in mind, just one. So the *de dicto* reading faces a serious dilemma: Either it has the usual *de dicto* syntax (prefixing thc necessity operator to a single assertoric proposition), in which case qualitative conversion is (*pace* Aristotle) invalid, or it saves that sort of conversion by appeal to a formulation in terms of conjunctions of *de dicto* propositions that (a) falls outside Aristotle's own object-language syntax and (b) blocks the usual procedure for term conversion.

What is worse, one cannot sidestep these problems even by the familiar appeal to a vacillation between modal *dicta* and predicates, for *neither* fits what Aristotle says about qualitative conversion. We just saw this for the case of modal *dicta.* As for modal predicates, the universal affirmative

(9) *ppA a B*

would entail that *no* B is either a necessary-*A* or a necessary-non-*A*. But this is plainly not equivalent to the universal negative:

(10) *ppA e B.*

What (10) says is that possibly-and-possibly-not-*A* applies to no *B*, which entails that *every B* is either a necessary-*A* or a necessary-non-*A*. So here the alternation between modal *dictum* and modal predicate breaks down completely; *neither* can give a plausible expression of what Aristotle had in mind. By contrast (as we shall verify at length in Chapter 5), the *cop* approach yields a reading on which these qualitative conversions are all well formed and valid.

This result is significant, since our goal is not to cast Aristotle's statements in terms of some already familiar approach, whether that be modern modal predicate logic or a traditional modal *dictum*/modal predicate framework, and then reconstruct Aristotle's arguments in those terms (by carrying out proofs in S5, say, or by way of modal predicate formulations, as with the examples given in the preceding section), but rather to think through Aristotle's modal logic – from his fundamental principles to his discussion of difficult local problems – in the way Aristotle himself did. (This obviously does not preclude the possibility of our detecting errors on Aristotle's part or, at a later stage, of considering the possibility of "translating" Aristotle's logic into some other system.) Analogously, for certain mathematical problems, one might program a calculator to arrive at the same answers as humans. But such a program might fail to reveal how humans conceive the problem or how they arrive at their solutions.

## 2.5. STRONG *COP* VS. *DE DICTO*

We have now seen that strong *cop* is to be distinguished on various formal grounds from *de dicto* modality. But the main reason these two readings might have appeared to be essentially the same in the first place is that each can be seen as asserting a relation between properties or universals or some such intensional entities. Nevertheless, one may see plainly that the two must be kept apart even at this level, from the simple fact that many *de dicto* truths are strong *cop* falsehoods. For example, 'nec: all bachelors are unmarried' and 'nec: all nearsighted philosophers are myopic' are both true. But since being a bachelor (a nearsighted philosopher) does not belong to the essence or what-it-is-to-be of any unmarried person (myopic person), the strong *cop* versions of these statements are false. The key point is that strong *cop* involves Aristotelian essentialism (in the requirement that *A* belong to the essence of all *B*'s), whereas these *de dicto* truths might as well be true by virtue of the meanings of the terms, or in some other way that carries no commitment at all to the notion of philosophers or anything else having essences.

Moreover, even many *de dicto* truths now widely termed "essentialist" will be strong *cop* falsehoods. If any property of a thing which that thing could not lack and still exist is to be considered an essential property of that thing, then *being self-identical* will be an essential property of everything. But self-identity is not part of what-it-is-to-be for a horse, so that 'self-identical $N_s$ all Horse' is false. (Because self-identity is not part of the essence of anything that is a horse, even the weak *cop* '$A$ $N_w$ all $B$' will be false.) Aristotle's essentialism is broad enough, of course, to include essences of things other than substances. As the *Topics* points out, there will be a *ti esti* for qualities, quantities, and so forth, as well as for substances. And the *Categories* foreshadows this idea in recognizing a strong "said of" relation among genera and species of non-substantial kinds of things. But this still falls far short of grounding such "essentialist" truths as 'every horse is identical with itself'.

There is also a more general reason for preferring a *cop* syntax over *de dicto* modalities, one that may be illustrated by the difference between a possible-worlds semantics and Aristotle's semantics of genus, species, and so forth. Where one defines necessary truth as truth in all possible worlds, and where truth is, at ground level, a property of non-modal sentences, it will be entirely natural, if one wants one's syntax to reflect one's semantics, to make whole sentences the basic subjects of necessity. After all, there is, as it were, no modality "inside" the ground-level sentences, but only the predication of a (non-modal) property of a subject (as in 'Theaetetus flies'). On a possible-worlds semantics, necessity is then interpreted not as a necessary application of predicate to subject, but as a non-modal sentence's holding in all possible worlds. From within this sort of modal semantics a *de dicto* construction can seem entirely fitting and proper, even if it is not the only possible option.

Now, in Aristotle's view, truth is a property of sentences or beliefs [e.g., *Categories* (*Cat.*) 4a22ff.], so one might suppose it natural enough to speak of necessary truth (or being necessarily true) as also a property of sentences. But even if Aristotle had been in the habit of speaking of necessary truth (which he was not), one still would have to ask about the *grounds* of the (necessary) truth of a sentence. And a proposition such as '$A$ $N_s$ $a$ $B$' will be true, if it is true, because $A$ is, say, a genus or differentia or proprium of the species $B$. So if one wishes to say in this case that '$A$ all $B$' holds in all possible worlds, this will be *because* $A$ is, say, a genus of the species $B$. The possible-world semantics no longer gives the primary interpretation of necessity (as it would for a "realist" about possible worlds), but becomes at best a kind of picturesque intellectual aid – as indeed it is regarded even by some (non-realist) possible-world semanticists.

Aristotle's own approach is reflected in many passages, some of which we may note briefly here. He speaks, for example, in the *Topics* of "all *A*'s being essentially *B*" in a slightly different idiom from anything we have discussed so far, but still using a modal copula: *estin A hoper B* ('*A just is B*' or '*A* is essentially *B*'). The *hoper* functions here as an intensifier attached to the copula. As Brunschwig[10] points out, this construction occurs repeatedly in the *Topics* to express the fact that *A* is a kind of *B* (*esti A B tis*), that *A* is a species of the genus *B*. (Brunschwig lists *Topics* 120a23sq., 122b19, 26sq., 123a, 124a18, 125a29, 126a21, 128a35; cf. the similar use of the expression at *Post. An.* 83a24–30.) So, too, the *Categories* uses the expressions 'is in' (*en*) and 'is said of' (*legetai*) to express two basic sorts of relations between predicate and subject, one where *P* inheres (as an accident) in *S*, and one where *P* is definitionally applicable to *S* (as, e.g., genus to species). (This seems in fact to be the forerunner of Aristotle's claim in the *Posterior Analytics* that predicates apply to their subjects either necessarily or accidentally. But that is a subtle issue we may set aside here.) Other passages have already been discussed (Section 2.1 on Aristotle's introduction of modal syllogisms and conversions) or will be discussed in detail later on (e.g., Section 2.6 on the "four predicables" of the *Topics*). But the common thread is that the syntax always reflects (1) a predicational relationship between predicate and subject terms and (2) an underlying thought that there are only a few ways in which a predicate can be related to its subject(s). Thus, even if we were to isolate in thought a definitional aspect of strong *cop* statements – which would get us as close as possible to a basis for a *de dicto* reading – there would still be decisive grounds for representing Aristotle's thought by use of modal copulae.

In sum, there is ample reason to steer clear of the modal *dictum*/modal predicate dichotomy, or the traditional *de dicto*/*de re* distinction in any of its more common forms, when trying to understand Aristotle's modal logic "from the inside": Aside from their severe syntactical defects, neither is an appropriate reflection of Aristotle's modal semantics. Moreover, Section 2.9 will develop the further point, critical for understanding how Aristotle might have failed to distinguish two readings of his modal propositions, that strong and weak *cop* share a common essentialist root for which there is no obvious counterpart on the side of modal *dicta* and predicates.

## 2.6. THE FOUR PREDICABLES AS SYLLOGISTIC TERMS

Aristotle's essentialist semantics has appeared occasionally in Sections 2.1–2.5. It is worth noting how the essence/accident distinction would fit with the *Topics'* discussion of the "four predicables," especially since there are some explicit indications in the *Prior Analytics* that syllogistic terms will be drawn from the predicables of *Topics* I. Three central points are forcefully stated in the *Topics* and explicitly confirmed in the *Prior Analytics:*

> Arguments (*logoi*) arise from premises (*protaseis*), and syllogisms (*sullogismoi*) are concerned with problems (*problēmata*). But every premise and every problem indicates either genus or proprium or accident (*genos ē idion ē sumbebēkos*). For the differentia (*diaphora*), being generic, should be classed along with the genus. Of the propria, some signify the what-it-is-to-be (*to ti ēn einai*), some do not . . . let the one part be called the "definition" (*horos*), the remaining part be called, in accordance with the usage customary in these cases, the "proprium" (*idion*). (*Topics* 101b18–24)

The first point, then, is that the subjects and predicates of the dialectical "premises" and "problems" of the *Topics* – and, I would suggest, of the *Prior Analytics* (for which more direct evidence is given just below) – are, in relation to one another, genera, species, propria, or accidents.[11]

*Topics* I.9 makes the second point, that the predicate terms of well-formed propositions are predicated of their subjects in one or another of ten ways, giving ten "categories" of predication (what-it-is, quality, quantity, etc.):

> Next it is necessary to distinguish the kinds of categories (*genē tōn katēgoriōn*) in which are found the four ⟨predicables⟩ mentioned above. These are ten in number: what-it-is, quantity, quality. . . . For the accident and the genus and the proprium and the definition is always in one of these categories. For all premises formed from these signify either what-it-is or quality or quantity or some other of these categories. (*Topics* I.9, 103b20–27)

There is a connection also with the *Categories* in that the subjects and predicables involved in any predication serving as a syllogistic premise or conclusion will be drawn from one or another of that work's ten kinds of things that there are (substances, qualities, quantities, etc.).[12]

Finally, the third point: Each of the four predicables will belong, or fail to belong, to a given subject in one of two ways: necessarily or contingently. This is implicit in the *Topics,* because not only genera and species but also *propria* belong necessarily to their subjects (102a17–30), whereas the only other sort of predicate is related accidentally to its subject.[13] *Post. An.* A.6 makes the same point:

> Now if demonstrative understanding depends on necessary principles (for what one understands cannot be otherwise), and what belongs to the objects in themselves is necessary (*ta de kath' hauta huparchonta anangkaia*) . . . it is evident that demonstrative deduction will depend on things of this sort; for everything belongs either in this way or incidentally, and what is incidental is not necessary (*hapan gar ē houtōs huparchein ē kata sumbebēkos, ta de sumbebēkota ouk anangkaia*). (74b5–12)[14]

Regarding accidents, the passage just quoted says only that "what is incidental is not necessary." This is true as far as it goes, although of course one must add that what is incidental is also not impossible (i.e., not necessarily inapplicable to that of which it is an accident). This presumably was intended, however – with 'not necessary' meant to cover both 'not necessarily applying' and 'not necessarily not applying' – but it is in any case explicit in various passages, including the "official" definition of (two-way) possibility at *Pr. An.* A.13, 32a18–20:

> I call "to be possible" and "what is possible" that which is not necessary and which, being assumed to obtain, results in nothing impossible (*legō d' endechesthai kai to endechomenon, hou mē ontos anangkaiou, tethentos d' huparchein, ouden estai dia tout' adunaton*).

Or, as the *Topics'* preferred definition has it, the accidental is that which may apply and may not apply to one and the same thing (*sumbebēkos de estin . . . ho endechetai huparchein hotōioun heni kai tōi autōi kai mē huparchein,* 102b4–7; cf. 120b34, referring back to 102b4–7: *sumbebēkos elegomen ho endechetai huparchein tini kai mē*).[15]

This metaphysical background is not announced within the chapters of the *Prior Analytics* that develop the logical system formally. One might well infer it, however, from Aristotle's examples of syllogistic propositions: The ones quoted here (e.g., Animal/Human, White/Snow, White/Human) are entirely typical. But later chapters of the *Prior Analytics* bring these matters into the foreground:

> One must select the premises in each case as follows: first, set down the subject and the definitions and all the *idia* of the thing; then set down everything which follows the thing [are necessarily implied by the subject] and again, those which the thing follows [which necessarily imply the subject], and the things that cannot belong to it. . . . Among those which follow the thing, one must distinguish all those included in the definition (*en tōi*

> *ti esti*), and all those predicated as *idia* and as accidents (*hōs sumbebēkota*). (A.27, 43b1–8)

These remarks are part of Aristotle's advice in A.27ff. on how to equip oneself for the ready construction of syllogisms. In the latter half of book A and in book B, he discusses a long series of metalogical questions, always utilizing the same sorts of examples, and occasionally making explicit their connection to those four basic types of predicables. Thus, in discussing the derivation of a true conclusion from false premises (B.2), Aristotle gives the following example:

> And if *BC* ['*B a C*' where, e.g., *B* = man, *C* = footed] is not wholly false but in part only, even so, that conclusion ['*A a C*', let *A* = animal] may be true. For nothing prevents *A* from belonging to the whole of *B* and of *C*, while *B* belongs to some *C*, as a genus to the species and differentia. . . . (54b3–7)

And again, in considering an instance of Ferio with true major, false minor, and true conclusion:

> Similarly if the proposition *AB* is negative ['*A e B*']. For it is possible that *A* should belong to *B*, and not to some *C*, while *B* belongs to no *C*, as a genus to the species of another genus and to the accident of its own species. (55a11–16)

Aristotle then proposes the concrete example: *A* = animal, *B* = number, and *C* = white, B middle.

Such explicit appeals to the various relations among genus, species, differentiae, and accident reinforce the general suggestion of the passage from chapter 27 quoted earlier, and of Aristotle's concrete examples, that his syllogistic was intended to express and reason categorically about the sorts of terms, and predicative relations between terms, central to the essentialist metaphysics of the *Categories* and *Topics*.

This not only advances our integration of Aristotle's modal logic with his essentialism but also makes it possible to link the modal logic with a specific level of metaphysical analysis, for one may abstract from the details of the *Prior Analytics* with its concrete examples of Human, Animal, Crow, White, Black, and so on, to various levels of generality. In all cases the modal system deals with relations of predication (*katēgoreisthai*) or applying to (*huparchein*), because its syllogisms will consist of propositions predicating one thing of another.[16] These relations subdivide into two types of cases: necessarily applying and two-way possibly applying. These two general divisions, in turn, cover, at the next level down, the four relations in which the "four (or five?) predicables" – genus (with differentia?), species, *idion,* and accident – may stand to their subjects, where the first three necessarily apply to their subjects. These four, finally, could be subdivided further in light of Aristotle's ten categories of predication.

The modal logic actually developed in chapters 3 and 8–22 is pitched formally at the level at which the relations of necessarily and two-way possibly applying emerge. It is these general ways of applying whose logical interrelations Aristotle methodically plots in chapters 8–22. Descent to the next lower level would require introduction of notation to distinguish applying necessarily as genus from applying necessarily as species, and so forth, as well as applying as an accident. For the lowest level, one could, in principle, distinguish among applying as an essence, as a (non-essential) quantity, quality, and so forth. That is, one could mark syntactically expressions for all the various ways in which a subject may be essentially or accidentally determined.

But it seems, for the main dialectical and scientific purposes Aristotle had in mind, and which emerge in the *Topics, Sophistical Refutations,* and *Posterior Analytics,* that genus (differentia), species, and *idion* can be grouped together under the heading of 'necessarily applies', and the rest under 'accidentally applies'. So, for example, for the testing of proposed definitions and of attributions of *idia* to subjects, one should look to see if the predicate–subject relation is really necessary rather than accidental. Or again, for scientific demonstration of a *per se* link between predicate and subject of one's conclusion (where it seems that the predicate might be genus, species, or *idion* of the subject), one needs two propositions of necessity in the premises. And we may note once more that Aristotle evidently did not feel compelled, for these or any other purposes, to investigate systematically syllogisms with one-way possibility premises.

Even so, he failed to mark one sort of distinction that comes into play already at the level of necessary vs. contingent relations, namely, that between two sorts of readings of propositions of necessity and (one-way and) two-way possibility. And we shall see that for certain important purposes, including scientific demonstration, the distinction is critical. But for now we turn to the formulation of those readings.

## 2.7. TWO READINGS OF THE NECESSITY PROPOSITION

The distinction between a stronger and a weaker version of Aristotle's universal affirmative necessity propositions corresponds to that between *A* belonging to the essence of all *B*'s and *A* not only belonging to the essence of the *B*'s but also having an essential connection with the nature or essence of *B* itself. Again, Aristotle's example letting *A* stand for Animal, and *B* for Human, illustrates the stronger reading. In his examples in which *A* stands for Animal, and *B* for White, in a situation in which all white things are animals, the weaker reading is true but the stronger false; Animal does belong to the essence of all white things, but it is not entailed by their being white. Using these notions, we can now spell out the stronger and weaker versions of all four necessity propositions:

(1) $A\ N_w\ a\ B$: *A* applies necessarily to every *B*.
(2) $A\ N_w\ i\ B$: *A* applies necessarily to some *B*.
(3) $A\ N_w\ e\ B$: For every *B, A* is contrary to something applying necessarily to that *B*.
(4) $A\ N_w\ o\ B$: A is contrary to something applying necessarily to some *B*.[17]

These cover the four weak necessity propositions. To determine whether or not they are true, we must first identify the actual *B*'s, then identify their essence (which may or may not include *B*), and finally consider the relation of *A* to that essence. (By '*A* being contrary to *X*' – where *X* stands, for example, for something included in the essence of all *B*'s – I mean that *A* cannot belong to anything to which *X* belongs.)[18]

For the strong versions of these, we have (1)–(4) *plus* (5)–(8), respectively:

(5) $A\ N_s\ a\ B$: *A* is included in the definition of *B*; or, being *A* is part of what it is to be a *B*.

Thus strong $A_n$ = (1 & 5).[19]

For strong $I_n$:

(6) For some *C*, $B\ N_s\ a\ C$ and $A\ N_s\ a\ C$.

This guarantees that *A* applies with strong *cop* necessity to something that is properly *B* rather than merely to some subset of the *B*'s. It also guarantees that *A* and *B* are both included in some common essentialist tree, as it were. Nonetheless, there are some special conditions discussed in the next section under which (6) does not entail that there be an essential link between *A* and *B* themselves.

For the strong negative *cop* propositions $E_n$ and $O_n$, respectively, we need the following:

(7) $A\ N_s\ e\ B$: *A* is contrary to something that is part of what it is to be a *B*.

So strong $E_n$ = (3 & 7).

(8) $A\ N_s\ o\ B$: For some $C$, $B\ N_s\ a\ C$ and $A\ N_s\ e\ C$

So strong $O_n$ = (4 & 8). Definition (8) covers such cases as

'Human $N_s$ *o* Animal' (e.g., Horse)

while excluding

'Human $N_s$ *o* Brown' (where some brown thing is a horse, and horses are accidentally brown).

The idea behind (8) is that the "some *B*" to which *A* is, in the strong *cop* sense, necessarily inapplicable is not just any set of *B*'s whose essence is incompatible with *A,* but some essential subdivision of *B* whose essence is incompatible with *A.*

One final addition: Aristotle's propositions of necessity cover the application of *idia,* as well as genera and species, to a subject. But because in Aristotelian science one would want to explain a thing's propria by reference to its essence in a narrow sense (that signified in the definition of the thing), and to set out the explanation in syllogistic form, one would, in order to take account of this, read '$A\ N_s$ all *B*' as '*A* applies to *B kath' hauto*', where the latter is taken to include *idion*-species relations as well as genus-species (and other) relations. Thus, 'Capable of learning grammar $N_s$ all Human' would be true even though the predicate term is not part of the definition of the subject term. One could then speak of a property belonging to a subject "essentially," but now in a broader sense covering not only items mentioned in the thing's definition but also all properties present because of the thing's essence.

Aristotle may well have intended, as many commentators have suggested, that an *idion* apply to its species *kath' hauto* in the second sense of *kath' hauto* defined in *Post. An.* A.4:

> One thing belongs to another in itself if it belongs to it in what it is . . . and they belong in the account which says what they are. . . . Those also belong *kath' hauto* which belong to things and those things are part of the formula of the being of the former, as straight and curved belong to line, and odd and even belong to number. (73a37–40)

It is reasonable to suppose that Capable of Learning Grammar would belong *kath' hauto* to Human (or to Rational), because Human (or, it may be, Rational) presumably would be a part of the essence of the former as such. Hereafter, I shall assume that *idia* are related to their species – and to the genera and differentiae of their species – by strong *cop* family ties, and sometimes speak of them as applying to their subjects "essentially" in the broad sense just indicated. The weak *cop* definitions will accommodate *idia* just as they stand; the strong *cop* definitions are easily amended: (5) will read '*A* is entailed by *B*', rather than '*A* is included in the definition of *B*'; (7) will read '*A* is incompatible with something entailed by *B*', and so forth. The Appendix contains fuller statements of these amended versions.

And let us remember in all that follows that '$A\ N_w\ a\ B$' as defined here *does not preclude an essential tie between* A *and* B *themselves* (as is explicitly asserted in '$A\ N_s\ a\ B$'): It leaves that question open, merely asserting that *A* applies necessarily to everything that is a *B*.

## 2.8. TWO NOTES ON ARISTOTLE'S CONCRETE TERMS

To some extent our definitions have been framed so as to accommodate the essentialist principles underlying Aristotle's use of such examples as 'Animal necessarily applies to some White' along with 'Animal necessarily applies to all Human'. But other examples involving such properties as White and Black raise at least three interesting issues. First, the use of White as applying accidentally to humans but necessarily to swans and snow (as Black applies necessarily to pitch and ravens) complicates the task of defining the basic types of modal propositions. Second, an apparent "inconsistency" in his examples – which include 'Black *N a* Raven' as well as 'White *pp a* Animal' – may lead one to wonder how careful Aristotle has been in general about constructing his examples.[20] Third, his examples occasionally show the same sort of alternation between two readings of modality that we have been discussing in connection with his treatment of conversions and syllogisms. (Discussion of the third point is reserved for Chapter 4, Section 4.2.)

As for the first question, we need to decide whether or not to take seriously the idea that White applies necessarily to swans or to any other subject. Such examples might in fact be dismissed as not representing true necessity statements at all. One could argue that Aristotle had his reasons for thinking that any predicate would relate either accidentally or necessarily to its subjects, but not one way to some, and the other way to another. For example, letting White be, say, a differentia of Swan would violate Aristotle's strictures on division, because White would not entail all the essential ancestors of Swan. In keeping with this suggestion, one could regard White as what came to be called an "inseparable accident" (the phrase is not found in Aristotle) of swans: It always inheres in the matter of swans but is no part of their essence, so that swans can be thought of apart from their whiteness. There is a long medieval tradition of debate over this point. Zabarella, for example, following Porphyry, says that "although a raven never exists without blackness, we still can imagine that a raven does not possess blackness. Hence, all propositions which are not *per se* are such that the predicate may possibly not belong to the subject." Although Black might apply with some low grade of necessity to ravens, Zabarella's considered opinion is that in fact, and in Aristotle's own view, this is not really necessity at all.[21] But aside from the lack of evidence for conceivability as a criterion of possibility, and of finding the requisite matter/form distinction in the *Organon,* a strict division between (invariably) essential versus (invariably) accidental properties would be difficult to establish, and in fact demonstrably false if one allows a further example from the (admittedly later) *Metaphysics* H: spatial location, which is accidental to many subjects, is part of the *essence* of a threshold (1043b8–10).

Even if a case could be made that there are in Aristotle's view no such ambidextrous predicates, one would then have to explain why he used these examples in the *Prior Analytics.* This is probably the easier part, for certainly Aristotle is not committed to the actual truth of his examples even when they are used as premises in a counterexample. Of course, one sometimes makes a point of selecting statements that will seem obviously true (and ones that are obviously false for the conclusion of the counterexample). But one need only provide premises that the audience is willing to suppose could be true (at the same time as the premises were false). And certainly one might select concrete terms with the audience in mind. One thinks here of Aristotle's "Numbers are substances" (27a17–19): Even if he thought this to be in fact impossible, he could still use it as a premise in a counterexample, given an appropriate audience. In this he would be no different from the rest of us.

A second option would be to take these examples seriously as Aristotelian necessities. There are various ways this might be done. Perhaps the simplest would be to introduce a new term wherever White is taken as a differentia, genus, or proprium of some species: White-bird, perhaps, on the model of Polypod-terrestrial and Polypod-aquatic, which, as David Balme points out, could be used to avoid the problem of cross-division.[22] This is not without its own problems, but would preserve a *per se* link between the terms of statements like 'White-bird *a/i* Swan'.

Or, third, we could allow 'White $N_s$ *i* Bird', with the problems that would raise for division; this would also give the odd result that *A* (e.g., White) and *B* (Bird) could themselves be part of a common essential tree (that of *C*, Swan), even if there were no *per se* link between them.

For present purposes I will assume that Aristotle would favor Balme's solution: This will avoid adding pesky qualifications (which, however, the reader is free to supply) involving the use of 'white'.

The resolution of the "inconsistent examples" problem has already been touched upon. Because Aristotle need not draw exclusively on his own convictions about the true state of the world, nothing prevents him from using what are, from his point of view, counterfactual examples. And unless his auditors are *very* dull, they are hardly going to be confused by considering in one place a possible situation in which there are ravens and these are all necessarily black, and in another a situation in which there are no ravens, and all animals are two-way possibly white. What he does need to avoid is putting two inconsistent premises into a single counterexample.

## 2.9. AN IMPORTANT MORAL

These two formulations (weak and strong *cop*) have arisen from two sorts of situations in which a given predicate will belong of necessity to all those things to which some subject term applies. Both sorts of situations are implicit in Aristotle's essentialist metaphysics, and it is for this reason that both are represented in the discussions and examples in *Prior Analytics* A (although among the examples the first type predominates). It follows with regard to attempts to save Aristotle from logical ambiguity[23] – that is, to devise a representation of his modal logic without incorporating two readings of necessity propositions – that these are one and all mistaken, even if understandable. If Aristotle's logic is even so much as to express the very facts about which he needs and wants to reason, then it will have to be able to express and distinguish between these two sorts of situations. Bringing out these two readings (one *requiring* a *per se* link between *A* and *B*, one not), making clear their similarities and differences, and investigating how both might be consistently integrated into the logical system (as, again, Aristotle himself failed to do) are in fact necessary for understanding exactly what Aristotle was trying to do and how it came about that his own efforts were incomplete.

## 2.10. INTENSIONAL RELATIONS AND THE UNITY OF THE TWO *COP* READINGS

When thinking in terms of a *de re/de dicto* distinction, one tends to suppose, as the Latin phrases may suggest, either that the former makes a statement about *things* [e.g., that certain things have some (modal) property], and the latter about propositions (that some proposition is a necessary truth), or that the one concerns a relation between terms and things, and the other an intensional relation between two terms (or natures) – or, perhaps, that one says something about things in and of themselves, and the other about things only under a certain description. But from the Aristotelian point of view, all these proposals are seriously misleading in that they obscure the nature of the relation between his two sorts of necessity propositions, and with it a further element of unity between them. If we ask what, at bottom, accounts for the truth of an Aristotelian weak *cop* necessity proposition, which is the sort of statement many commentators would want to read *de re* (e.g., 'Animal $N_w$ *i* White', where some white things are humans), it is that the essence of some of the things that are in fact white is itself essentially linked to the nature Animal. And this puts the emphasis on the *essence* (Human) of the "things" (Socrates, Coriscus) designated by the subject term (White), rather than on those individual things as such. It thereby helps bring out the fact that what makes the weak *cop* statement true is that two *natures* *A* and *C* (Animal, say, and Human) are essentially linked in a certain way – in this case, as genus to species – and that the logical subject term *B* (White) of our initial weak *cop* sentence happens to apply to some actual things that are essentially *C* (Human). Or, somewhat more formally, and equivalent to the definition given earlier:

'*A* $N_w$ all *B*' is true iff
(1) for every *B* there is some *C* such that *C* applies essentially to that *B* and
(2) being *A* is entailed by being *C*[24]

Conditions (1) and (2) guarantee that *A* belongs to the essence of everything that is a *B*, but they leave open the questions whether or not *A* has any essential relation to *B* itself and whether *B* is accidental or essential to the *B*'s.

By way of contrast, consider strong *cop*:

'$A\ N_s$ all $B$' is true iff
(1) $A$ applies essentially to all the $B$'s and
(2) being $A$ is entailed by being $B$

Notice that *both* sets of truth conditions, strong and weak, involve not only (i) the notion of some nature or predicate being part of the essence of certain designated object(s) but also, and perhaps surprisingly, (ii) an essential link between two natures or terms. The latter is the element that one might have supposed was introduced only by a strong *cop* reading – analogously to the way *de dicto* might be thought to introduce intensional relations between terms or natures. Conversely, one might have expected that the essential term-thing link, as opposed to a term-term connection, was what distinguished a weak *cop* proposition, just as *de re* propositions are thought to predicate a property of a thing rather than of a thing-under-a-description. But in fact, *both cop* readings entail *both* sorts of components.[25] The key difference between them lies rather in weak *cop*'s allowing for the subject term's *accidentally* applying to its designata, along with its being accidentally related to the predicate term.[26]

## 2.11. CONVERSION OF NECESSITY PROPOSITIONS

These results concerning Aristotle's essentialist semantics call for a different approach than those surveyed earlier in regard to the conversion problem for necessity statements. One cannot approach the question with a very broadly conceived notion of logical necessity or possibility and then apply recognized *de dicto* principles (e.g., 'If *p* strictly entails *q*, then nec: *p* entails nec: *q*.') either to obtain a blanket proof of conversion of *A, E,* and *I* necessity and one-way possibility propositions or to show that Aristotle must have had in mind in *Pr. An.* A.3 a *de dicto* reading of modal propositions. Nor can one simply alternate, where necessary, between this and the modal predicate reading on grounds that obtaining the logical results claimed by Aristotle entitles one to conclude that he should be read with modal *dictum* here, and modal predicate there.

But with those props gone, and given the ambiguity, even within a *cop* approach, between strong and weak *cop* necessities (and their counterparts for other modalities), what are we to do about establishing the *starting points* of the formal system – about determining which syllogisms or conversion principles are rightly taken as valid and primary? The answer is that we can consult the underlying relations among genus, species, accident, and proprium for which the modal system is supposed to provide a

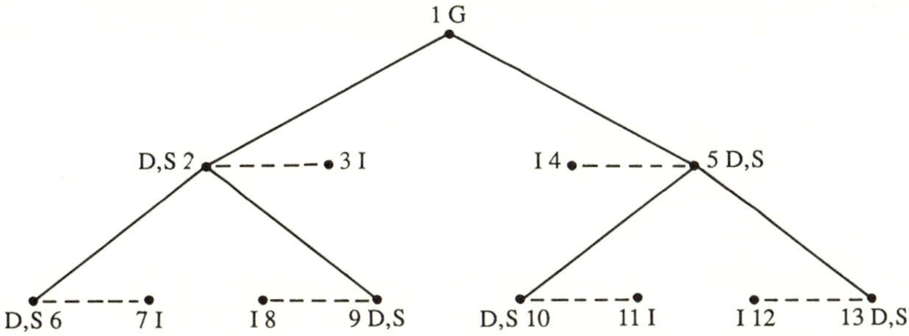


Figure 2.1. G, genus; D, differentia; S, species; I, *idion*

logical calculus. This would be analogous to a contemporary realist about possible worlds inventing from scratch a formal logic, plain and modal, for the purpose of expressing and reasoning deductively about what there is. He might, to start with, come up with a *de dicto* syntax (for reasons touched on earlier). Then he would need to ask himself, among other things, whether or not various principles should be laid down as axiomatic. For example, if it is necessary that *p,* is it necessary that it is necessary that *p?* If it is possible that *p,* is it necessary that it is possible that *p?* And so on. As our possible-worlds realist said yes to certain of these questions, he would add appropriate axioms. Depending on his intuitions in these cases, he could, in effect, have decided that S4 or S5 or some other system is the "logic of being," as it were.

The general situation with Aristotle is analogous, but complicated from the outset by the fact that when one looks critically at his conversion principles, one can see, in light of the underlying semantics of accidental and necessary properties of subjects, that each principle has two natural Aristotelian readings, only one of which converts. But the semantic considerations by means of which this discovery is made are obviously at the same time the means by which one may separate the Aristotelian sheep from the goats – that is, disambiguate, and then test for validity, his modal conversions.

The conversion proofs for strong *cop* $A_n$, $I_n$, and $E_n$ can be somewhat cumbersome to spell out in full.[27] Nonetheless, a simple diagram (Figure 2.1) should show very quickly why these propositions convert. (Notice that most nodes are occupied by both a differentia and a species. Where a species has more than one differentia – as is certainly the case for living things later on in Aristotle's thought[28] – these are all represented by a single 'D'. Also, I have placed each differentia on the same level as the species it differentiates. Depending on how one resolves some nasty questions about differentiae, one might want to place them higher than their corresponding species, but lower than the next genus up. If so, the following remarks could easily be modified in an appropriate way. All *idia* of a given species are represented by a single 'I' located on the same level as its species and connected to it by a horizontal line.)

The four strong *cop* propositions can now be represented in terms of the diagram as follows:

> '$A\ N_s\ a\ B$' is true iff there is a continuous path from node $A$ to node $B$ that does not move upward. (One can go down or sideways, or back and forth sideways, or from one letter to another at the same location, only never upward. Call this a "non-ascending path," or, for short, a "path.")
> '$A\ N_s\ i\ B$' is true iff there is some $C$ such that there is a path from $B$ to $C$ and from $A$ to $C$.
> '$A\ N_s\ e\ B$' is true iff there is no path from $A$ to $B$ or vice versa.
> '$A\ N_s\ o\ B$' is true iff there is some node $C$ such that
> (i) there is a path from $B$ to $C$
> and (ii) $A\ N_s\ e\ C$.

Now to conversion: First, does '$A\ N_s\ a\ B$' entail '$B\ N_s\ i\ A$'? There are many possibilities to consider. If $A\ N_s\ a\ B$, then either

> (1) $A$ is higher than $B$ on some path from $A$ to $B$
> or (2) $A$ is on the same level as $B$ and there is a path from $A$ to $B$.

If (1), then $A$ could be a genus, and $B$ any subspecies of that genus, or any subspecies of a subspecies . . . of $A$, or any *idion* or differentia of any such subspecies. Or $A$ could be an *idion* or differentia, and $B$ some lower species on a common path with $A$, or any differentia or *idion* of any such subspecies.

If (2), $A$ could be related to $B$ as

> a species to one of its own differentiae
> or a species to one of its own *idia*
> or a differentia to its own species
> or an *idion* to its own species
> or a differentia to any *idion* of the same species
> or an *idion* to any differentia of the same species
> or a differentia to any other differentia of the same species
> or an *idion* to any other *idion* of the same species

It would be laborious to work through each possibility, showing in each case that '$B\ N_s\ i\ A$' will hold. Fortunately it is immediately obvious that it will hold. For if (1), $A$ is on a *higher* level than $B$, and there is a path from $A$ to $B$, then we need only find some $C$ such that there is a path from $A$ to $C$ and a path from $B$ to $C$. But we can always take as an appropriate $C$ any letter lower than $B$ and on a path from $B$. The only case in which this would not work would be that in which there was nothing lower than $B$ (on a path from $B$). But if that were the case, then $C$ could be a differentia or species or *idion* on the same level with $B$. (Whichever one of these $B$ may be, let $C$ be one of the others.) Either way, we shall have a $C$ that is both on a path from $B$ (so that $B\ N_s\ a\ C$) and on a path from $A$ (so that $A\ N_s\ a\ C$). To verify that there is a path from $A$ to $C$, we used the obviously correct assumption that 'being on a path from' is transitive.

Second, we must consider the case in which $A$ is on the *same level* as $B$ and there is a path from $A$ to $B$. Here we can pick as $C$ anything lower than $B$ that is on a path from $B$ or anything other than $B$ and $A$ that is on the same level as $B$ and is on a path from $B$. Either way, '$B\ N_s\ i\ A$' will hold.

In sum, if $A\ N_s\ a\ B$, then $B\ N_s\ i\ A$; that is, strong *cop* $A_n$ converts "to a particular."

The proof for strong *cop* $I_n$ is too similar to need separate discussion.

If the universal negative '$A\ N_s\ e\ B$' is true, then in terms of the diagram, one must either jump a gap (as in going directly from node 3 to 4 in Figure 2.1) or move along solid lines upward and downward to get from $A$ to $B$. These are the only ways to get from $A$ to $B$ if, as stated in the definition of '$A\ N_s\ e\ B$', $A$ and $B$ do not lie on any common path. But both of these conditions are obviously symmetrical. So given '$A\ N_s\ e\ B$', it holds that '$B\ N_s\ e\ A$'. We can get the same result even more directly, but by a less scenic route: If there is no path from $A$ to $B$ or vice versa, then there is no path from $B$ to $A$ or vice versa. So strong *cop* $E_n$ converts.

Strong *cop* $O_n$ does not convert (i.e., '$A\ N_s\ o\ B$' does not entail '$B\ N_s\ o\ A$'). For a counterexample, suppose that $A$ and $C$ are coordinate subspecies of the genus $B$. Then it will be true that $A\ N_s\ o\ B$, for $C$ will be "some $B$" such that $B\ N_s\ a\ C$ and $A\ N_s\ e\ C$ (e.g., let $A$ = 'S' of node 6 = Human, $B$ = node 1 = Animal, $C$ = 'S' of node 10 = Horse), but not that $B\ N_s\ o\ A$.

So all strong *cop* necessity propositions convert in the way Aristotle says his necessity propositions convert. It is important also that for anyone thinking in terms of Aristotelian species-genus trees, it will be immediately obvious, despite the large number of pairs of items relating *per se* to one another, that these propositions will convert as Aristotle describes.

To any readers who may be uncomfortable with these proofs, carried out as they are with pictorial aids – or, alternatively, via lengthy expositions in a natural language – rather than as deductions within a familiar and trusted formal system, I would say first that if we assume certain Aristotelian principles concerning the small number of possible predicative relations between terms, it is possible to determine quite rigorously, given that *A* relates to *B* in a certain way, whether or not *B* relates to *A* in a specified way. Once the relevant principles about species, genus, and so forth, are laid down, and Aristotle's semantically ambiguous modal statements are disambiguated, the job of testing various conversions is in fact rather mechanical. The important point is that these proofs do show that *there can be no counterexamples* to certain of Aristotle's conversions (interpreted in specified ways).

Second, in verifying the valid conversions on strictly Aristotelian (semantic) grounds, we secure important building blocks for a formal model of Aristotle's modal syllogistic. As we obtain various "complete" modal syllogisms to go along with those conversions, we shall, in effect, build up a model that is based directly on Aristotelian principles and within which we can construct deductions reflecting Aristotle's text step by step.

## 2.12. *DE DICTO* CONVERSION AS PARASITIC ON STRONG *COP*

Finally, one can go further than simply replacing *de dicto,* as it were, with strong *cop* necessity, for in fact the conversion of strong *cop* $A_n$, $I_n$ and $E_n$ will *explain* the conversion of the corresponding *de dicto* statements. Although many strong *cop* statements whose *de dicto* counterparts are true will themselves be false, it is clear that the truth of the former entails the truth of the latter.

One might put it this way: *A* being true "by (Aristotelian) definition" of all *B*'s to whose essence it belongs ($A\ N_s$ all *B*) entails the *necessity* of the *dictum* that *A* applies to all *B*'s (nec: *A* all *B*). Therefore, because a *strong cop* proposition validly converts, so will its *de dicto* counterpart. That is, a strong *cop* statement to the effect that '$A\ N_s$ all *B*' will entail the *de dicto* statement 'nec: *A* all *B*' and also (*via* the strong *cop* conversion proved earlier) '$B\ N_s$ some *A*'. But '$B\ N_s$ some *A*' entails, *via* its definitional component, the corresponding *de dicto* truth 'nec: *B* some *A*'. In this way one can show by using the conversion of strong *cop* necessity (along with the assumption that strong *cop* definitional propositions entail the corresponding *de dicto* necessary propositions) that in the case of any *de dicto* counterpart to a strong *cop* statement that does convert, if the original *de dicto* statement is assumed true, then its converse must also be assumed true. However, its conversion will be, from the Aristotelian point of view, a surface phenomenon, for it ('nec: Animal all two-footed', for example), as well as its converse ('nec: two-footed some Animal'), will derive from underlying strong *cop* propositions ('Animal $N_s$ all two-footed' and its converse, 'two-footed $N_s$ some Animal'). The crucial conversion is the one at the strong *cop* level, where Aristotelian concepts of genus, species, and so forth, and their relationships to one another come into play. This is not to deny the validity of the modern principle that if *p* strictly entails *q*, then nec: *p* entails nec: *q*, by which one can directly prove the conversion of *de dicto* necessities. It is only to say that (a) it is no accident that this principle gives end results, at least as regards conversion, exactly parallel to our results for Aristotle's strong *cop* necessity and (b) it does so without correctly representing Aristotle's notion of (strong *cop*) necessity, or revealing why such propositions convert as they do, or how they are related logically to other modal propositions of the system.

With these results about conversion in hand, we are now in a position to consider the pure necessity and mixed assertoric/necessity syllogisms of *Pr. An.* chapters 8 and 9, respectively.[29]

# *Chapter 3*

# Syllogisms with two necessity premises

Aristotle's discussion of modal syllogisms opens in chapter 8 with those composed entirely of necessity premises and conclusions. His larger ground plan for discussion of modal syllogisms cannot fairly be deemed whimsical: He proceeds methodically to those with one necessity premise and one plain premise in the first, second, and third figures (ch. 9, 10, and 11), and then, chapter by chapter, to all the various premise permutations that result from substituting at least one necessity or contingency proposition for the premise(s) of a valid assertoric syllogism:

| | Figure | | |
|---|---|---|---|
| Premise combination | 1 | 2 | 3 |
| Necessity/Necessity | 8[a] | 8 | 8 |
| Plain/Necessity | 9 | 10 | 11 |
| Two-way/Two-way | 14 | 17 | 20 |
| Plain/Two-way | 15 | 18 | 21 |
| Necessity/Two-way | 16 | 19 | 22 |

[a]Chapter number.

The treatment of pure necessity syllogisms is extremely compressed. We may begin by translating the whole of Aristotle's brief chapter 8, and then look more closely at some of the details of the text and their implications. Although there is always a certain value in discovering merely which syllogisms work and which do not, we shall be primarily occupied throughout with a series of highly engaging related issues: (1) an unexpected violation of the age-old *peiorem* rule (a rule first formulated by Theophrastus, not by Aristotle), which says that the strength of the conclusion (including its modal strength) must always be no greater than that of the weaker premise – a violation unexpected not because rules are rules but because it seems unlikely on the face of it that a strong *cop* proposition should be derivable from anything less than two strong *cop* premises; (2) the role of existential import in validating necessity syllogisms; (3) an ingenious modal proof technique of Aristotle's that unfortunately turns out to be invalid; (4) the particular use of *ekthesis* in validating Baroco and Bocardo *NNN*, moods that cannot be established *via* conversion or *reductio ad impossibile* proofs; (5) an important general feature of *ekthesis* proof that was first noted if not by Aristotle then by Alexander; (6) the surprisingly complex matter of the bearing of chapter 8 on the theory of scientific demonstration elaborated in the *Posterior Analytics*. First, the text as we have it:

29b29 Since to apply (*huparchein*), and to necessarily apply, and to possibly apply are different (for there are many things which apply, but which do not apply of necessity, and others which neither apply of necessity nor apply at all, but still possibly apply) it is clear that the syllogisms from these will be different, and the terms will not be alike, one ⟨syllogism⟩ being from necessary ⟨premises⟩, another from assertoric ones, another from possible ones.

29b36 With regard to necessary ⟨premises⟩ things are virtually the same as with those of belonging. For if the terms are placed in the same way in the premise of belonging (*en . . . tōi huparchein*) and in that of necessarily belonging (*tōi ex anagkēs huparchein*) or necessarily not belonging, there will or will not be a syllogism ⟨in the same way⟩, except that they will differ through the addition of 'necessarily belonging' or '⟨necessarily⟩ not belonging' to the terms.

30a2 For the ⟨universal⟩ negative will convert in the same way, and "being in the whole of" and "⟨applying⟩ to all" will be defined in the same way. In the others [outside the first figure], then, the conclusion
a5 will be proved necessary in the same way as with the syllogisms of belonging, via conversion.

30a6 In the middle figure, however, when the universal ⟨premise⟩ is affirmative and the particular negative [Baroco *NNN*], and again in the third, when the universal is affirmative and the particular negative [Bocardo *NNN*], the proof will not be the same, but one must set out ⟨that⟩ to which each ⟨of the predicates in the particular negative premises⟩ does not belong, and construct the syllogism about this. For it will be necessary in the cases of these ⟨things set out⟩. And if it is necessary of that which has been set out (*kata tou ektethentos*), then it is necessary also of some of that ⟨from which the selection was made⟩. For that which is set out just is some of that
a13 (*to gar ektethen hoper ekeino ti estin*). And each of these syllogisms comes about in its own figure.[1]

# *3 Syllogisms with two necessity premises*

## 3.1. THE GENERAL PARALLEL TO ASSERTORIC SYLLOGISMS

The entirety of chapter 8 is part of a larger Aristotelian strategy of building upon his results in chapters 4–7 with regard to plain syllogisms.[2] So Aristotle writes not only that the modal premises and conclusions will be just like the plain ones, except that 'necessarily belongs' (rather than simply 'belongs') is added to the terms, but also that the pure necessity syllogisms themselves will be valid or invalid just where their plain counterparts were. Moreover, one will validate second- and third-figure syllogisms in the same way one did their plain counterparts – by conversion of $A_n$, $I_n$, and $E_n$ (where before, one converted plain *A, I*, and *E*) to effect a "reduction" to the perfect moods of the first figure. Aristotle notes two exceptions: Baroco *NNN* and Bocardo *NNN* will not be proved by *reductio* arguments, as with their plain counterparts;[3] instead, one must use the proof technique called *ekthesis.* Finally, it is implicitly clear that the four first-figure necessity syllogisms will be "perfect" or "complete" (just as with plain Barbara, Celarent, Darii, Ferio), while the rest will be validated – and this part is explicit – *via* appropriate conversions or, as just remarked, in two cases, by *ekthesis* proofs. Again, by "complete" or "perfect" (*teleios*) Aristotle means not only that a syllogism is valid but also that its validity is evident or obvious (*phaneros*) on the basis of the premises as given; "nothing further is needed . . . to make the necessity [of the conclusion's following] obvious" (see 24b22–26). In all of these matters, Aristotle adheres as closely as possible to the procedures of chapters 4–7 on plain syllogistic.

A final, terminological, point: Aristotle observes throughout the modal chapters a distinction between a conclusion's following of necessity from the premises and the conclusion itself being a necessity proposition. That is his consistent *practice;* he also states the point explicitly in chapter 10 in terms of a conclusion's "being necessary if these [premises] are the case" (*tinōn ontōn anangkaion*) as opposed to its being "necessary without qualification" (*anangkaion haplōs*, 30b31–33, 38–40).[4] (The former expresses the notion of logical consequence reflected already in the general definition of a syllogism in *Pr. An.* A.1. Aristotle treats the notion as what we would call a "primitive": He gives many examples of it, some of them "obvious," but no explicit definition.[5]) Of course, the distinction between strong and weak *cop* is a different distinction from that between "being necessary if the premises are the case" and "being necessary without qualification"; the weak vs. strong *cop* distinction applies *within* the category of propositions "necessary without qualification."

## 3.2. FIRST-FIGURE SYLLOGISMS

### *3.2.1. Weak* cop

The first-figure pure weak *cop* necessity moods are all valid, given – as Aristotle remarks – the definition of "belonging (necessarily) to all" and "to none." Consider weak *cop* Barbara: If *A* necessarily applies to everything to which *B* applies, and *B* necessarily applies to everything to which *C* applies, then obviously *A* will necessarily apply to every *C*. Or, if every *C* is necessarily a *B*, and if every *B* is necessarily an *A*, then every *C* is necessarily an *A*. (It may have occurred to the reader that the conclusion will follow even if the minor premise is a plain rather than a necessity statement. But we shall let that notorious sleeping dog lie for one more chapter.) The validity of Celarent $N_wN_wN_w$,[6] Darii $N_wN_wN_w$, and Ferio $N_wN_wN_w$ is equally obvious.

### *3.2.2. Strong* cop

In the first figure, two strong *cop* premises will – again obviously – entail a strong *cop* conclusion:

$$\frac{\begin{array}{c} A\ N_s\ a\ B \\ B\ N_s\ a\ C \end{array}}{A\ N_s\ a\ C}$$

Here the definitional component of the premises will guarantee that the terms *A* and *B, B* and *C* – hence *A* and *C* – occupy places in a common definitional tree (e.g., *A* = Living thing, *B* = Animal, *C* = Human). The only technical point of interest is that the definition of '$A\ N_s$ all *B*' given in Chapter 2 needs to be slightly loosened. With the particular substitution of terms just suggested, *A* (Living thing) will not be part of the definition *per genus et differentiam* of *C* (Human), nor will their common highest genus, Substance. One wants to reply that *A* and *C* are nonetheless members of the same definitional hierarchy or chain, of which, in the present example, Human is the lowest link, and *that* is what we really want to capture – for example, in "scientific" contexts – by certain strong *cop* syllogisms. This basic intuition is perfectly sound and can be implemented logically by appeal to what would nowadays be called an "ancestral" relation, namely, that *A* must be included in the definition of *C*, or in the definition of something in the definition of *C*, or in the definition of something in the definition of something in the definition of *C*, and so on, so that *C* is a (definitional) "descendant" of *A*. Introducing this into the definition of strong *cop* necessity allows us to express the fact that every member of an Aristotelian definitional hierarchy will have a positive definitional link, immediate or otherwise, to all the subordinate and superordinate members of that same hierarchy, even though a given term or its definition will not necessarily overlap with the definitions of all other terms in the same hierarchy. Aristotle achieved this top-to-bottom linkage in the *Categories* by making the "said of" relation transitive (1b10–11).

This makes possible the construction of an extended *sorites* (to use the traditional term: a chain of linked two-premise syllogisms) out of pure strong *cop* propositions in a way exactly parallel to that for weak *cop* and assertoric sorites. This may be of only occasional import with the latter sorts of propositions, but insofar as demonstrative understanding is based on sorites composed of strong *cop* propositions, it will be crucial for setting out the results of Aristotelian science.

Strong *cop* Celarent is also valid and perfect:

$$\frac{\begin{array}{c} A\,N_s\,e\,B \\ B\,N_s\,a\,C \end{array}}{A\,N_s\,e\,C}$$

By the minor premise, *B* is part of the essence of each *C* as such. And by the major premise, *A* itself is incompatible with *B* itself. Thus *A* is incompatible with something (*B*) that is part of the essence of each *C* as such – thus also with *C* itself. Hence, by the definition of the strong *cop* universal negative, $A\,N_s\,e\,C$. [In the language of the four predicables, *A* will be something incompatible with all *B*'s as such (by the first premise) and hence also with any genus, species, differentia, or proprium that itself entails *B*, including (by the second premise) *C*. Thus, $A\,N_s\,e\,C$.] Validation of Darii and Ferio $N_s\,N_s\,N_s$ will follow along obvious lines.

In sum, all of Aristotle's allegedly complete pure necessity syllogisms are indeed valid (and complete or perfect) on either a strong or weak *cop* reading.

## 3.3. STRONG *COP* AND SCIENTIFIC DEMONSTRATION

Aristotle's discussion in *Post. An.* A.4–10 of the component statements of scientific demonstrations makes it clear that in a large range of cases, though perhaps not all, these will be strong *cop* assertions, and the scientific mood par excellence will be a pure strong *cop* syllogism in Barbara. The principal reason for the latter is that demonstration will aim primarily, although not exclusively, at establishing universal affirmative conclusions, and Barbara is the only syllogism by which this can be accomplished.[7] That scientific premises and conclusions will be strong *cop* rather than weak *cop* necessities follows from the fact that their terms must be related *per se* if the premises are to *explain* the *per se* connection asserted in the conclusion. These *per se* connections include, as we noted earlier, those between a species and its propria – and those among the propria themselves – as well as the definitional ones among genus, differentia, and species. By contrast, propositions in which weak *cop* necessity holds, but strong *cop* fails (e.g., Cat $N_w$ all White Thing on the Mat), will in fact contain terms bearing only an ''accidental'' relationship to one another, and so will not figure in (explanatory) scientific demonstrations.

One welcome implication of this is that the weak *cop* premise (e.g., 'Animal $N_w$ all White'), which does *not* convert, and which therefore cannot be used in Aristotle's validations of syllogisms *via* reduction to the first figure, simply cannot arise in a strictly scientific context. On the other hand, as was shown in Chapter 2, strong *cop* $A_n$, $I_n$, and $E_n$ do convert, so that, insofar as one wants to use scientific demonstrations in the second or third figure, these can be validated *via* Aristotelian conversion proofs.

These observations can also help us avoid a common oversimplification of Aristotle's conditions on scientific demonstration. It is frequently said (by Aristotle, among others, e.g., *Post. An.* A.6, 74b26–30) that the basic scientific demonstration is one containing two necessity premises and concluding validly to a *per se* relation between its extreme terms. But the following syllogism will not constitute a scientific demonstration even though it is valid and consists entirely of true propositions of necessity:

Animal *N* all Cat
Cat *N* all White Thing on the Mat
---
Animal *N* all White Thing on the Mat

Even in a situation in which its premises and conclusion are true, the *terms* of the conclusion are nonetheless accidentally related. Nor can the premises say *why* the conclusion *must* be true: for the minor premise (like the conclusion!) may happen to have been false. Again, the point is that even some of Aristotle's own examples of true propositions of necessity could not be part of any ''scientific'' (*apodeictikē*) demonstration. (For this reason I have avoided the common but misleading practice of calling necessity propositions in general "apodeictic propositions." Aristotle himself uses *anangkē* or some cognate to describe his propositions of necessity.) Thus the appropriate necessary condition on such demonstration would not be simply "only propositions of necessity," but rather "only strong *cop* propositions of necessity." (Again, scientific demonstration involving two-way possibility will be examined in Chapter 6.)

## 3.4. THE SURPRISING STRENGTH OF SOME FIRST-FIGURE MIXED *COP* MOODS AND THEIR RELATION TO SCIENTIFIC DEMONSTRATION

It may seem obvious that a strong *cop* conclusion requires two strong *cop* premises. After all, how could one infer an essential link between major and minor terms when one or the other is linked only accidentally to the middle? But just here the modal system has a surprise in store. Certainly, to recall to mind our late friends the modal *dictum* and modal predicate readings, *any* modal *dictum*, modal predicate/modal *dictum* syllogism in Barbara will be invalid:

*N*(Animal all Human)
*n*Human all White Thing at 10 Downing Street

*N*(Animal all White Thing at 10 Downing Street)

In the possible situation in which Churchill is the only white thing at 10 Downing Street, both premises will be true, and the conclusion false. One can readily supply counterexamples also to the mixed syllogism with modal predicate major and *de dicto* minor.

The same holds, as one would expect, for strong *cop*, weak *cop*/strong *cop* moods:

Animal $N_s$ all Human
Human $N_w$ all White Thing in the White House

Animal $N_s$ all White Thing in the White House

Again, it is quite possible that both premises be true and the conclusion false. The same goes for Celarent, Darii, and Ferio $N_s,N_w/N_s$.

But consider Barbara with weak *cop* major and strong *cop* minor:

Animal $N_w$ all White
White $N_s$ all Snow

Animal $N_s$ all Snow

(Let us suppose, for the sake of the argument, that the minor premise is true and also put aside momentarily our proviso [Section 2.8] concerning such terms as 'White'.) Notice that these same terms will show invalid the corresponding modal predicate, modal *dictum*/modal *dictum* mood. And in the *cop* version just formulated, the conclusion is false. But *can* the premises both be true (thus completing a counterexample)? A moment's reflection shows that they cannot, for Aristotle's universal affirmatives presuppose that there do exist some objects to which their subject terms refer. (Because *I* propositions carry existential import, and convert, and *A* propositions convert *kata meros* to *I*'s, then *A* propositions must also carry existential import for both terms.) But if there is snow, then there are white things (*via* the second premise). And if there are white things, then there are animals (by the first premise). But then the premises are already incompatible with one another, for even if we weaken both premises to their plain assertoric versions, they imply (*via plain* Darii, whose validity is obvious) that some snow is (an) animal, which is impossible. So this particular modal version of Barbara cannot be invalidated by this set of concrete terms. (This is not yet, of course, to show the syllogism valid.)

Similarly, if one were to give the same existential import to universal *de dicto* and modal predicate necessity premises, then the premises of

*n*Animal all White
*N*(White all Snow)
―――――――――――
*N*(Animal all Snow)

*could* not any longer be true together. If there is snow, hence (by the minor premise) something white, hence (by the major) some Necessary-animal, it follows that Necessary-animal applies to all Snow, which is impossible. This is not necessarily to say that this syllogism is *valid* either, but only that one promising counterexample fails, once we grant the premises existential import.

As to the validity of our weak, strong/strong *cop* syllogism, one still supposes that a counterexample can be found. This looks like one (in a possible situation in which all white things are dogs):

Dog $N_w$ all White
White $N_s$ all Antique White
―――――――――――
Dog $N_s$ all Antique White

Clearly, there is a problem with this example. The first premise is true only if White functions as an adjective, so that the subjects of which Dog is predicated are all things that are colored white. By contrast, White must be taken as a noun in the second premise (if that premise is to come out true): The kind of white color called Antique White is not itself a white-colored entity. The premise says, rather, that Antique White is a kind of white color. So, in effect, we have two terms corresponding to the word 'white': 'thing colored white' in the major premise, and 'white color' in the minor. Thus the syllogism as a whole, having four terms, is ill-formed. We could correct for this by changing either term, but that would render one premise or the other false. So this proposed counterexample fails, too.

These failures are entirely predictable once one thinks through the situation in semantic terms – here, Aristotelian metaphysical terms – for then one sees that the syllogism is in fact valid, and why this should be so. (In principle, one might think of validating the syllogism by a *reductio* argument. But the reducing syllogism, starting with the contradictory of the conclusion of the syllogism to be reduced, would not be expressible in the Aristotelian apparatus available to us.) If we have a true proposition '$B\ N_s$ all $C$', then $B$ and $C$ must both be included in the essence of the subject ($C$) they introduce, whether these be substances or such non-substances as kinds of color or particular instances of color (e.g., 'Animal $N_s$ *a* Human', 'Color $N_s$ *a* White Color', 'White Color $N_s$ *a* Colonial White Color'). Thus, given $B\ N_s\ a\ C$, $B$ will belong to the *logos* of the essence of the $C$'s, and $B$ and $C$ will both belong to that of the $C$'s. Turning to the major premise, if it is also true that $A\ N_w$ all $B$, then $A$ will belong to the essence of the $B$'s, and hence (given the minor premise) of the $C$'s, and so will belong to the *logos* stating $C$'s essence. So now not only do $B$ and $C$ belong to a common definitional tree – that stating the essence of the $C$'s as such – but $A$ and $C$ also belong to a common definitional chain (because they both belong to the definition of the essence of the $C$'s). So now not only do $A$ and $C$ belong to a common definitional tree, but given the universal affirmative nature of the premises, it will hold that $A\ N_s$ all $C$. (This proof presupposes the obviously correct principle that 'belonging to the same definitional tree' is transitive.) Recall also that with *idia* included among the things predicated *kath' hauto*, we could speak of an essential chain rather than of a narrower "definitional" chain.)

The key difference from the case of the invalid Barbara $N_sN_wN_s$

$$\frac{\begin{array}{c} A\ N_s \text{ all } B \\ B\ N_w \text{ all } C \end{array}}{A\ N_s \text{ all } C}$$

is that although in that case the premises do show that $A$ is included in the definition of the (essence of the) $C$'s, they do not show that $C$ is so included. For all those premises say, $C$ might be related only accidentally to the $C$'s. This is not precluded by the truth of the weak *cop* '$B\ N_w$ all $C$', as it is by '$B\ N_s$ all $C$': Let $A$ = Animal, $B$ = Horse, and $C$ = Brown, where all brown things are horses and horses are accidentally brown. So the critical factor in establishing that $A$ and $C$ are connected *per se* is that they belong to a common definitional chain (that of the $C$'s). The unexpected result is that this can be shown in the first figure even without appeal to two strong *cop* premises – so long as the strong *cop* premise is the minor, and the weak *cop* is the major premise.

Treatment of the remaining three moods of this figure will now go smoothly. Celarent, Darii, and Ferio with strong *cop* major and weak minor are all invalid; with weak *cop* major and strong minor they are, like Barbara, valid. Although these valid $N_wN_sN_s$ syllogisms are of interest for the insight they provide into certain surprising connections between Aristotle's essentialist semantics and his modal syllogistic, they are still, despite their strong *cop* conclusion and two necessity premises, inadequate for the purpose of scientific demonstration. The reason, once again, is that the weak *cop* premise does not guarantee anything more than an accidental relation between its terms. Thus in none of these syllogisms, valid or invalid, can the links between the middle term and the two extreme terms give a principled explanation of why the major and minor terms are essentially related. But to conclude this section on a more positive note, we may take the valid mixed moods as showing that systematic disambiguation of Aristotle's propositions of necessity need not lead to two separate modal systems: The underlying semantics has shown how strong and weak *cop* premises can be combined to produce valid arguments.

## 3.5. SECOND-FIGURE SYLLOGISMS

### 3.5.1. *Weak* cop

Putting aside Baroco for the moment, we have three weak *cop* syllogisms:

| Cesare $N_wN_wN_w$ | Camestres $N_wN_wN_w$ | Festino $N_wN_wN_w$ |
|---|---|---|
| $B\ N\ e\ A$ | $B\ N\ a\ A$ | $B\ N\ e\ A$ |
| $B\ N\ a\ C$ | $B\ N\ e\ C$ | $B\ N\ i\ C$ |
| $A\ N\ e\ C$ | $A\ N\ e\ C$ | $A\ N\ o\ C$ |

Although conversion proofs will not work here (because no weak *cop* proposition converts), it may seem intuitively obvious that these are all valid: If, as in Cesare $N_w N_w N_w$, *B* is incompatible with something in the essence of the *A*'s, and *B* necessarily belongs to something in the essence of the *C*'s, it would seem to follow that no *C* could possibly be an *A*, for then at least one thing (some *C* that could be an *A*) would have a nature or essence that was both compatible with and incompatible with *B*.

There is a mistake, however, in that beguiling argument. What the premises of Cesare, for example, do establish is that there is, in the essence of each *A*, something incompatible with *B* and that there is, in the essence of each *C*, something that entails *B*. And this shows that for any actual individuals *a* and *c*, there is something in the essence of that particular *a* that is incompatible with something in the essence of that particular *c*. But this does not show that *A itself* is incompatible with anything in the essence of any individual *c*, nor that *C itself* is incompatible with anything in the essence of any individual *a*. Let the middle term *B* = Human, and let *A* = White Thing in the Barn, and *C* = Wakeful Thing, in a situation in which all white things in the barn are horses and all wakeful things are rational animals:

Human *N e* White Thing in the Barn (horses)
Human *N a* Wakeful Thing (rational animal)

---

White Thing in the Barn *N e* Wakeful Thing

Then both premises will be true: None of the things (horses) that are in fact *A*'s (white things in the barn) could possibly be *B* (human), and all the things that are in fact awake (rational animals) are necessarily *B* (human). But *A* itself (White Thing in the Barn) is still two-way possibly applicable to the things (rational animals) that are in fact *C* (Wakeful Thing), and *C* itself is two-way possibly applicable to all the things (humans) that are in fact *A*. Finally, *A* and *C* are compatible with one another. These possibilities are all left open by the fact that in all these moods the terms *A* and *C* appear in weak *cop* premises, and only in the logical subject position, so that they may, for all the premises say, apply only incidentally to their subjects. Here consideration of the underlying essentialist semantics shows clearly why, despite their logically tempting appearances, these syllogisms are invalid. Similar considerations (and counterexamples) apply to the rest of the weak *cop* second-figure syllogisms.

### 3.5.2. *Strong* cop

Aristotle says of these moods merely that they are valid on the basis of reduction, via conversions, to the first figure – the sole exception being Baroco, for which one needs an *ekthesis* proof. On a strong *cop* reading, things do work out pretty much that way: Camestres, Cesare, and Festino $N_sN_sN_s$ are all valid, because all reduce to the first figure *via* appropriate conversions (shown valid in Chapter 2) of $E_n$ and $A_n$. Baroco we shall take up later (Section 3.7), along with the *ekthesis* proof for Bocardo.

### 3.5.3. *Mixed strong/weak* cop

The situation with mixed *cop* syllogisms of the second figure is interestingly different from what we encountered in the first figure. With Camestres, there would be two arrangements to check for the possibility of deriving an $N_s$ conclusion from a mixed strong/weak premise pair. These are again worth looking at, for what they show about the interplay between Aristotle's modal logic and his metaphysics of genus, species, and so forth:

$$\frac{B\,N_s\,a\,A \quad B\,N_w\,e\,C}{A\,N_s\,e\,C} \qquad \frac{B\,N_w\,a\,A \quad B\,N_s\,e\,C}{A\,N_s\,e\,C}$$

But letting $B$ (middle) = Animal,
$A$ = Human,
$C$ = White Thing on the Mat,
in a situation in which all white things on the mat are cloaks, we have:

$$\frac{\text{Animal}\,N_s\,a\,\textit{Human} \quad \text{Animal}\,N_w\,e\,\text{White}\ldots}{\text{Human}\,N_s\,e\,\text{White}}$$

In that possible situation the premises are true and the conclusion is false.

But letting $B$ = Human,
$A$ = White Thing on the Mat,
$C$ = Horse,
in a situation in which all white things on the mat are human, we have:

$$\frac{\textit{Human}\ \ N_w\,a\,\text{White}\ldots \quad \text{Human}\ \ N_s\,e\,\text{Horse}}{\text{White}\ldots\ N_s\,e\,\text{Horse}}$$

In that possible situation the premises are true and the conclusion is false.

So both are invalid. Similar counterexamples will show the remaining second-figure hybrids with strong *cop* conclusions invalid.

But why is it that unlike the first figure, the second figure yields no strong *cop* conclusion on any combination of strong with weak *cop* premises? Recall first-figure Celarent $N_wN_sN_s$:

$$\frac{\begin{array}{c} A\,N_w\,e\,B \\ B\,N_s\,a\,C \end{array}}{A\,N_s\,e\,C}$$

Our reasoning established, in effect, that *A* was essentially incompatible with whatever was in the essence of the *C*'s and that *C* was in the essence of the *C*'s. Thus, given the premises, the connection between *A* and *C* could not have been accidental; rather, it was one of essential exclusion. And for Barbara we were able to show that *A* and *C* must both be in the definitional tree of the *C*'s. Thus in both cases (and with Darii and Ferio) we were able to establish that neither *A* nor *C* was accidentally related to the *C*'s: In one case, both were necessarily entailed by the definition of the *C*'s; in the other, one was entailed, and the other excluded.

But in all our *second-figure* premise pairs, the placement of the terms is such that nothing precludes whichever one of the extreme terms (*A, C*) is included in the weak *cop* premise from being an accidental property of all its denotata. Thus there is no way to show either that they are both included in the definition of any common subject or that one is included in some definition with which the other is incompatible. For example, in Cesare $N_wN_sN_s$ ($B\,N_w$ no *A*, $B\,N_s$ all *C*), the major term *A* may, for all the premises say, be only accidentally applicable to both the *A*'s and the *C*'s, and to *C* itself; with Cesare $N_sN_wN_s$ ($B\,N_s$ no *A*, $B\,N_w$ all *C*), *C* may be an accident of the *C*'s and related only accidentally to *A;* in Camestres $N_wN_sN_s$ ($B\,N_w$ all *A*, $B\,N_s$ no *C*), *A* may be an accident of the *A*'s and of the *C*'s; and so on. In none of these cases will it be possible to bring *A* and *C* themselves into any essential relation, whether of incompatibility or entailment, to one another.[8]

## 3.6. THE THIRD FIGURE AND THE EVEN MORE SURPRISING STRENGTH OF SOME WEAK *COP* PREMISES

In the third figure (putting aside Bocardo for the moment), all the syllogisms recognized by Aristotle are valid and can be proved valid in the way Aristotle wants, on a *strong cop* reading. These all reduce to the first figure *via* conversion of an $I_n$ or $A_n$ premise. Disamis $N_sN_sN_s$ requires conversion of the conclusion as well, but this is in order because both premises are strong, and hence entail a strong $I_n$ conclusion, which will convert.

On a weak *cop* reading, these syllogisms cannot be validated *via* conversion (again, no weak necessity proposition converts), but unlike their second-figure counterparts, they are in fact valid. Because one cannot appeal, with Aristotle, to conversion proofs, one must turn to proofs by *ekthesis* or by *reductio ad impossibile.* Such proofs can be easily supplied. Let us consider just one example, Darapti $N_wN_wN_w$:

$$\begin{array}{ll} (1) & A\,N_w\,a\,B \\ (2) & \underline{C\,N_w\,a\,B} \\ (3) & A\,N_w\,i\;C \end{array}$$

Suppose $A\;p\;e\;C$ ($A$ one-way possibly fails to apply to every $C$), the contradictory of (3), and combine this with (2):

$$\begin{array}{l} A\;P\;\;e\,C \\ \underline{C\,N_w\,a\,B} \\ A\;P\;\;e\,B \end{array}$$

This syllogism is obviously valid, and its conclusion contradicts (1). So we now have a *reductio* proof of the validity of Darapti $N_wN_wN_w$. Notice that this sort of proof is not available within Aristotle's system as he left it, because, once again, he did not treat syllogisms with one-way possibility premises. An *ekthesis* proof will work just as well, however, and can be carried out within the system: By (1), $A$ necessarily applies to every $B$, including the individual *b;* by (2), $C$ necessarily applies to every $B$, including *b;* hence there is some $C$ (namely, *b*) to which $A$ necessarily applies (i.e., $A\,N_w\;i\;C$). This parallels Aristotle's *ekthesis* proof for plain Darapti (28a23–26). (There he also gave conversion and *reductio* proofs; here only *ekthesis* would work.) Notice that in both cases the ekthetic proof must set out an individual $B$*:* Setting out a group of $B$'s to all of which both $A$ and $C$ applied would only give us another case of Darapti, the mood to be established. (We shall return to this point in Section 3.7.) Similar proofs could be supplied for the other valid third-figure weak *cop* necessity moods.

We are now prepared to contemplate the most astounding and unheard-of feat ever performed by a pair of weak *cop* necessity premises – an exploit possible only in the third figure – namely, entailment of a strong *cop* conclusion.

Consider, if you will, the amazing Darapti $N_wN_wN_s$:

$$\begin{array}{l} (1)\ A\ N_w\ a\ B \\ (2)\ \underline{C\ N_w\ a\ B} \\ (3)\ A\ N_s\ i\ C \end{array}$$

This deduction cannot be validated within Aristotle's system, either by converting a premise, because neither converts, or by use of a *reductio*, which would require a syllogism with a one-way possibility premise. For the benefit of those who (understandably!) doubt that two weak premises could ever entail a strong conclusion, we must once again consult Aristotle's semantics: (1) tells us that $A$ is part of the essence of the $B$'s (whether or not $B$ itself is); (2) tells us that $C$ also is part of the essence of the $B$'s. But if the natures $A$ and $C$ themselves are both part of the essence of the $B$'s, then $A$ and $C$ will belong to a common definitional chain, namely, that of the essence of the $B$'s. This may hold even if $A$ and $C$ do not belong to the definition of the nature $B$ itself, and $B$ itself does not belong to the essence of the $B$'s (e.g., let $A$ = Animal, $C$ = Human, and $B$ = White, where all white things are humans). But if $A$ and $C$ belong to a common definitional tree, then no matter which may be higher on the tree, it will hold that $A\ N_s$ some $C$.

By contrast, the *first*-figure Barbara $N_wN_wN_s$

$$\begin{array}{l} A\ N_w\ a\ B \\ \underline{B\ N_w\ a\ C} \\ A\ N_s\ a\ C \end{array}$$

is, as one might intuitively have expected, invalid. But because we have seen that such intuitions can go wrong, let us see exactly why this pair of weak premises cannot entail the stated conclusion. For all the premises tell us, $C$ might relate incidentally to the $C$'s, and also to $A$ and to $B$. So there are no grounds at all for placing *any* two of our terms in a common definitional tree. Thus this mood should be liable to counterexamples. And so it is: Let $A$ = Animal, $B$ = Human, and $C$ = Walking, in a situation in which all things walking are humans: $A\ N_w\ a\ C$ will follow, but not $A\ N_s\ a\ C$.

However (another surprise), the premises just considered do entail weak $A\ N\ a\ C$, which combines with the minor premise $B\ N\ a\ C$ to give $A\ N_s\ i\ B$ by Darapti $N_wN_wN_s$, the third-figure mood shown valid earlier. So we can, by auxiliary use of a third-figure mood, obtain a strong *cop A–B* conclusion from the premises of Barbara $N_wN_wN_s$, if not the standard $A$–$C$ conclusion. This is indeed surprising, for it is not at all obvious that the premise pair

$$
\begin{array}{rl}
& (1)\ A\ N_w\ a\ B \\
& (2)\ \underline{B\ N_w\ a\ C} \\
\text{should entail} & (3)\ A\ N_s\ i\ B
\end{array}
$$

Premise (1) tells us only that $A$ belongs to the essence of each $B$, *not* that $A$ and $B$ have any essential connection. And how could the additional information given in (2) that $B$ belongs to the essence of every $C$ allow us to conclude that $A$ and $B$ do after all have an essential link to one another? We have just shown that this does follow, using Barbara $N_wN_wN_w$ and Darapti $N_wN_wN_s$; so one might simply accept the result and try to grow accustomed to the face of this new "Barbari." But again one could consult Aristotle's essentialism. The conclusion asserts that $A$ belongs to the essence of some $B$ and that $A$ and $B$ are linked (with our proviso about White; without it we get '$A\ N_s\ i\ B$' but not necessarily a *per se* connection between $A$ and $B$). This will hold just in case $A$ and $B$ are included in some common essential path. (For this term, see Section 2.11.) To show that that is the case, the premises must show that there is some $E$ to whose essence both $A$ and $B$ belong. Now the premises show immediately [by (2)] that $B$ belongs to the essence of every $C$, including $c$. They also show, since they entail (by Barbara $N_wN_wN_w$) '$A\ N_w\ a\ C$', that $A$ belongs to the essence of every $C$, including $c$. So both $A$ and $B$ belong to the essence of $c$. Therefore they will belong to a common essential path; hence $A\ N_s\ i\ B$. But, as remarked earlier, there is no need to add this syllogism as a "starting point" of the system: The conclusion was reached by successive use of Barbara $N_wN_wN_w$ and Darapti $N_wN_wN_s$.

These considerations point to some important general principles encountered earlier and covering all three figures: First, a premise pair will entail that some pair of terms $A$ and $C$ are related in a strong *cop* manner if and only if they entail either that $A$ and $C$ themselves belong to a common essential path or that one term is necessarily excluded from a path in which the other is included. This can come about with *mixed* strong/weak premise pairs, and even pairs of weak premises, as well as with pairs of strong premises. In every case our derivation of a strong *cop* conclusion has been in accord with that principle, whether the principle applied directly and obviously, as in the pure strong *cop* cases, or not at all obviously, as to certain mixed cases and to third-figure moods with two weak *cop* premises.

Second, for the same sorts of reasons as applied in cases of mixed strong/weak premise pairs, the valid $N_wN_wN_s$ moods will fail to qualify as scientific demonstrations: Although they do manage to conclude validly to propositions asserting a positive or negative essential relation between their extreme terms, and although their premises are all propositions of necessity, they do not *explain* that strong link. So there is no reason from that quarter to investigate these combinations further.

## 3.7. THE *EKTHESIS* PROOFS FOR BAROCO AND BOCARDO

As Aristotle says, the necessitated versions of Baroco and Bocardo cannot be proved by reduction to the first figure *via* conversion of one or the other premise. Because the particular negative premise does not convert, only the universal affirmative could be converted in such a proof. But it would convert to a particular affirmative, leaving us with two particular premises, which prove nothing in any figure. These moods cannot, within Aristotle's system, be proved by *reductio* arguments either, for such a proof would use, as one premise, the negation of '*A N o C*' (i.e., '*A p* all *C*'). But Aristotle never investigates the mixed necessity/one-way possibility premise pairs he would need here. Rather, he says, one must use *ekthesis.* On an unampliated weak or strong *cop* reading, Aristotle's reasoning for Baroco can be reconstructed as follows:

(1) *B N a A*
(2) *B N o C*

To prove: (3) *A N o C*

Notice that in the second figure (as opposed to our earlier *ekthesis* proof for third-figure Darapti $N_wN_wN_w$) we cannot choose an individual for "setting out": Picking out either an *A* or a *C* to which *B* necessarily applies or fails to apply obviously will not help. Here we must select some subset of the *C*'s and then (as Aristotle remarks) make use of an auxiliary syllogism. If (2) is true, then for some *D* (designating an appropriate subset of *C*), so will '*B N e D*' hold.

Then, by Camestres $A_nE_nE_n$, we have

(1) *B N a A*
(2) *B N e D*

(3) *A N e D*

And if '*A N e D*' is true, then so is '*A N o C*', for, as Aristotle would say, *D* just is some *C.* This style of proof fails on the weak *cop* reading, however, because it appeals to the invalid Camestres $N_wN_wN_w$.

Leaving aside the failure of this particular proof, it turns out that pure weak *cop* Baroco is in fact invalid. Let *B* (middle) = Animal, *A* = White Thing in the Agora, and *C* = In the Agora, in a situation in which all white things in the Agora are humans and some plant is in the Agora. Then we have *B N a A* and *B N o C*. But it might be the case, given all we have specified in the premises, that all things in the Agora are one-way possibly white, so that *A N o C* is false.

On a strong *cop* reading, on the other hand, Baroco $A_nE_nE_n$ is valid; moreover, Aristotle's *ekthesis* proof will go through using strong *cop* Camestres.[9]

Third-figure Bocardo can be validated using Aristotle's proof on *either cop* reading. For the weak *cop* version, we have

(1) *A N o B*
(2) *C N a B*
To prove: (3) *A N o C*

Premise (1) entails *A N e D* for some reading of *D*. This, combined with *C N a D*, gives, *via* Felapton $N_wN_wN_w$ of the third figure (which, unlike Camestres $N_wN_wN_w$, is valid), the desired *A N o C*. The strong *cop ekthesis* proof is exactly similar. As Aristotle remarks, the *ekthesis* proof is carried out, for both Baroco and Bocardo, by use of a syllogism from the same figure as the one being validated. It may be added, however, that on a weak necessity reading, third-figure Bocardo can also be validated by an *ekthesis* proof setting out an individual *B*.

*Ekthetic* proof in general is a large, complex topic. Although a comprehensive discussion would be out of place here,[10] my account of the proofs for pure necessity Baroco and Bocardo call for some further remarks about how my own view differs from Patzig's well-known discussion. With regard to Aristotle's early ekthetic proof of the conversion of plain *I* propositions (25a16–17; this is embedded in a *reductio* proof for conversion of plain *E* propositions), I believe (*pace* Patzig) that Aristotle may well have had in mind the "setting out" of an individual (e.g., Socrates). Patzig objects to this because he believes that appeal to a concrete individual can provide a counterexample by which to *invalidate* an inference, but cannot validate a syllogism.[11] His application of this idea to the present passage seems to me mistaken; in fact, the proof would simply use, in effect, the rule of existential generalization from first-order predicate logic. (On the other hand, Patzig is right to reject the view of Alexander that appeal to an individual here would involve the *imagination* rather than a strictly logical procedure.[12])

Nor should it be thought a problem for my view that Aristotle's syllogistic propositions use only terms to which quantifiers can attach. The topic of singular terms is another vexed issue going beyond the concerns of modal logic proper. However, we may set aside the whole question of whether or not and how singular terms might fit into Aristotle's system – a question for which his use of 'Pittakos' at *Pr. An.* B.27, his use of 'the moon' at *Post. An.* 89b17 and 93a37, and the somewhat puzzling use of 'Mikkalos' and 'Aristomenes' in *Pr. An.* A.33, where he in fact entertains the possibility of premises with quantifiers attached to proper names, would be relevant.[13] An ekthetic proof setting out an individual (as opposed to one setting out a subset of some term) will simply not use any auxiliary syllogism. Thus it will not, on that ground at least, require quantifiable terms. At the same time, this view also has the advantage that otherwise[14] Aristotle apparently would prove certain conversions by use of syllogisms that he later would validate by use of those same conversions. One might observe, finally, that even if his ekthetic proofs setting out individuals did go beyond the system he devised by implicitly using singular terms, that would be only one additional case in which Aristotle's insight into a logical situation exceeded the capacity of his formal system to express that insight.[15] This is hardly to be wondered at, given the frequent subtlety of Aristotle's observations and the limitations of his formal system.

Patzig himself thinks that Aristotle bases ekthetic proof in general on a principle not enunciated until chapter 28:

$$A\,i\,B \leftrightarrow (\exists C)\,(A\,a\,C\;\&\;B\,a\,C)$$

He may be right, but his position would be more plausible if this principle had been mentioned earlier in the *Prior Analytics*, preferably in the context of the proof of the plain *I* conversion, or in the context of some other proof for which (on Patzig's view) it is essential.

The question has implications for a related issue. I tend to agree with Alexander that there is an interesting distinction (probably noticed by Aristotle himself) between those ekthetic proofs that do use a syllogism and those that do not. As mentioned earlier, where a proof can proceed by setting out an *individual*, as, for example, with plain *I* conversion and plain third-figure Darapti (the first application of *ekthesis* to a syllogism), there is no need for any auxiliary syllogism. (Moreover, as we saw in the preceding section, setting out a *D* that signified some subset of the middle term would in this case beg the question, because our auxiliary syllogism would just be Darapti again.) And of course Aristotle makes no mention of a syllogism being used in either of those proofs.

The first ekthetic proof *actually given* by Aristotle that *cannot* proceed by setting out an individual is the one for Baroco *NNN* in *Pr. An.* A.8 (30a6–14):

$$\frac{\begin{array}{c} A\,N\,a\,B \\ A\,N\,o\,C \end{array}}{B\,N\,o\,C}$$

Clearly, setting out an individual *C* will not help. Rather, we must set out a subset *D* of *C*, then make use of the fact that *A N e D* by combining it with the premise *A N a B* to get pure necessity Cesare:

$$\frac{\begin{array}{c} A\,N\,e\,D \\ A\,N\,a\,B \end{array}}{D\,N\,e\,B}$$

Converting the conclusion gives *B N e D*; and since *D* just is some of *C*, we get *B N o C*. (Whether or not that conversion is valid is obviously beside the present point.) Here, for the first time, Aristotle speaks of a syllogism being used in an ekthetic proof (30a10) and remarks that Baroco and Bocardo will each be "proved using a syllogism in its own figure" (30a13–14): Baroco uses Cesare; Bocardo uses Felapton. (Aristotle has economically combined his treatment of the two proofs; in fact, Bocardo *NNN could* also be proved without any auxiliary syllogism by setting out an individual.) Certainly there is much more to be said about *ekthesis*. But it seems to me plausible that Aristotle's earliest ekthetic proofs set out individuals and hence do not need or mention any auxiliary syllogisms, or the formula from *Pr. An.* 28 adduced by Patzig.

How, finally, do our pure weak and strong *cop* syllogisms compare with Aristotle's own claims? The strong *cop* moods are all valid, as Aristotle claims, and moreover can all be proved by the sorts of proofs he gives. The weak *cop* moods are valid in the first figure (where, of course, no proof is given), but are invalid in the second figure. In the third figure they are again valid, but, with one exception, cannot be validated by the sort of proof Aristotle gives: *Ekthesis* still works for Bocardo, but Aristotle's conversion proofs for the rest break down because weak *cop* propositions of necessity do not convert. Unfortunately, this does not add anything to our meager evidence for what sort of necessity Aristotle might have had in mind here in A.8, for he gives no concrete examples in this chapter, and he does not pause to discuss these syllogisms individually at all, except to remark on the need for *ekthesis* with regard to Baroco and Bocardo. Instead, he simply says that because propositions of necessity convert in the same way as their plain counterparts, conversion proofs will work here exactly as they did there.

## *Chapter 4*

# Mixed syllogisms: one assertoric and one necessity premise

### 4.1. THE TWO BARBARAS: ARISTOTLE'S POSITION AND ITS CRITICS

Over the centuries, the principal test case – one might say the principal battleground – for interpretations of Aristotle's modal syllogistic, and indeed for the question of the very viability of the system, has been the "two Barbaras." Aristotle begins chapter 9 of *Pr. An.* A with the remark "It sometimes happens, even when only one of the premises is a necessary proposition that the conclusion must be a necessary proposition; only not whichever premise it may happen to be, but the one having to do with the major term" (30a15–17). Thus,

$$\begin{array}{l} A\,N \text{ all } B \\ \underline{B \text{ all } \quad C} \\ A\,N \text{ all } C \end{array}$$

is, according to Aristotle, valid, but not

$$\begin{array}{l} A \text{ all } \quad B \\ \underline{B\,N \text{ all } C} \\ A\,N \text{ all } C \end{array}$$

Theophrastus had already objected (as reported by Alexander, *In Aristotelis Analyticorum*, 124.8–127.16) that both Barbaras were invalid, maintaining that the strength of the conclusion must follow that of the weaker premise;[1] that is, the link asserted in the conclusion between major and minor terms – *A* and *C* – can be no stronger than the link between major and middle – *A* and *B* – or between middle and minor – *B* and *C*: A chain is no stronger than its weakest link. As Theophrastus put it, if the bond between minor and middle – *C* and *B* – is not necessary (referring to the first Barbara, the one Aristotle considers valid), then those two terms are "separable from one another." By contrast, the major and middle – $A$ and $B$ – are *necessarily* joined together. So if $B$ were separated from $C$, so would $A$ be separated from $C$ right along with it: $A$ and $B$ are indissolubly joined and so would break away from $C$ together. Thus there is no necessary connection between $A$ and $C$.

Theophrastus also gives an example of a syllogism set up, so he says, just like the allegedly valid first Barbara, but itself obviously not valid (124.24–30):

(1) Everything that walks necessarily moves $(A_n)$

And it might be true that

(2) Every human is walking $(A)$

But it could not be true that

(3) Every human is *necessarily* in motion $(A_n)$

Theophrastus' claim is, in effect, that this argument has exactly the same logical form as the first Barbara, but is clearly invalid, which shows that the first Barbara is also invalid.

Ross also declares both syllogisms invalid, appealing more directly to metaphysical concepts. As he puts it, Aristotle's premises need to show that every $C$ is $A$ "by a permanent necessity of its [$C$'s] own nature, [but] all they do show is only that so long as all $C$ is $B$, it is $A$, not by a permanent necessity of its own nature, but by a temporary necessity arising from its temporarily sharing in the nature of $B$."[2]

On the other side, we find Łukasiewicz upholding the validity of both syllogisms. He proposes the following analogy for the first Barbara:

(1) Every $B$ is connected by a wire to some $A$
(2) Every $C$ is a $B$

(3) Every $C$ is connected by a wire to some $A$

The analogy is supposed to illustrate the principle that "whatever is true in some way of every $B$ is also true in the same way of every $C$, if every $C$ is a $B$."[3]

For the other Barbara, one would have

(1) Every $B$ is an $A$
(2) Every $C$ is connected by a wire to some $B$

(3) Every $C$ is connected by a wire to some $A$

The guiding principle here is that "if every *B* is an *A*, then if every *C* is connected in any way whatever with a *B*, it must be connected with an *A* in just the same way. . . . This seems to be obvious."[4]

Albrecht Becker and Peter Geach consider each syllogism from the *de dicto* and (modal predicate) *de re* points of view, concluding that neither is valid *de dicto*, but also that Barbara *NAN* is valid *de re.*[5] As Geach observes, "From 'necessarily: *A* (every *B*)' and '*B* (every *C*)' there certainly does not follow 'necessarily: *A* (all *C*)' unless 'nec. *A* (every –)' is read as '*nA* (every –)'. . . . [Then the conclusion is validly reached] simply by applying Barbara . . . with a modalized major term, '*nA*'. . . ."[6]

Becker and Geach are on the right track in trying to pin down the interpretation of Aristotle's necessity propositions before pronouncing judgment on the issue of validity. And they are right about the invalidity of both syllogisms read *de dicto*. Let *A* = Awake, *B* = Animal, and *C* = Human. Clearly, for the second Barbara we could have it true as a matter of fact that every animal is awake and necessarily true that every human is an animal; but 'every Animal is Awake' would *not* be a *necessary* truth. Thus Barbara *ANN* read *de dicto* is invalid. To show Barbara *NAN* read *de dicto* invalid, let *A* = Animal, *B* = Human, and *C* = White Thing in the Conference Room. We could then have 'nec: every Human is an Animal' and 'every White Thing in the Conference Room is Human' true, but 'nec: every White Thing in the Conference Room is an Animal' false.

Becker, Geach, and numerous others are also right about the validity of Barbara *NAN* read with a modalized major term (although in Chapter 2 we saw other grounds for resisting the modal predicate approach). Finally, it must be added that the modal predicate version of Barbara *ANN* is invalid:

$$\frac{\begin{array}{c} A \text{ all } B \\ nB \text{ all } C \end{array}}{nA \text{ all } C}$$

This syllogism does not even have a middle term, for *B* and *nB* are two distinct terms. One could readily remove this flaw, however, by use of the premise '*B* all *nB*', or '*B* applies to everything to which necessary-*B* applies'. (The minor premise tells us that there are some necessary-*B*'s.) Then we can construct two syllogisms in Barbara, each with a correct disposition of terms:

| | |
|---|---|
| (1) $A$ all $B$ | (3) $A$ all $nB$ |
| (2) $B$ all $nB$ | (4) $nB$ all $C$ |
| (3) $A$ all $nB$ | (5) $A$ all $C$ |

Now we do have a pair of well-formed syllogisms in Barbara, and ones that together provide a link between *A* and *C;* but they still cannot give the desired link between *nA* and *C.* To show that any further efforts to save this second modal predicate Barbara would be futile, let *A* = White, *nB* = necessary-Animal, and *C* = Human, in a possible setup in which all necessary-animals are white. The premises will then be true, but not the desired conclusion 'necessary-White *a* Animal'.

Thus, from the logical point of view, the modalized predicate reading does give the results Aristotle claims (Barbara *NAN* valid, *ANN* invalid), whereas the *de dicto* reading does not. For this reason, these first-figure "mixed" moods of *Pr. An.* A.9 have seemed, in terms of the older modal *dictum* vs. modal predicate debate, to demand a *de re* interpretation.[7]

If these last few points may be regarded as settled, let us return to the unsettling claims, and supporting examples, of Theophrastus. As for Theophrastus' reasoning, if, for example, in saying that Walking and Human are "separable" he means that it is not a *necessary truth* that any humans be walking or that any walking things be human, then he is right. (In effect, he correctly denies the *de dicto* reading of the proposition.) But if he thinks that because walking possibly fails to apply to all humans (which is true enough), then Human is possibly inapplicable to everything to which Walking applies, he is wrong – at least where these statements are read *de re.* In the possible situation in which some walking things are human, it will be true that Walking is possibly false of, or "separable from," all humans, because all walking humans are accidentally walking. But it is false that Human is possibly false of (or separable from) all walking things, because in the situation imagined, some of the things that are walking are necessarily human. (Compare: The quality White may be separable from all things that are human without it being the case that *being human* is separable from all those things, e.g., Socrates, Coriscus, that are in fact white.)

The initially plausible metaphor of the chain that is no stronger than its weakest link (supplied earlier as a friendly gloss on Theophrastus' stated position) is, for this reason, misleading. It suggests a picture of *A*'s, *B*'s, and *C*'s as three distinct groups linked one to the next by some sort of weaker or stronger logical joint. On this picture it is quite natural to suppose that the resultant link between *A*'s and *C*'s cannot possibly be any stronger than either that between *A*'s and *B*'s or that between *B*'s and *C*'s.

Now we just saw that *A* can apply *necessarily* to all *C*'s, even where *C* does not apply necessarily to any *A*. So the symmetricality of the "separability" relation between links of a chain (if *C* is physically separable from *B*, then *B* is separable from *C*) does not obtain when applied generally to predicates and their subjects in modal contexts. It would be better, to counter the chain metaphor and the picture of *A*'s, *B*'s, and *C*'s as distinct items needing to be joined, to represent Barbara *NAN* by stressing the *identity* of each *C* with some *B* or other:

(1) *A* necessarily applies to everything that is in fact identical with some *B* or other
(2) Every *C* is in fact identical with some *B* or other

(3) *A* necessarily applies to every *C*

The validity of this argument is clear enough, and also serves to highlight Aristotle's own reasoning: Because the *C*'s simply are some of the *B*'s, and *A* necessarily belongs to every *B, A* will necessarily belong to every *C* (30a21–23).

Theophrastus' purported counterexample,

All that walks necessarily moves
Every human is walking

Every human is necessarily moving

fails because it is not necessarily of the same logical form as the first Barbara: Whether it has the same form or not depends on how one construes the structure of the modal propositions involved. The first premise is true only if read *de dicto.* Because on that reading the conclusion is false, we do have a counterexample to the validity of this mood read *de dicto* throughout. On a weak *cop* or modal predicate interpretation, however, the first premise is false. Hence, although on these readings the conclusion is false, we do not have a case of true premises with false conclusion. So Theophrastus' example works only against the *de dicto* reading of the syllogism.

One can also now see why Ross was troubled. Probably what he means in saying that *C* is not *A* "by a permanent necessity of its own nature" is that it is not *necessarily true* that *C* is *A* (that all White are Human) – or, perhaps, it is not *qua C* that *C* is *A*, or *A* is not part of what-it-is-to-be-a-*C*. In this he is right. But he overlooks the fact that *A* can belong to the essence of all *C*'s, hence belong necessarily to all *C*'s, without belonging to them *qua C*'s, without being any part of what-it-is-to-be-a-*C*. Where *A* = Human and *C* = White, and all white things are human beings,

*A does* in fact belong to all the *C*'s "by a permanent necessity of their own nature" – only the latter are here picked out by a property (*C*) that is not part of their own nature. The pertinent fact is that however these objects are picked out, by color or weight or location – by whatever accidental or essential property – *if* they are all human beings, then Animal belongs necessarily to them.

Łukasiewicz's reasoning is considerably more beguiling. But the metaphor of "being connected by a wire" must be employed with care. To preserve the idea of *A* belonging necessarily to all *B*, he speaks of every *B* connecting to "some *A*" by a wire. If, then, every *C* is a *B*, it obviously follows that the *C*'s all connect by a wire to some *A*. My central objection is that it is very odd – and in any case un-Aristotelian – to model '*A* (some predicable) necessarily applies to all *B*' by a relationship between *individual B*'s and *A*'s, as if an individual *B* will be necessarily *A* just in case there is some individual *A* to which it attaches in the right way. If we want to think in terms of Łukasiewicz's wire analogy,[8] it would seem more appropriate to model '*A* necessarily applies to all *B*' as '*A* attaches by a wire to every *B*'. Then for Barbara *NAN* we would have

*A* attaches by a wire to everything that is a *B*
Every *C* is a *B*
______________________
*A* attaches by a wire to everything that is a *C*

And this does seem obviously valid.[9]

The premises of Barbara *ANN* would read

Every *B* is an *A*
*B* attaches by a wire to every *C*
______________________

But from this it does not follow that *A* attaches by a wire to any *C*'s – or, for that matter, to anything at all. (If *B* attaching by a wire to every *C* entails *B* attaching to every *C*, and that in turn implies that every *C* is a *B*, then we can validate plain Barbara, but still not Barbara *ANN.*)

So Łukasiewicz's wire analogy, if set up in a way that plausibly models '*A* necessarily applies to all *B*', does not make both Barbaras valid, but – in agreement with Aristotle and the modal predicate reading – makes the first valid and the second invalid.

In sum, it is clear that neither Barbara *NAN* nor *ANN* is valid when read *de dicto:* When read *de re* with modalized predicates, the former is valid, and the latter invalid. In addition, three more or less informal attempts to show *both* Barbaras valid (Łukasiewicz), or *both* invalid (Theophrastus, Ross), have been refuted.

## 4.2. MIXED ASSERTORIC/*COP* NECESSITY SYLLOGISMS

How do things stand with regard to our two *cop* readings of Barbara? And how do different readings fare with regard to other mixed syllogisms of the first figure, and with those of the second and third figures? These questions can now be answered with little ado, because most of the issues involved have already been addressed.

### 4.2.1. First figure

The (unampliated) weak *cop* syllogisms work out just as Aristotle says they should:

(1) *A* necessarily applies to all *B*
(2) *B* applies to all *C*

therefore *A* necessarily applies to all *C*

If *A* necessarily applies to everything to which *B* applies, and *B* applies to all *C*, then *A* necessarily applies to all *C*. By contrast, consider Barbara *ANN:*

(1) *A* applies to all *B*
(2) *B* necessarily applies to all *C*

therefore *A* necessarily applies to all *C*

The premises do imply that *A* applies to all *C:* If *A* applies to everything to which *B* applies, and *B* necessarily applies to all *C*, then *A* applies to all *C*. But there should be nothing here to tempt one to the conclusion that *A necessarily* applies to all *C*. For those who are tempted, Aristotle's own counterexample should help them to resist. Let *A* = Moving, *B* = Animal, and *C* = Human, and suppose that *A* all *B* (every Animal is Moving) and that *B N* all *C* (Animal necessarily applies to every Human Being). These premises may be true, but the desired conclusion (Moving N all Human, or every Human is necessarily Moving) is nonetheless false. The reasoning is sufficiently like that discussed earlier in the chapter, especially in connection with Łukasiewicz's treatment of these arguments, that no more need be said here.

Both syllogisms are invalid, however, on a strong *cop* reading. For Barbara *NAN*, let *A* = Animal, *B* = Human, and *C* = White, where all white things are humans. The strong *cop* conclusion (Animal $N_s a$ White) would be false even though the premises would, in the same situation as described earlier, be true. For strong Barbara *ANN*, let $A$ = White, $B$ = Animal, and $C$ = Human, with all animals being white and some humans being accidentally white.

Results for the other three first-figure moods (Celarent, Ferio, Darii) exactly parallel those for weak and strong Barbara and are easily established on the basis of our discussion to this point. Thus the weak *cop* reading yields the four perfect first-figure mixed moods Aristotle wants, whereas the strong *cop* versions are all invalid.

### 4.2.2. *Second figure*

*4.2.2.1. Invalidity overlooked.* Aristotle's treatment of the second figure is notable chiefly for the error (shared by many commentators) of regarding several of these moods as valid, but also for the large number of housekeeping chores it presents. Aristotle wishes to reduce Cesare *NAN* and Camestres *ANN* to the first-figure mood Celarent *NAN* (as their plain counterparts had been reduced to plain Celarent in chapter 5) *via* conversion of an $E_n$ premise:

| Cesare *NAN* | | Camestres *ANN* | |
|---|---|---|---|
| $B\ N\ e\ A \rightarrow$ | $A\ N\ e\ B$ | $B\ a\ A$ | $C\ N\ e\ B$ |
| $B\ a\ C \rightarrow$ | $B\ a\ C$ | $B\ N\ e\ C$ | $B\ a\ A$ |
| $A\ N\ e\ C$ | $A\ N\ e\ C$ | $A\ N\ e\ C$ | $C\ N\ e\ A \rightarrow A\ N\ e\ C$ |

On a strong *cop* reading, these conversions are valid. Unfortunately, the weak *cop* reading is needed to make the reducing syllogism, first-figure Celarent *NAN* in both cases, come out valid. But on that reading the conversions are no longer valid. Worse yet, not only are Aristotle's conversion proofs thereby ruled out, but also on a weak *cop* reading both these second-figure moods are in fact invalid. Let $B$ (middle) = Human, $A$ = Asleep, and $C$ = Eating, where all sleeping things are horses and all eating things are humans. In this possible setup (on a weak reading), '$B\ N\ e\ A$' and '$B\ a\ C$' would be true, but not '$A\ N\ e\ C$'. So Cesare *NAN* is invalid on a weak *cop* reading. For Camestres *ANN*, let the terms be the same in a situation in which all eating things are horses and all sleeping things are humans; again the mood is invalid.

The reason these syllogisms, unlike their first-figure brethren, and contrary to Aristotle's opinion, come out invalid is that the predicate terms of their conclusions (*A*) are so situated in the premises that we know very little about their relations to their own designata. Thus in Cesare *NAN, A* may be only accidentally related to any and all of the *A*'s. In fact, the premises give us no assurance that *A* applies necessarily to anything, including the *C*'s. Camestres *ANN* is invalid for the same reason.[10]

*4.2.2.2. Aristotle's invalidation of Camestres NAN: a too-clever proof and a curious counterexample.* Aristotle's interesting argument at 30b24 for the invalidity of Camestres *NAN* fails on a weak *cop* reading with or without ampliation:

> Further, if the conclusion is necessary, it follows that *C* will be necessarily inapplicable to some *A*. For if *B* is necessarily inapplicable to all *C*, then *C* will be necessarily inapplicable to all *B*. But *B* must belong to some *A*, if *A* belongs of necessity to all *B*. Thus *C* must be inapplicable to some *A*. But nothing prevents *A* from being selected such that it is possibly applicable to all *C*. (30b24–31)

Here is the syllogism in question:

$$\frac{\begin{array}{l} A\ N\ a\ B \\ A\ e\quad C \end{array}}{B\ N\ e\ C}$$

Aristotle's argument is the following:

(i) If $A_nEE_n$ in figure 2 is valid, then '*A N a B*' and '*A e C*' entail '*B N e C*' and thus '*C N e B*'.
(ii) If $E_nI_nO_n$ in figure 1 is valid, then '*C N e B*' and '*A N a B*' (which converts to '*B N* some *A*') entail '*C N o A*'.
(iii) So if $A_nEE_n$ in figure 2 and $E_nI_nO_n$ in figure 1 are both valid, then '*A N a B*' and '*A e C*' entail '*C N o A*'.
(iv) But '*A N a B*' and '*A e C*' do not yield '*C N o A*'.
(v) So $A_nEE_n$-1 and $E_nI_nO_n$-2 cannot both be valid.
(vi) Since $E_nI_nO_n$ is obviously valid in figure 1, $A_nEE_n$ of figure 2 must be rejected.

The fly in this ointment now appears only too grossly. From '*A N a B*' and '*A e C*' one can derive only the weak version of '*B N e C*'; but this version does not convert to '*C N e B*' as step (i) requires. [On an ampliated version of the argument, we would have ampliated $E_n$ in step (i) (*B N e C p*'), which would convert on either a weak or strong reading. But then in step (ii) we would have to convert '*A N a B p*', which does not convert.]

So on a weak *cop* reading, the mood cannot be shown invalid by the argument Aristotle proposes.

But as Aristotle says, this mood can be shown invalid "by terms," letting $A$ (middle) = Animal, $B$ = Human, and $C$ = White (30b31–40). Then the plain conclusion 'Human no White' will follow from the premises but will not itself be necessary, for a human may become white. To Aristotle's use of this particular example, Peter Geach understandably objects that

> to get a counterexample to 'no white thing is possibly-a-man', one would need . . . a possible set-up in which an identifiable white thing was first not a man and then a man, and this is not at all the same thing [as a setup in which some man is first not white and then white].[11]

Notice, however, that Aristotle's counterexample *does* work for any reading of the $E_n$ conclusion on which that conclusion *is* convertible (i.e., *de dicto*, ampliated *de re*, ampliated weak *cop*, and strong *cop* with or without ampliation), for if a pair of premises yield 'nec: Human no White' on any of *these* readings, it will yield 'nec: White no Human' on that same reading. The case of a human becoming white would then show that this latter proposition ('nec: White no Human') is false on any of those readings. Thus the (true) premises of $A_nEE_n$ do not yield either 'nec: White (no Human)' or its converse on any reading on which that proposition is convertible. Still, the most important point is that Geach's misgivings about the counterexample are appropriate where the $E_n$ conclusion is taken in any sense in which it does not convert – that is, with unampliated modal predicate, which Geach in fact had in mind, and unampliated weak *cop*, which is needed to make Aristotle's foundational first-figure moods valid.

It is interesting that we still lack any demonstration that the unampliated weak *cop* version of Camestres *NAN* is invalid. Could it possibly be valid? Well, perhaps Aristotle's example *can* work even if his reason for proposing it (i.e., that a human might become white, 30b37) does not suffice on a weak *cop* reading. All we need is a white thing that is not human but which may become human. Perhaps some menses is white (or pale) and is potentially human. Or if we change the terms from Human to Oak, and from White to Brown, it may be that a brown acorn, which is not an oak, but which is possibly an oak, will fill the bill. It is far from clear, however, that Aristotle would have had any such thing in mind. For one thing, it would introduce potentiality for substantial coming-to-be in a way nowhere explicitly involved in *Pr. An.* A.1–22, either in its examples or in its discussion of possibility and contingency. Moreover, even in this very passage, Aristotle, instead of pointing however obliquely to the doctrine of substantial coming-to-be, reverts to the fact that "some human may become white."

Still, these issues could all be avoided, and the mood invalidated, as Aristotle suggests, "by terms," if we simply let $A$ = Animal, $B$ = White, and $C$ = Cloak ($A$ middle) in a possible setup in which all white things are (necessarily) animals such as swans, no cloak is an animal, and at least some cloak is possibly white.

### *4.2.3. Third figure*

The third figure is relatively tidy as it stands. Here it is worthwhile to mention only some small slips on Aristotle's part and to add one point about the effects of ampliation. First, Darapti *ANN* will give the weak *cop* conclusion '*C N* some *A*' regardless whether the minor premise is strong or weak necessity. But that conclusion will not convert, as Aristotle desires, to give the conclusion '*A N* some *C*'. For the same reason, Disamis *ANN* will yield '*C N* some *A*' but not '*A N* some *C*'.

Among the valid third-figure moods, $A_nAI_n$, $E_nAO_n$, $A_nII_n$, and $E_nIO_n$ will, *via* conversion of *plain A* or *I* premises, reduce to valid first-figure moods ($A_nII_n$, $E_nIO_n$, $A_nII_n$, and $E_nIO_n$, respectively).

Aristotle has probably slipped in rejecting $O_nAO_n$; he recognizes it as valid elsewhere (34a38–40), and it is valid on a weak *cop* reading, as can be quickly shown by *ekthesis:*

$$\begin{array}{ll} (1) & A\ N\ o\ B \\ (2) & \underline{C\ a\ B} \\ & A\ N\ o\ C \end{array}$$

Let the "some $B$" of (1) = $b$. Then we have '*A N*-not *b*' and '*C* applies to *b*', which entail '*A N*-not something (namely *b*) to which *C* applies'.[12]

In sum, the results for mixed *cop* syllogisms of A.9 differ in several ways from Aristotle's claims. On a strong *cop* reading, all these moods are invalid in all figures, so one gets disagreement with all his claims of validity and – as it happens – agreement wherever he claims invalidity. Most commentators have understandably wanted to read these arguments *de re*, where results are exactly parallel to those for our weak *cop* necessity, yielding the four perfect first-figure syllogisms Aristotle wants. But notice that even on these last two readings there is agreement with Aristotle only in the first figure.

In the second figure, Aristotle declares valid three moods (Festino and Cesare *NAN*, Camestres *ANN*) whose weak *cop* versions are invalid. This divergence is due in each case to his use of illegitimate conversion principles. In the third figure, one does get Darapti, Felapton, and Datisi *NAN*, as Aristotle says, but these must be proved by *ekthesis* rather than by the conversion proofs he gives. In the two cases where Aristotle claims validity, but weak *cop* gives an invalid argument (Darapti and Disamis *ANN*), it is worth noting that one does get a necessary *C–A* conclusion, but not the *A–C* conclusion claimed by Aristotle. He infers the latter from the former by appeal, once again, to an illegitimate conversion.

Finally, Aristotle rejects two third-figure moods (Disamis and Bocardo *NAN*) that are in fact valid with weak *cop* necessity. Both can be proved in an obvious way using *ekthesis*. Aristotle overlooks this because he thinks he has counterexamples at hand for both of them. On inspection, however, it turns out that both counterexamples depend on a strong *cop* reading.

| Against Disamis *NAN* (31b31–33) | Against Bocardo *NAN* (32a4–5) |
|---|---|
| Two-footed *N i* Animal | Two-footed *N o* Animal |
| Awake *a* Animal | Moving *a* Animal |
| Two-footed *N i* Awake | Two-footed *N o* Animal |

Aristotle says in both places that the premises are true but the necessity version (if not the plain version) of the conclusion is false. As just noted, both moods are easily validated by *ekthesis* on a weak *cop* reading, and the counterexamples obviously fail on that reading. It is just as obvious, however, that read with strong *cop* necessity, all our premises will be true, but both conclusions false, so that the counterexamples do work on this reading.

This leads to an interesting point mentioned in passing earlier (Chapter 2, Section 2.8) concerning Aristotle's examples in general. Taken in isolation, all the necessity propositions of chapters 9–11 (and of 8–22, for that matter) will of course satisfy the definition of weak *cop* necessity, and most of them clearly qualify also as strong *cop* ('Animal *N a* Human': 30a29ff., 30b33ff., 31b40–41; 'Animal *N a* Horse': 31b4–8; 'Two-footed *N i/o* Animal': 31b28–31, 31b31–33, 32a4–5), and the rest *might* be strong *cop* ('Animal *N i/o* White': 30b5–6, 31a15–17, 32a1–4) (we shall return to these less clear-cut cases in a moment). Taken in context (i.e., as they function within a given counterinstance), most of them can be read either way, in the sense that the counterexample works on either reading. But in the two cases under discussion, Aristotle's counterinstances work if and only if one uses strong *cop* necessity. By contrast, there are no examples so far that must be read with weak *cop* only, because examples such as 'Animal *N i* White' can be true with strong *cop* necessity if we take White as essential to swans. On the other hand, we had doubts earlier (Chapter 2, Section 2.8) about granting this, and if we do not grant it, we shall have to read examples of this sort with weak *cop* necessity. And because, if we may peek ahead to chapter 17, Aristotle discusses 'White *pp a* Human' and its converse 'Human *pp i* White' in a way that does require a weak *cop* reading, one might prefer to read these Animal/White (Human/White) examples from chapters 9–11 with weak *cop*. But these considerations are hardly decisive. Perhaps the most judicious course would be to restrict oneself to the conservative but not uninteresting point that the necessity propositions in Aristotle's counterexamples are liable to the same basic semantic ambiguity found in connection with conversion principles and the premises and conclusions of syllogisms.

## 4.3. THE TWO BARBARAS AND A CLOSE LOOK AT SOME UNIVOCAL READINGS

I have, thus far, emphasized several significant differences, syntactic and semantic, between a modal copula approach and one in terms of an alternation between modal *dictum* and modal predicate. But from the wider perspective, my general position is, in distinguishing weak from strong *cop* necessity, closer to the latter in one important respect than to a reading that denies any ambiguity in Aristotle's system. Some time ago, Nicholas Rescher[13] and Storrs McCall[14] sought to provide, in one case informally, and in the other formally, an interpretation preserving virtually all of Aristotle's claims about validity or invalidity in particular cases, but without invoking any alternation between two different conceptions of modality. More recently, Wolfgang Wieland, Jeroen van Rijen, Fred Johnson, and Paul Thom have all, in various ways, attempted to string this bow. Such interpreters can be most appropriately taken up at this point because the "two Barbaras" of *Pr. An.* A.9 provide a kind of touchstone for all of them. We begin with McCall and Rescher (McCall considers his own views a completion and formalization of Rescher's informal presentation[15]), with the discussion falling into two parts: first, McCall's treatment of the two Barbaras, by which he wishes to clear the way for his own positive interpretation of Aristotle's modal logic;

second, the positive Rescher-McCall account of Aristotle's modal syllogistic.

### *4.3.1. McCall on the two Barbaras*

McCall's general strategy is to put to the test various proposed ways of distinguishing between the two Barbaras with regard to validity. He says, first of all, that Aristotle's own comments on behalf of Barbara *NAN* (in McCall's translation, "*A* necessarily belongs to every *B*, and since *C* is one of the *B*s, it is clear that for *C* also the positive . . . relation to *A* will hold necessarily," 30a21–23) are "not . . . conclusive: I do not see, for example, how Aristotle's restatement of the premises will serve to show the *invalidity* of Barbara *XLL*" (emphasis added).[16] This opening sally is rather odd, because Aristotle does not propose to show anything about Barbara *ANN* by "restatement of the premises": He gives two arguments (discussed in detail later in Section 4.3.2), one a variation on his usual sort of *reductio* argument, designed to show that impossible results follow from the assumption that Barbara *ANN* is valid, the other consisting in a concrete substitution instance of Barbara *ANN* on which the premises are clearly true but the conclusion false.

More important, however, are McCall's comments on Aristotle's actual practice. Before turning to the arguments given by Aristotle himself, McCall considers the possible applicability of the Aristotelian "suggestion (30b4) that from the denial of the conclusion of a valid syllogism there must result an impossibility, i.e., a proposition inconsistent with the premises,"[17] or, in other words, that any valid syllogism can be shown valid by a *reductio ad impossibile* argument. McCall argues that if that principle is true, then *both Barbaras are equally valid.* For Barbara *NAN* (the one Aristotle says is valid), McCall formulates the following *reductio* proof:

| | | | |
|---|---|---|---|
| If | (1) | all *B* is necessarily *A* | (*A N* all *B*) |
| and | (2) | all *C* is *B*, | (*B* all *C*) |
| then | (3) | all *C* is necessarily *A*, | (*A N* all *C*) |
| for suppose | (4) | some *C* is not necessarily *A*, | (*A P*-not some *C*) |
| then, since | (2) | all *C* is *B*, | (*B* all *C*) |
| therefore | (5) | some *B* is not necessarily *A*, which contradicts premise (1). | |

Before proceeding to Barbara *ANN*, McCall lodges the objection that this *reductio* uses Bocardo *PAP* [in deriving (5) from (4) and (2)], and because *that* syllogistic form stands in need of validation just as much as Barbara *NAN*, this overall argument for the latter is unsatisfactory.[18] But in fact, Bocardo *PAP* can be validated by an *ekthesis* proof:

| | |
|---|---|
| Some *C*, let us say *c*, is not necessarily *A* | [from (4)] |
| Every *C*, including *c*, is a *B* | [from (2)] |

Therefore something that is a *B* (i.e., *c*) is not necessarily an *A* [which is (5)]. So Bocardo *PAP* is valid, and the *reductio* proof proposed by McCall for Barbara *NAN* [(1)–(5)] is perfectly in order.

Still, McCall also maintains that with regard to Barbara *ANN*, which Aristotle considers invalid, a *reductio* argument exactly analogous to the one just given will in fact show the syllogism *valid*, so that on this approach (i.e., validation via *reductio*) there is no distinction between the two Barbaras as to validity, even if one grants that first *reductio* argument. For Barbara *ANN*, one has

| | | | |
|---|---|---|---|
| If | (1) | all *B* is *A* | (*A* all *B*) |
| and | (2) | all *C* is necessarily *B*, | (*B N* all *C*) |
| then | (3) | all *C* is necessarily *A*, | (*A N* all *C*) |
| for suppose | (4) | possibly some *C* is not *A*, | (*A P*-not some *C*) |
| then, since | (1) | all *B* is *A*, | (*A* all *B*) |
| it follows that | (5) | possibly some *C* is not *B*. | (*B P*-not some *C*) |

But this contradicts premise (2). "Here," says McCall, "the *reductio* syllogism, Baroco *APP* [by which we obtain (5) from (1) and (4)], seems on the face of it just as valid as Bocardo *PAP*, and for this reason the method of proof by *reductio ad absurdum* cannot be used to distinguish between Barbaras *LXL* and *XLL*."[19] But the *reductio* syllogism Baroco *APP*, whether or not it appears at first sight just as valid as Bocardo *PAP*, is demonstrably invalid. Let *A* (middle) = Moving, *B* = Animal, and *C* = Human. It is quite possible that all animals are in fact in motion and that some human is possibly not in motion; but it is false that some human is possibly not an animal.

The interesting result is not so much the defense of Aristotle against McCall, but the positive discovery, made in response to McCall's inventive challenge to Aristotle, that this method of proof by *reductio does* distinguish between the two Barbaras as to validity and draws the distinction exactly in accord with the claims of Aristotle, as opposed to Theophrastus, Łukasiewicz, Ross, and others. Notice, finally, that the method can force a distinction without circularity, because the validation of Bocardo *PAP*, used here to validate Barbara *NAN* by *reductio*, was itself achieved independently of Barbara *NAN* by an *ekthesis* proof. (By contrast, Aristotle's validation of *assertoric* Bocardo appealed to assertoric Barbara; however, he could have used *ekthesis* there as well.)

### *4.3.2. Aristotle's arguments for the invalidity of Barbara ANN*

Let us turn now to the first of the two arguments actually used by Aristotle to *invalidate* Barbara *ANN* and to remarks by McCall and Jaakko Hintikka on Aristotle's argument. Aristotle first reasons (30a25–28) that if Barbara *ANN* were valid, "it would follow, through the first and the third figures, that *A* necessarily belonged to some *B*. But this is false. For the *B*'s to which *A* applies may be such that *A* possibly fails to apply to them." The reasoning seems to be the following:[20]

| | |
|---|---|
| If (1) | all *B* is *A* |
| and (2) | all *C* is necessarily *B*, |
| then (3) | all *C* is necessarily *A* (by Barbara *ANN*). |
| But (4) | some *B* is *C* [by conversion of the weakened, assertoric version of (2)]. |
| Hence (5) | some *B* is necessarily *A* [via Darii *NAN*, with (3) and (4) as premises]. |
| But | (5) clearly goes beyond anything stated or implied in (1) and (2). |

As Hintikka indicates,[21] the more general structure of the argument is this:

| | |
|---|---|
| If | mood (a) (Barbara *ANN*) is valid, then (1) and (2) together entail (3). |
| If | mood (b) (Darii *NAN*) is valid, then (3) and (4) together entail (5). |
| So | if moods (a) and (b) are both valid, then (1) and (2) together entail (5). |
| But | (1) and (2) do not entail (5). |
| And | mood (b) is valid (Darii *NAN* is a perfect syllogism). |
| Therefore | mood (a) (Barbara *ANN*) is invalid. |

The argument works because, as Aristotle says, the *B*'s to which *A* belongs [as asserted in (1)] may be such that *A* possibly fails to apply to them [i.e., such that (5) is false]. This is consistent with what the first premise tells us, and certainly nothing about *B* necessarily applying to all *C* (the second premise) rules out *A* being possibly false of all *B*.

Aristotle's second argument consists in showing the mood invalid "by terms" (30a28–32): "let $A$ = Motion [moving], $B$ = Animal, $C$ = Human. For human is necessarily animal, but animal is not of necessity in motion, nor is human." (That is, with these terms it is possible to have 'Moving all Animal' and 'Animal *N a* Human' true, but 'Moving *N a* Human' false.) This counterexample confirms the result of the previous argument and would stand by itself as a refutation of Barbara *ANN*.

Hintikka suggests, however, that there may be a flaw in that *reductio* proof after all. The argument, in effect, came down to a choice between Darii *NAN* and Barbara *ANN*. But is it so obvious that Darii *NAN* must be preferred over Barbara *ANN?* Hintikka has constructed an ingenious argument parallel to Aristotle's to show that if Barbara *ANA* (note the *plain* conclusion) is valid, then Aristotle should in fact reject the first-figure mood Darii *NAN:*

(i) If mood (a) (Barbara *ANA*) is valid, then $p$ (i.e., '*A* all *B*') and $q$ ('*B N* all *C*') yield '*A* all *C*', and hence by conversion yield $r$ ('*C* some *A*').

(ii) If mood (b) (Darii *NAN*) is valid, then $q$ ('*B N* all *C*') and $r$ ('*C* some *A*') yield '*B N* some *A*', which by conversion yields $s$ ('*A N* some *B*').

(iii) But $p$ ('*A* all *B*') and $q$ ('*B N* all *C*') do not yield $s$ ('*A N* some *B*').

Given (iii), it follows that Barbara *ANA* and Darii *NAN* cannot both be valid. But because the former is indubitably valid, the latter must be rejected. And if Darii *NAN* is rejected, then the previous argument against Barbara *ANN* collapses. Peter Geach has rightly pointed out, however,[22] that step (ii) of Hintikka's argument depends on conversion of $I_n$ and that on the *de re* reading, which is in any case needed (i.e., within the confines of the *de re/de dicto* distinction accepted by both Geach and Hintikka) to make the syllogisms of chapter 9 turn out valid, $I_n$ does not convert. (Recall that exactly the same situation obtains for the weak *cop* reading of $I_n$.) So one is not forced by acceptance of Barbara *ANA* to reject Darii *NAN*. Rather, one must reject either Darii *NAN* or the conversion of $I_n$. On either the modalized predicate *de re* reading or the weak *cop* reading, it is clear both that conversion of $I_n$ must go and that Darii *NAN* is valid.

But Hintikka anticipated that response and replied as follows:

> . . . we should hesitate before rejecting any rule of conversion Aristotle uses. Since they are usually his most important tools, rejecting them would mean abandoning all hope of understanding what he actually had in mind in developing his syllogistic.[23]

This response, although understandable, seems to me both an oversimplification and a *non sequitur.* It oversimplifies because as soon as one realizes that $A_n$, $I_n$, and so forth, can be read in (at least!) two different, if related, ways (we have actually canvassed four ways – or eight, if one adds the ampliated versions of *de re, de dicto*, strong *cop*, and weak *cop* – or really twelve, since each may be ampliated in two ways), one also realizes that rejection or acceptance of $I_n$ conversion is a complex matter, not something one settles across the board in a single decision. So, for example, $I_n$ conversion is valid on the unampliated *de dicto* and strong *cop* readings, even though it *must* be rejected when read *de re* or weak *cop* (without ampliation). Second, Hintikka's conclusion does not follow: Although the rejection of $I_n$ conversion (on certain readings) is the rejection of a very important logical tool, its rejection hardly entails that we must abandon all hope of understanding Aristotle's own approach to modal syllogistic. On the contrary, it is at least a coherent working hypothesis that there are two basic modal conceptions at work in Aristotle's approach. Hintikka himself recognizes this and suggests that Aristotle failed to distinguish two readings because his own modal propositions can so easily be read in two different ways. But if it is plausible that there are implicitly two readings at work in Aristotle's modal syllogistic, and if we work through the logical implications of each reading, we may find (in fact, do find) that the best explanation of how Aristotle's system arose does involve the supposition that he appealed to principles of conversion that are valid only if read one of those ways rather than the other. Thus a blanket refusal to question any of Aristotle's conversion principles may in fact obstruct the understanding of his logic.

Returning to Aristotle's invalidation of Barbara *ANN*, let it be noted that he rightly says that the argument may proceed also through the third figure. What he has in mind will be the following (with the numbers standing for the same proposition types as at the beginning of this section):

If Barbara *ANN* is valid, then (1) ('*A* all *B*') and (2) ('*B N a C*') entail (3) ('*A N a C*').
If Darapti *NAN* is valid, then (2) and (3) entail (5) ('*A N i B*').
So if Barbara *ANN* and Darapti *NAN* are both valid, (1) and (2) entail (5).
But (1) and (2) do not entail (5).

So either Barbara *ANN* or Darapti *NAN* is invalid. Because the latter is easily verified (either by *ekthesis* or by reduction, via conversion of the assertoric minor premise, to Darii *NAN*), Barbara *ANN* must be rejected.

It must finally be said, however, that this gives Aristotle the benefit of a certain doubt. Because he had, at this point, verified to his own satisfaction the pure necessity versions of Darii and Darapti (in ch. 8), but not, as yet, their mixed varieties (Darii and Darapti *NAN*), it is most likely that the moods he had in mind when he said that the argument "can work through the first and third figures" were the pure necessity moods Darii and Darapti *NNN*. Probably, then, he had in mind the use of conversion of (2) ('*B N* all *C*') – rather than conversion of the plain '*B* all *C*' – to get (4) ('*C N* some *B*') in the first version of the proof (given at the beginning of this section). But this conversion will not work on any reading of necessity that will make either Darii *NNN* or *NAN* come out valid. Similarly, although the second version's use of Darapti *NNN* ("through the third figure," as set out just above) would not need to use any conversions at all, Aristotle's validation of the mood in chapter 8 via reduction to Darii *NNN did* use conversion of $A_n$. So one might question Aristotle's right to appeal to that mood here. This is not of great importance, however, for we have seen that Aristotle could, in the first version, simply weaken (2) ('*B N* all *C*') to the assertoric '*B* all *C*', then validly convert that to get (4) ('*C* some *B*'). And the second version, the one working through the third figure, could use Darapti *NAN* (rather than Darapti *NNN*), which is validated by reduction to Darii *NAN* via conversion of the *assertoric* minor premise. Moreover, both Darii and Darapti *NNN* can be validated by *ekthesis*, even if not *via* conversion – a fact Aristotle presumably would have discovered if he had seen the need to look any further than his proposed conversion proofs.

In sum, on a weak *cop* reading, Aristotle's results are correct, and the general form of argument he devised can be used to validate them. Some steps must themselves be justified, however, in ways different from those he actually used (i.e., in ways not appealing to conversion of $I_n$). But such remedies were entirely open to him.

### *4.3.3. Rescher and Averroës' rule*

There remain several interpretations that not only resist any attempt to place two readings on Aristotle's necessity propositions but also claim to provide an intuitive Aristotelian basis for a modal system that is consistent and that captures virtually all of Aristotle's own logical results while using only one reading of modal propositions. Our first representative of this approach comes from Nicholas Rescher, who deplores the position of Becker (that Aristotle unknowingly alternated between *de re* and *de dicto* readings of necessity propositions) as "desperate and farfetched."[24] He offers instead a "wholly new approach" that eschews any "straightforward, direct application of the methods of modern symbolic logic."[25] Rescher's approach consists of several main elements that I shall set forth at some length, because, Rescher's own disclaimers notwithstanding, his results are very similar in spirit and in many details to modern interpretations in terms of *de dicto* modality:

1. The modality of a statement is to be conceived "not as an explicit integral facet of the statement itself – as is done in construing 'All S is necessarily P' as '(x) (Sx → Nec: Px)' or as '(x) Nec: (Sx → Px)' – but rather as a way of according a certain status to the 'ordinary' proposition 'All S is P'." The sign of modality thus "categorizes" propositions "from the outside" in such a way that one cannot tell what status a proposition has just by looking at the statement in which it is expressed (e.g., 'All S is P'). Thus, "ordinary, unmodalized categorical propositions A, E, I, O can be qualified by four modalities: A (assertory, 'is actual'), P (problematic, 'is possible'), N (apodeictic, 'is necessary'), and C (contingent, 'is neither necessary nor impossible')."[26]

2. The logic of such propositions so classified as to modal status is then given *via* rules of inference pertaining to conversion, negation, and so on. Rescher's rules, in effect, simply encapsulate Aristotle's own pronouncements in *Pr. An.* A.3, 8, and elsewhere. ($A_n$ converts to $I_n$, $I_n$ to $I_n$, etc., through all the modalities; $A_n$ is the contradictory of $I_p$, etc.)[27]

3. Rules for syllogistic modal inference: First-figure rules have "nothing to do with the *type* (A, E, I, or O) of conclusion drawn, but [enter in] only in determining the modal status of the conclusion."[28] In the "basic" cases the modality of the conclusion follows that of the major premise, as asserted by Averroës' principle: *modus conclusionis sequetur modum propositionis maioris.*[29] The meaning of this rule is then given on the basis of a scale of the strength of modal propositions (*N, A, P, C*) and the following two rules:

(a) If the modality of the major premise is *not* stronger than that of the minor premise, then the modality of the conclusion simply follows that of the major.
(b) If the modality of the major premise *is* stronger than that of the minor premise, then the modality of the conclusion is always stronger than that of the minor premise (but need not be as strong as that of the major premise).

"It is this rule assigning the determining role to the major premise which represents – in contrast to Theophrastus' *peiorem* rule [that the modality of the conclusion can be no stronger than that of the weaker premise] – Aristotle's basic intuition into the logic of modal syllogistic inference.''[30]

4. The final step is to give proofs for validating moods in figures other than the first; this is carried out via term conversion, qualitative conversion, *reductio ad impossibile*, and *ekthesis*. But it is Rescher's treatment of the first figure, and above all the central rule just quoted, that will be our chief concern here.

First, we may observe that this approach does harmonize with Aristotle's text insofar as it yields results similar to his. In fact, the rules quoted here by and large simply follow Aristotle's own endorsement of various conversion rules and syllogistic moods. Second, the basic rule in which Rescher's approach culminates will validate Barbara *NAN* while ruling out Barbara *ANN*. Since in Barbara *NAN* the modality of the major premise is stronger than that of the minor, the conclusion's modality will be stronger than that of the assertoric minor. But the only status stronger than assertoric is that of necessary. By contrast, since in Barbara *ANN* the major is not stronger than the minor, the modality of the conclusion will simply follow that of the major, which means that these premises will yield a plain conclusion, but not one whose status is that of necessary.

Still, the interpretation has several serious drawbacks. First, Rescher does not seem to have made clear the difference between his external classification of propositions as to modal status and good old-fashioned *de dicto* modality. One might imagine that his external classification is in itself consistent with a *cop* reading: The necessity operator simply classifies something as, say, a necessity proposition, leaving open the question of the internal structure of such statements. However, Rescher does not want to introduce modality anywhere *except* as an external classifier. So it becomes unclear whether his assigning this or that modal status to a plain proposition taken as a whole really does differ from *de dicto* modality. In any case, the point of this approach was to get away from ''straightforward, direct'' application to Aristotle of modern symbolic modal logic. But this is a vain hope, for merely in adopting such modal relations as the contradictoriness of '*N: A* all *B*' to '*P: A* not some *B*', and the like, one includes these ''status-classified'' structures within a symbolic modal calculus.

More important, there is a serious internal problem with that modal calculus. Rescher's primary Averroist rule depends on a modal scale of strength from necessity down through actuality, then through (one-way) possibility to contingency. But this scale is nowhere accepted by Aristotle, nor should it be. Of course, it does make sense to say that necessity entails actuality, which entails ⟨one-way⟩ possibility, but not vice versa, and this does give a scale $N \rightarrow A \rightarrow P$. But it does not make sense to rank *C* "below" *P* or below *A*, or below *N* for that matter. On the contrary, because *C* entails *P*, but not vice versa, *C* should rate *above P* on this scale. Also, *A* does not entail *C*, nor does *C* entail *A*. So neither can be rated higher than the other on a scale of logical strength. Moreover, *N* does not entail *C* but in fact *excludes* it, as Aristotle says, just as *C* excludes *N*. So there is no means of ranking these two, either, on Rescher's scale of modal strength. Because a great many of Aristotle's modal syllogisms involve some mixture of *C* premises with assertoric or necessary ones, it is difficult to see how the rule can be "Aristotle's basic intuition into the logic of modal syllogistic inference."

Third, that "intuitive" rule, taken in itself, is rather thin on intuitive content. In and of itself it corresponds to no rule or intuition anywhere formulated or even hinted at by Aristotle. Rescher does, however, try to provide an underlying rationale by connecting the rule to Aristotelian ideas about scientific demonstration, a connection that he says will let us "see why Aristotle taught that the modality of the major premise can strengthen that of the conclusion above that of the minor."[31] The basic idea is that in a syllogism such as Barbara *NAN*,

(a) the major premise lays down a necessary rule of some sort,
(b) the minor describes some special case that has been shown by observation or induction to fall under this rule, so that
(c) the conclusion is justified that this special case necessarily conforms to the rule.

The paradigm of such reasoning is as follows:

*Law* (necessary rule): All *B*'s are *A*'s.
All twinkling things are distant.
*Special case* (observation): All *C*'s are *B*'s.
All stars are twinkling things.
*Explained consequence* (necessary result): All *C*'s are *A*'s.
All stars are distant.[32]

Thus, if *A* necessarily attaches to everything that is a *B*, and the *C*'s are just one special case of things that are *B*, then *A* necessarily attaches to all the *C*'s. Here the modality of the conclusion can be "upgraded" over that of the minor, and this will be so wherever the major asserts some necessary connection (affirmative or negative) of *A* to all the *B*'s, and the minor asserts that (some or all) *C*'s are just a special case of those things that are *B*'s.

In one way this "special-case" idea is thoroughly Aristotelian. After all, Aristotle's own comment on Barbara *NAN* – that because *A* applies necessarily to all the *B*'s, and all the *C*'s are among the *B*'s, *A* applies necessarily to all the *C*'s – in effect simply says that the *C*'s are a special case of the *B*'s and hence, like all the *B*'s, are applied to necessarily by *A*. But this has nothing in particular to do with scientific demonstration. On the contrary, because the minor premise of Barbara *NAN* is merely assertoric, it may be, for all we know, possibly false. But if so, it cannot give an explanation of why there is a necessary link between *A* and *C*.

Putting aside the matter of scientific demonstration, however, we must still evaluate Rescher's proposal that the special-case idea is the central intuition behind Aristotle's modal logic as a whole. As already remarked, it does tally with Aristotle's own comments on first-figure moods. But it is not so easy to apply the special-case principle beyond the first figure. Consider the second-figure mood Cesare *NAN:*

| | |
|---|---|
| *B N* no *A* | *(All A*'s are necessarily not *B*'s, or no *A* is possibly a *B*) |
| *B* all *C* | *(All C*'s are *B*) |
| *A N* no *C* | *(No C*'s are possibly *A*'s) |

I see no intuitive ground for saying that the minor premise "describes a special case" of the relationship asserted in the major. One reply might be that the second- and third-figure moods can be "reduced" to the first figure *via* conversion of $I_n$, $A_n$, or $E_n$, or by *reductio* arguments using first-figure syllogisms. Thus they may be regarded as entailing their conclusions *by* entailing, in a way that may not be immediately obvious, a form of reasoning (i.e., a first-figure syllogism) that does, after all, fit the special-case idea. This retreats a bit from the direct application of that intuitive principle in the case of the first figure: One must at least combine that principle with an appropriate conversion rule or, where *reductio ad impossibile* comes into play, rely on a combination of the special-case idea with other, equally crucial principles underlying the very use of the *reductio* technique itself. And let us not forget Baroco and Bacardo *NNN*, which must, at least within the confines of Aristotle's system, be proved by *ekthesis.*

But this still will not do, even if we simply help ourselves to the *reductio* technique and also close our eyes to the problem with Baroco and Bocardo. There still remains a central and very stubborn problem: One cannot, without further justification, make free use of the conversion of $I_n$, $A_n$, and $E_n$, which Aristotle uses to validate most moods outside the first figure. Because concrete counterexamples clearly demonstrate that these are at least problematic, these principles must themselves be justified.

(This was discussed at length earlier, in Chapter 2.) And the special-case approach does not provide any such justification. How does '*B N* some *A*' describe a special case of '*A N* some *B*', or '*B N* no *A*' describe a special case of '*A N* no *B*'? Yet such conversion rules are needed to bring the majority of second- and third-figure moods under the aegis of the special-case principle. So it seems that Aristotle's modal syllogistic relies heavily on basic elements that cannot be rationalized by appeal to the concept of a special case.

Finally, certain syllogisms in the second and third figures pose an additional fundamental problem. As Storrs McCall points out, it is not always clear that the minor premise will be a special case of the major:

> In the second and third figure there cannot be any such pat formula as this, for the pairs Cesare *LXL* and Camestres *XLL*, Datisi *LXL* and Disamis *XLL*, indicate that the minor can be at one time the "special case" of the major [Cesare *LXL*, Datisi *LXL*, where the major is apodeictic], at another time its "general rule" [Camestres *XLL*, Disamis *XLL*, where the minor is apodeictic]. If we wish to uphold the principle that the modality of the conclusion follows that of the general rule, we must have a way of determining, in all cases, which premise *is* the general rule.[33]

This is closely related to the problem just noted: There one had to worry that there was no evident reason, with many second- and third-figure syllogisms, to regard *either* premise as giving a special case of the other; now it turns out that even if we accept that every premise pair contains one general rule and one special case, we lack a principled means of establishing in each case which is which.

So, in sum, let us say (1) that Rescher's rule for determining the status of modal conclusions does not correspond to any principles discussed or endorsed by Aristotle, but only contributes to a system that pretty much duplicates Aristotle's end results, (2) that that system achieves those results by appeal to a scale of modal strength that is also not to be found in Aristotle and that is seriously flawed on logical grounds, (3) that the further proposal concerning special cases works well for certain first-figure syllogisms (including Barbara *NAN*) but must not be confused with scientific demonstration nor, therefore, justified by direct appeal to Aristotle's views on science, (4) that in any case the proposal would have to be generously supplemented (if it is to provide the sought-after intuitive basis of Aristotle's modal logic) by a discussion of the foundations of Aristotle's conversion rules and of his proofs by *reductio* and by *ekthesis*, and (5) that even assuming each premise pair to consist of a general rule and a special case, one lacks a way to decide on non-arbitrary grounds which premise is the general rule and which is the special case.

### *4.3.4. McCall and distribution*

We have already examined McCall's approach to the two Barbaras. As for his more general approach, his central positive suggestion lies in his use of the traditional concept of distribution to preserve the special-case element of Rescher's general approach and to build on that in a more formal way. McCall's basic principle, generalizable to all three figures, now becomes that "a premise in which the middle term is distributed serves as the *general rule* of which the other premise is a *special case.*"[34] And where this is so, the modality of the conclusion can match that of the general rule, even if that is greater than the modality of the premise containing the special case. (Again, Theophrastus' *peiorem* rule is rejected.) For example, in

All *B*'s are necessarily *A*
All C's are *B*'s
______________
All *C*'s are necessarily *A*

the middle term *B* is distributed in the major premise but not in the minor. Thus we have a special case of the special-case conception, and that is why the modality of the conclusion here follows the major premise. Where the middle term is distributed in *both* premises, each, on McCall's view, may be considered a special case of the other. This validates Darapti *LXL* and Darapti *LXX*, both of which Aristotle had declared valid, but which appeared problematic on Rescher's approach.

One advantage is that now one can deal *directly* and forcefully with Cesare and the rest of that rough crowd from the second and third figures: Because the middle term is distributed in the major (necessity) premise, the minor asserts a special case of the major, so that the conclusion may be a necessity proposition and not just an assertoric one. Unfortunately, as McCall points out, third-figure Bocardo *ANN* is still a problem, as are Baroco *ANN* (second figure) and Felapton *ANN* (third figure). They still do not fit the pattern, because they are, according to Aristotle, invalid, but do qualify as valid under the stated "distribution" rule. To handle these special cases, McCall appends two special clauses: "(a) a universal premise cannot be the 'special case' of a particular," which rules out Baroco *ANN*, and "(b) a negative premise cannot be the 'special case' of an affirmative premise," which rules out Felapton and Bocardo *ANN*.[35]

This new proposal would answer the objection raised by McCall himself to Rescher's general rule. But it leaves other problems untouched and raises new ones of its own. First, the doctrine of distribution itself, even as applied to plain syllogisms, is post-Aristotelian and is unlikely to provide the intuitive key to Aristotle's own conception of what he was trying to do. Second, even if we accept the distribution approach, we must, I think, leave behind the special-case idea. At least, if there is an underlying connection, if the distribution rule captures and generalizes the special-case conception as McCall claims, it is far from evident exactly how it does so. (How intuitive is it, for example, to regard '*A N a B*' and '*C a B*' – the premises of Darapti *LXL* – as special cases of one another?) This is a serious issue, because the special-case idea was supposed to supply the fruitful intuition behind the more general distribution rule and so provide a way of obtaining Aristotelian results in an Aristotelian way.

Third, it is important to recognize that those exceptions to the rule (Baroco, Felapton, Bocardo *ANN*) stand as a substantial challenge to the interpretation, especially because it is put forward as an intuitive foundation of Aristotle's modal logic. How, without peeking at the results claimed by Aristotle for various syllogisms, would McCall arrive at his exception clauses? Put another way, when the special-case approach and the distribution approach clash, as with those three moods for which special conditions (which amount to amplifications of the special-case rule) were devised, how does one intuitively justify favoring one approach over the other? Without simply setting up things so that they turn out the way Aristotle wants, how does one justify McCall's declaration that certain items are not special cases of others when in fact the distribution principle says that they are?

A fourth difficulty concerns the relation of distribution to conversion. Although he does acknowledge that the "[conversion] laws themselves require intuitive justification,"[36] McCall seems not to have considered critically his assumption that in modal propositions the same terms are distributed as in their plain counterparts. For example, in '*A* no *B*', we know that *A* and *B* are distributed because the proposition tells us, of *every B*, that it is not an *A* and, of *every A*, that it is not a *B*. But notice that it is precisely because the statement entails its converse ('*B* no *A*') that it manages to tell us something about every *B* and also about every *A*. What, then, of '*A* necessarily fails to apply to every *B*'? *B* is distributed, because we know, of every B, that it is necessarily not an *A*. We could say that it is *modally* distributed, in that we know, of each *B*, that *A necessarily* does not apply to it. Moreover this is crucial to the upgrading of the modality of the conclusion. But what about *A*? McCall assumes that it, too, is distributed, and it is: from '*A N e B*' we can infer '*A e B*', and this converts to '*B e A*'. But *A* is not here modally distributed, and this is what will be needed, in some contexts, to assign the conclusion its proper modality. To get *A* modally distributed we would in effect have to assume that $E_n$ converts. But this begs a major, and by now familiar, question. In simply assuming in effect that both terms of such propositions are modally distributed, McCall's system does not so much solve the problems surrounding this conversion as ignore them.

The same problem confronts McCall's axiomatization of Aristotle's system, even though it is internally coherent and contains an argument for conversion of $E_n$.[37] Aside from primitive symbols, rules of formation, and rules of inference, this calculus includes among its axioms the four first-figure mixed moods recognized as valid by Aristotle (Barbara *NAN*, etc.) and the "law of conversion of the apodeictic *I*-premise."[38] Thus, conversion of $I_n$ "must be assumed without proof,"[39] and is asserted by axiom 11.[40] Meanwhile, conversion of $E_n$ is proved by use of Cesare *NAN* ('*B N* no *A* and *B* all *C*, therefore *A N* no *C*'), which is itself included as axiom 6. But this proof is circular, as we can see when we ask how things stand with the reducing syllogism, Cesare *NAN*. Returning to the root idea of the calculus, which the axioms are supposed to capture in a formal way, we might try to view Cesare *NAN* under the special-case rule. But as we saw, the idea has no *direct* intuitive application to this second-figure syllogism: It is illuminating and lends intuitive credibility to this argument form only when the major premise is converted so that we have Celarent of the first figure:

*A N* no *B* (the converse of '*B N* no *A*', Cesare's major premise)
*B* all *C*
―――――
*A N* no *C*

Here, with *first*-figure Celarent *NAN*, we do have a clear application of the special-case principle. Unfortunately, this justification of Cesare *NAN* requires the conversion of $E_n$, which is precisely the conversion law that Cesare was, in McCall's presentation, supposed to prove.

The only escape I can see from this circularity would be to do without conversion and to justify the syllogism directly via the distribution rule. *B* is the middle term, and if *B* is distributed in '*B N* no *A*', then the rule does apply and so yields a necessity proposition as conclusion. But as we have seen, *B* is *modally* distributed in '*B N* no *A*' only if the proposition validly converts to '*A N* no *B*'. Once again, $E_n$ conversion must be assumed in order to validate Cesare *NAN*, so that that mood cannot be used to validate the $E_n$ conversion.

So, in the end, McCall's axiomatization has, in effect, simply presupposed the conversion of $E_n$ and $I_n$. Speaking more generally, in simply taking over as axioms various conversions and primary-argument forms that Aristotle had declared valid, McCall has not really addressed the issue of semantic ambiguity drawn many centuries ago. He may have produced a system that is formally coherent, but he has not shown – nor begun to address the problem of showing – how this system plausibly expresses any particular conception(s) of necessity. Finally, McCall's own proposed intuitive justification (resting on traditional logic's doctrine of distribution) of the validity of Aristotle's mixed necessity/assertoric moods is, in itself, problematic and, as an interpretation of Aristotle, lacking in textual support.

### *4.3.5. Van Rijen, the homogeneity of terms, and improper predication*

Van Rijen's approach[41] resembles Rescher's in seeking the key to Aristotle's modal logic in his theory of scientific demonstration. And although he is in accord with McCall and others in wishing to avoid the introduction of two readings of individual modal propositions, van Rijen is less shy about proposing an interpretation whose results fail to tally perfectly with Aristotle's. He rightly observes that merely counting up the number of syllogisms preserved by a given interpretation is not a good measure of fidelity to Aristotle: The more important factor is that one attribute fewer *kinds* of errors, rather than fewer errors *simpliciter*, and that one be able to give a plausible explanation of any sort of error one does attribute to Aristotle.[42] His treatment of modal syllogistic is very selective, however, covering only the "apodeictic" moods of *Pr. An.* A.8–11, and these not in great detail. More important to him is the reconciliation of the *Prior* and *Posterior Analytics* on a certain point about the requirements of scientific demonstration, namely, that whereas in the latter an apodeictic conclusion requires two apodeictic premises, the former allows that in some cases an apodeictic conclusion is obtainable from one apodeictic and one assertoric premise (as in Barbara *NAN*).[43] This, in turn, serves his primary goal of giving a coherent reading of Aristotle's logic of modalities in *De Interpretatione* (on future contingencies) and *De Caelo* (on the relation between omnitemporality and necessity), as well as the *Analytics.*

His discussion of *De Caelo* and *De Interpretatione* is excellent, both for what it has to say about difficult details of key passages and for its attempt to include both works within a general framework of "absolute vs. hypothetical" necessity, where those relate ultimately to the essences and accidents of actual substances. His treatment of syllogisms involving necessity propositions seems to me less successful. The main points of that treatment are as follows:

(i) Demonstration within any particular Aristotelian science will take place against the background of a homogeneous subject matter. As *Post. An.* A.7–10 and A.28 maintain, "a 'science that is one' deals with a single subject or genus. Of this genus, the *kath' hauto* affections or attributes are studied."[44]

(ii) Any apodeictic proposition – which for van Rijen includes all of Aristotle's necessity propositions – contains two homogeneous terms. Terms *S* and *P* are homogeneous iff any thing *as named by S* and any thing *as named by P* both belong to some one genus or some homogeneous domain of discourse.[45]

(iii) Some of Aristotle's necessity propositions in the *Prior Analytics* are expressly instantiated by non-homogeneous pairs of terms (recall the weak *cop* examples we have discussed containing terms like White and Human): These must be disposed of, because they conflict with (ii). Van Rijen proposes that these are usually given by Aristotle in order to invalidate some mood by counterexample and that Aristotle is sometimes "nonchalant" about this, and so is liable to make a silly mistake occasionally.[46] Besides, these examples are suspect anyway, because they violate Aristotle's requirement (*Pr. An.* A.15) that premises be taken without temporal restriction.[47]

(iv) Now comes the critical confrontation with the two (mixed) Barbaras. The problem facing van Rijen is that on his interpretation, all necessity propositions must contain homogeneous terms, and it is far from clear how one will derive such a conclusion, as in Barbara *NAN*, from one necessity and one assertoric proposition. (The problem is parallel to that of deriving a *de dicto* conclusion from one *de dicto* premise and one assertoric premise, or of deriving a strong *cop* conclusion from one strong *cop* and one assertoric premise.) Van Rijen finds a solution in the distinction of *Post. An.* A.22 between proper or unqualified predication and predication *per accidens:*

> You can say truly 'the white (thing) is walking' or 'that big (thing) is a log' and also 'the log is big' or 'the man is walking'. Yet, there is a difference between the latter two statements and the former two. When I say 'the white (thing) is a log' I mean that that which happens to be white is a log, and not that the subject/substratum of wood is the white (thing). For it is not the case that being white or *qua* white (thing) it became wood; so it is only *per accidens*. But when I say 'the log is white' I do not mean that

> something else which happens to be a log is white; . . . but the log is the subject/substratum that became (white) not *qua* something else but *qua* log or *qua* this log.
>
> If we are to legislate, let us speak about *predication* in the latter case, and in the former not at all about predication or not about predication *simpliciter (haplōs)*, but *per accidens*. . . . Let us assume, then, that the predicate is always said *simpliciter* of that of which it is said, and not *per accidens*. For that is the way in which demonstrations demonstrate. (82a2ff., van Rijen's translation)

What van Rijen seems to want from this passage is the idea that within the context of scientific demonstration, at least, the *subject* term of one's propositions will signify (name) some entity of which the predicate is predicated *simpliciter*. Then he formulates two "Aristotelian demonstrational maxims":

> maxim 1: the domain of discourse of a demonstration must be homogeneous.
>
> maxim 2: the subject terms of all the premisses of a demonstrative deduction must be homogeneous with the name of the category or kind of elements of the homogeneous domain of discourse.[48]

Let us apply this now to Barbara *NAN:*

$$\frac{\begin{array}{l} A\ N\ \text{all}\ B \\ B\ \text{all}\ C \end{array}}{A\ N\ \text{all}\ C}$$

Because the first premise is "apodeictic," the terms *A* and *B* are (by maxim 1) homogeneous. And because by maxim 2 (assuming that Barbara *NAN* is supposed to be demonstrative) the subject term *C* of the second premise must be homogeneous with the background domain of discourse, *C* will be homogeneous with *A* and *B*. But then if assertoric '*A* all *C*' is true (which it will be, given the premises), and *A* and *C* are homogeneous terms, it is evident that '*A N* all *C*' will also be true. So Barbara *NAN* is valid.

By contrast, consider Barbara *ANN:*

$$\frac{\begin{array}{l} A\ \text{all}\ B \\ B\ N\ \text{all}\ C \end{array}}{A\ N\ \text{all}\ C}$$

Because '*B N* all *C*' is "apodeictic," we know by maxim 1 that *B* and *C* are homogeneous. And if we suppose that Barbara *ANN* is supposed to be demonstrative (as Barbara *NAN* was supposed to be), then by maxim 2 the subject term of the other premise (*B*) is "homogeneous with the name of the category or kind of elements of the homogeneous domain of discourse." If so, *B* will be homogeneous with the terms of the minor premise (*B* and *C*). But this does not tell us anything about the homogeneity of *A* with either *B* or *C*.[49] Thus we are unable to derive the apodeictic '*A N* all *C*'. So Barbara *NAN* is valid, and Barbara *ANN* is not, just as Aristotle says.

(v) Working through *Pr. An.* A.8–11 with these maxims in mind, one finds that all of Aristotle's rejected moods, except one ($AO_nO_n$-2), are in fact invalid. On the other hand, six moods declared valid by Aristotle are in fact invalid ($A_nAI_n$-3, $E_nAO_n$-3, $A_nII_n$-3, $E_nIO_n$-3, $AA_nI_n$-3, and $IA_nI_n$-3).

That seven moods do not come out right (just counting these three chapters, one may add) is said not to be so alarming, because they can be explained away. $AO_nO_n$-2 is declared invalid because of Aristotle's nonchalance in constructing a counterexample; it is nothing more than a "silly fault."[50] The rest are "all reducible to the same source and, moreover, readily explicible: the application of assertoric conversion rules may in some contexts interfere with maxim 2."[51]

I am afraid this solution cannot be correct. First, I see no evidence that Aristotle, in either the *Prior* or *Posterior Analytics*, wanted to use Barbara *NAN* for scientific demonstration. The problem with such a syllogism, according to principles laid down in the *Posterior Analytics*, is not that it is not valid (for it is valid), but that it is not scientifically demonstrative: Because the assertoric premise in each may not have been true, the premise pair cannot reveal *why* the extreme terms *must* stand in some essential relation to one another. Recall A.6, 74b26–30:

> That the deduction must depend on necessities is evident from this too: if, when there is a demonstration, a man who has not got an account of the reason why does not have understanding, and it might be that *A* belongs to *C* from necessity but that *B*, the middle term through which it was demonstrated, does not hold from necessity, then he does not know the reason why.[52]

So Aristotle is not looking, in *Post. An.* A.22, for a way to salvage this sort of mood for demonstrative purposes: It is clearly not salvageable for that purpose.

Why does van Rijen take Barbara *NAN* to be an apodeictic mood? Apparently just because it attempts to prove an "apodeictic" (i.e., scientific) conclusion. But I know of no evidence that Aristotle ever took Barbara *NAN* or its conclusion to be scientific in that strong sense. Van

Rijen even goes so far as to say that "a sentence '*B* belongs of necessity to all *A*' can only be true if '*A*' and '*B*' are homogeneous terms."[53] Although this would be true of the conclusion of a scientific demonstration, it is not true of necessity propositions in general. In fact, it conflicts, as van Rijen acknowledges, with several of Aristotle's own examples – examples that van Rijen must then discount in the manner previously cited. But as argued earlier, it is misguided to even try to explain away these examples, for they correspond to firmly held Aristotelian essentialist views.

To put much of this another way, although it may not be immediately obvious, van Rijen's maxim 2 does not rescue the distressed Barbara *NAN;* rather, it guarantees, by homogenization of the premises, that the syllogism will in effect be Barbara *NNN* – with all three propositions (in our terminology) strong *cop* necessities, no less. But Aristotle's Barbara *NAN*, as it appears in the *Prior Analytics* (and even in *Post. An.*, if that is the mood in view at 74b26–30), is *meant* to contain a genuinely plain or assertoric premise.

As for the alleged discrepancy between the *Prior* and *Posterior Analytics* that van Rijen's reading is supposed to resolve, I would suggest that there simply is no such discrepancy. The discrepancy arises only if certain mixed necessity-assertoric moods of the *Prior Analytics* are taken to be demonstrative in the strong sense discussed in the *Posterior Analytics*, where *two* necessity premises are expressly required. But because Barbara *NAN* is not supposed to be scientifically demonstrative, there is no conflict with that requirement.

### *4.3.6. Wieland and Pr. An. A.12: modalities* sui generis

Like McCall and others, Wieland recognizes that various currently familiar formal approaches to modal logic fail, when applied to Aristotle's claims, to yield a consistent modal system encompassing conversions, syllogisms, and so forth.[54] Like McCall, again, he insists that one must accept the basic principles of the system, on the ground that if we forsake those, we run the risk of losing our grip on Aristotle's own intentions.[55] Wieland stands almost alone, however, in choosing to honor, come what may, his understanding of Aristotle's statement in *Pr. An.* A.12 that "there will not be a syllogism of belonging (*tou huparchein*) unless both premises are of belonging (*en tōi huparchein*), whereas there will be a necessity syllogism when only one of the premises is a necessity proposition" (32a6–8). Aristotle has just discussed, in A.9–11, valid syllogisms with a necessity conclusion and only one necessity premise. Wieland takes a hard line on the first clause of the lines just quoted: If an assertoric conclusion requires two assertoric premises, then one cannot derive an assertoric conclusion from one necessity and one assertoric premise, even though one can draw a necessity conclusion from such a combination. By contrast, most commentators would take it for granted that Barbara *ANA* and *NAA* are valid, simply because the necessity premise in each case entails its assertoric counterpart, giving the assertoric Barbara *AAA*. (Similarly, in modern propositional modal logic, one has the principle 'if nec: *p*, then *p*'.) Barbara *NAA* would normally be accepted, given the validity of Barbara *NAN*, on grounds also that the latter's necessity conclusion entails the corresponding assertoric proposition.

Wieland freely accepts that his interpretation of A.12 requires the rejection of that commonly accepted "intermodal law"; indeed, his insistence on the point is precisely the most distinctive and interesting feature of his interpretation. In his view, the *reason* Aristotle would not accept that necessarily applying entails plain applying is that the former pertains to such apodeictic propositions as one finds in Aristotelian science – principles about which one cannot be in error, which are not true in the same sense as "ordinary" (*gewöhnlich*) or merely "factual" truths.[56] On his view, plain applying is not a weaker version of necessary application or belonging, nor is it a kind of genus ('Applying') of which necessary application is one species. Rather, these are two distinctive modalities that mutually exclude one another: They, along with two-way possibility, are all modalities *sui generis* and so are not entailed by one another and are not to be ranked in any scale of relative strength.[57]

This means, among other things, that Barbara *ANA* and *NAA* must be rejected, and Aristotle makes this explicit (according to Wieland) in the passage quoted earlier from chapter 12. On the other hand, Wieland acknowledges that Aristotle's own comments on this and several other mixed moods indicate that the premises do entail *some* conclusion. The key phrase, in Wieland's view, is "the conclusion is not necessary," where it seems clear (as he rightly argues, against Patzig[58]) that Aristotle means "the conclusion is not a necessity proposition." What these moods do entail, according to Wieland, is a "non-necessity conclusion"; that is, they entail only that something of some modality other than necessity does follow. Or, there is a "disjunctive" conclusion to the effect that some statement of this modality or that follows, only not a necessary conclusion. This novel sort of conclusion he symbolizes as $\sim (N)$, which is not to be confused with the familiar negation of a necessity proposition (which would simply be equivalent to a one-way possibility proposition).[59]

Although Wieland is quite right in attempting[60] to relate Aristotle's modal logic to his views on science, on definition, and on accidental vs. *per se* predication, there are serious problems with his interpretation of the modal system and thence with his manner of bringing together modal logic and Aristotelian science and metaphysics. The next few pages will develop three of these difficulties: (1) The ultimate implications of his treatment of Barbara *ANA* and *NAA* are very difficult to accept. (2) There is a much more likely interpretation of A.12 than his, and one on which it is not ruled out that necessarily applying will entail plain applying. (3) There is conclusive evidence that Aristotle did accept the implication from "necessarily applies" to (plain) "applies," even when the former occurs in the context of a truly "apodeictic" proposition. These points taken together refute the central tenet of Wieland's approach (that necessity, two-way and one-way possibility, and plain propositions each represent a modality *sui generis*), while at the same time leading us back, ultimately, to the *cop* reading advocated here.

First, as to Barbara *ANA*, the specific mood of which Aristotle says "the conclusion will not be necessary," Wieland never spells out what its "disjunctive" conclusion might be. But we can render the matter somewhat less vague by considering that there are only four sorts of propositions that could serve as disjuncts: *A, N, P*, and *PP*. We know that on Wieland's view of A.12 no assertoric conclusion follows, and Aristotle tells us that no necessity conclusion follows, either. This leaves *P* and *PP*, so that if there is a disjunctive conclusion, it presumably is '*P* or *PP*'. But on what logical basis can one derive '*P* or *PP*' (i.e., '*A P* all *C* or *A PP* all *C*') from the premises '*A* all *B*' and '*B N* all *C*' – especially when one is barred from using the intermodal laws that necessarily applying entails applying, and that this in turn implies one-way possibly applying? On what possible philosophical or logical grounds would anyone who ruled out both an assertoric conclusion and a necessity conclusion still want to assert that a '*P* or *PP*' conclusion followed? In the absence of any plausible response to this question, we already have something close to a *reductio ad absurdum* of Wieland's interpretation.

The principal textual source of that interpretation was A.12, which seemed to say that an assertoric conclusion required two assertoric premises and hence entailed the invalidity of Barbara *NAA* and *ANA*. The usual reading of the passage – based largely on a reluctance to give up the inference from '*A N* all *B*' to '*A* all *B*' – is that Aristotle here sets a *minimal* condition, so that an assertoric conclusion requires two premises that assert the actual application of some predicate to a subject. "Assertoric" in this broad sense thus includes plain (non-modal) propositions and necessity ones, but not one- or two-way possibility statements.[61] But just here there is a problem for *any* interpretation, in that, as Ross points out,[62] Aristotle will go on in chapters 16, 19, and 22 to argue that appropriate combinations of a necessity premise with a two-way possibility premise will entail an assertoric conclusion. In these moods, one has an assertoric conclusion but *no* assertoric premises in the narrow sense. Moreover, unless one wants to say that *PP* is logically at least as strong as *A* (i.e., entails *A*), then even the "minimal condition" interpretation of A.12 is defeated, because the mood will not have two premises of belonging, even in the broad sense. But we noted earlier that *PP* does not entail *A*.

These examples from chapters 16, 19, and 22 of mixed *N, PP/A* moods would in any case have to be reconciled with Aristotle's next remark, which together with the sentence quoted in the first paragraph of this section (32a6–8) constitutes the whole of the brief chapter 12:

> In both cases [those of moods with *A* or *N* conclusions], whether the conclusions are affirmative or negative, it is necessary for one premise or the other to be like the conclusion. By "like" (*homoion*) I mean, if assertoric, then assertoric, if necessary, then necessary. Thus, this much is clear, that the conclusion will be neither necessary nor assertoric if one has neither a necessary nor an assertoric premise. (32a8–14)

Because in Celarent *N, PP/A*, for example (36a7–17), neither premise has exactly the same modality as the conclusion, here one must (1) take "like" (and "assertoric") to mean "is at least as strong as" ("is at least assertoric"), or (2) retain a narrower sense of these terms and conclude that Aristotle later discovered some moods he had not anticipated as of A.12 (though he never went back to correct his earlier, premature statement), or (3) take this remark in A.12 as intended to apply only to results achieved "so far" (i.e., in chapters A.8–11 on syllogisms with plain or necessity premises). These options are not mutually exclusive. For example, the last two readings might both be true. Also, the first could well be true along with the third and the latter conjunct of the second. And, of course, one might take "assertoric" broadly in the first sentence of the chapter, but "like" more narrowly, then make appropriate decisions about (1)–(3). There are several plausible, and not merely logically possible, readings.

We probably shall never know the whole truth about this passage. But I think the old controversy about the broader "at least as strong as" reading of "assertoric" and of "like" can be partially settled and that the portion of the passage just quoted (32a9–14) can be seen as making an interesting and correct claim that it cannot make on the narrower reading. It seems to me that "assertoric" (*en tōi huparchein*) must be read in the broader sense in the first sentence of the chapter, because Aristotle does clearly recognize in chapters 9–11 the validity of certain moods with one assertoric and one necessity premise but an *assertoric* conclusion (e.g., Camestres, Baroco, Felapton *NAA;* the evidence for this will be given shortly). That being so, his summary statement in the first sentence of A.12, *even if limited to chapters 8–11*, must mean "assertoric" in the sense of "asserts actual application" – where actual application obviously includes necessary application.

This does not reconcile the first sentence of A.12 with the later *N, PP/A* syllogisms, which is why I think that the second conjunct of (2) or (3) – or both – is probably also true. Still, we should remember that the point uppermost in Aristotle's mind probably was to emphasize (in the first sentence of the chapter) the surprising result (already contested by Theophrastus and Eudemus and still controversial today) that in certain circumstances only one, rather than two, necessity premise is needed to derive a necessity conclusion.[63]

The second sentence of A.12 then looks at the results of chapters 8–11 from a slightly different point of view, emphasizing now a point of *similarity* between those syllogisms having an assertoric conclusion and those having a necessity conclusion: In both cases it holds that one premise or the other (or both) will have to be "like" the conclusion. Recalling that Aristotle has explicitly recognized four main types of syllogisms in these chapters (*A, A/A; N, A/A; N, A/N; N, N/N*), and assuming it is these he has in mind, his statement will be true with "like" taken in either a narrow or a broad sense. In the narrow sense, a premise will be like an assertoric, as opposed to a necessity, conclusion, just in case it is itself assertoric as opposed to necessary. And only a necessity premise will be like a necessity conclusion. In the broad sense, a necessity premise would be like an assertoric conclusion, because it asserts all that the conclusion asserts (and perhaps more), namely, the actual application of some predicate to a subject; but an assertoric premise will not be like a necessity conclusion because it does not assert all that the conclusion asserts, namely, the necessary application of some predicate to a subject. Put another way, being (with respect to modality) "like" a proposition *P* requires being in the same modal category or categories as *P*, where there are (at this point) two modal categories: (1) that of asserting actual application and (2) that of asserting necessary application. Merely assertoric propositions belong to (1), while necessity propositions belong to (1), but also, and distinctively, to (2). Thus necessity propositions are "*en tōi huparchein*" and hence like merely assertoric ones; but the latter are not "*tou d' anangkaiou*," hence not like necessity ones. One could certainly find fault with this usage, but the question is whether Aristotle could have meant himself to have been understood in this way, and I do think he could have. Notice that if we do take "like" in a narrow sense, and also assume that Aristotle would recognize as valid certain *N, N/A* syllogisms (as I think he surely would), we must assume that he is not thinking of such syllogisms here, but only the four types he has actually recognized in A.8–11. If we take "like" more broadly, his statement is consistent with the validity of some *N, N/A* syllogisms, as well as with that of the *N, PP/A* syllogisms recognized later. But this does not show that he had any of these additional cases in mind, or that "like" should be taken broadly.

With all this as preamble, we may observe that the last sentence of A.12 may well be drawing out an important point that does hold for the entire system and that cannot be made on the narrower reading of "like": If we are going to conclude that some predicate *P* applies *necessarily* to some subject *S*, then we have to be told in the premises at least that *something* belongs necessarily to something. If we do not get that much in the premises, then there is no way we can infer that anything, including the major term *P*, belongs necessarily to anything else, including the minor term *S*. Similarly, if we do not know from the premises at least that something *applies*, whether plainly or necessarily, to something else, then we can never infer in the conclusion that anything actually applies to anything else. With regard to syllogisms having necessity conclusions, this means that although we can, surprising as it may seem, obtain a necessity conclusion using less than two necessity premises, we must still have at least one necessity premise. With syllogisms having merely assertoric conclusions, we need at least one premise asserting the (plain or necessary) application of some predicate to a subject. This is a weaker requirement than that expressed in Aristotle's preceding sentence [which required two (at least) assertoric premises for an assertoric conclusion] and would in fact be consistent with the later admission of *N, PP/A* moods (again, with "assertoric" taken broadly), even if Aristotle did not have them in mind here. This is less daring than certain principles formulated by Theophrastus and Averroës, but it does have the advantage of being true.

Whether Aristotle actually meant to express this principle, or merely meant to summarize in a more superficial way the results of A.8–11, is impossible to say. (To put the two options slightly differently: On the latter, he merely had in mind that the principle was true *a fortiori*, i.e., *because* things are as stated in the first sentence of the chapter; on the former, he grasped directly, as it were, the necessity of the principle.) The fact that he has just (incorrectly, by his own later lights) asserted in the first sentence of the chapter that assertoric conclusions require two (at least) assertoric premises does not show that he did not have the weaker (and correct) principle in mind in the last sentence of A.12. The discovery of valid *N, PP/A* moods (if there are any such valid moods; Aristotle's claims on their behalf in A.16, 19, and 22 will be examined in due course) only shows that a certain stronger principle regarding assertoric conclusions does not also hold.

To return now to Wieland, when Aristotle implies that certain mixed premises do entail some conclusion, but says that they do not entail a necessity conclusion, there is no bar to the common and natural assumption that in fact, and in Aristotle's view, what they do entail is an assertoric conclusion.

But does Aristotle *ever say* that such mixed premise pairs will give a plain conclusion? Wieland says no, and to a large extent he is right: In general, Aristotle tends to say that "the conclusion will not be necessary," without explicitly specifying what sort of conclusion follows. (See 30a24, 29, 35; 30b3, 9, 19, 23, 32; 31a4, 13, 16, 23, 38; 31b1, 3, 21, 26, 38, all cited by Wieland.[64]) But apparently Wieland misses the fact that one of these very passages does explicitly acknowledge an assertoric conclusion while rejecting a necessity one:

> Moreover if one selects terms it is possible to show that the conclusion [of Camestres *NAN*] is not necessary *simpliciter*, but is necessary, these things being the case (*ouk estin anangkaion haplōs, alla toutōn ontōn anangkaion*). For example let A = animal, B = human, C = white, and the premises be taken in the same way [i.e., as 'Animal *N* all Human', 'Animal no White']. For it is possible that animal applies to no white thing. Nor, then, will human apply to any white, but not of necessity [will it fail to apply]. For it is possible for a human to become white, only not so long as no white thing is an animal. Therefore, these things being the case the conclusion will necessarily follow, but it will not be necessary *simpliciter (hōste toutōn men ontōn anangkaion estai to sumperasma, haplōs de ouk anangkaion).* (30b31–40)

The opening and concluding clauses invoke the difference between a conclusion's being the necessary consequence of the premises (or "being necessary if these things are the case") and its not only following from the premises but also being itself a proposition of necessity (*anangkē haplōs*). This much would, I think, be uncontroversial. But Aristotle also affirms here the validity of Camestres *NAA*

$$\frac{\begin{array}{l} A\ N\ \text{all}\ B \\ A\ \text{no}\ C \end{array}}{B\ \text{no}\ C}$$

("the conclusion ['*B* no *C*'] is not necessary *simpliciter*, but is necessary, these things being the case") and even provides a substitution instance:

Animal *N* all human
Animal no white

---

Human no white

Camestres *NAN*, by contrast, he declares invalid, indicating that the same terms will constitute a counterexample. Thus, as he says, the conclusion ('Human no White') will follow, but it will not be necessary *simpliciter* (i.e., 'Human *N* no White' does not follow).

Further investigation reveals that Aristotle also *explicitly* affirms the validity of moods deriving an assertoric conclusion from one assertoric and one necessity proposition at 31a10–14 (Baroco *NAA*) and at 31a37–b4. (Felapton *NAA:* From '*A* no *C*' and '*B N* all *C*' it follows "that *A* fails to apply to some *B*, but not of necessity [does it fail to apply]. For it was shown in the first figure that if the negative premise is not itself necessary, then neither will the conclusion be necessary.") The placement of this last example is especially striking, because it occurs in chapter 11 and therefore only shortly precedes the passage from A.12 on which Wieland places so much weight.

However, it would be quite wrong to suppose that Wieland's approach represents nothing more than stubborn insistence on an eccentric reading of *Pr. An.* A.12. His positive philosophical rationale for denying that necessarily applying entails plain applying is, again, that the latter is a distinctive modality in its own right, rather than a genus of which the former is a subtype. Like contingency, plain belonging (or "ordinary" belonging or simple "factual" belonging) would exclude and be excluded by necessary application. At the same time, plain belonging cannot be identified with two-way possibility, because propositions expressing the latter modality have a different "microstructure" from assertoric ones, as is shown by the fact that some assertoric assertions are affirmative, and some negative, whereas *all* contingency propositions are, as Aristotle says, affirmative.[65] One could add that two-way possibly applying does not entail applying, so their respective propositional expressions can hardly be identified with one another.

This is an original and interesting proposal. But aside from the objectionable (and, I think, fatal) consequences we have already considered, it does not seem to square with Aristotle's metaphysics or, above all, with the theory of the four predicables. Aristotle says in the *Topics* (and this is assumed in the *Prior Analytics*, as we shall see in a moment) that there are four predicative relations in which a predicate (or predicable) may stand to a subject: It may be the subject's genus (with differentia), or essence, or proprium, or accident. In view of this, the most plausible construal of Wieland's plain applying would be that it means *applying, and* applying in the manner of an accident, as ''pale'' or ''in the Agora'' applies to Socrates. This makes it mutually incompatible with necessarily belonging, still distinct from two-way possibility, and naturally describable as ''everyday'' or ''ordinary'' or ''factual'' applying. So the positive assertoric statement '*A* all *B*' would entail that all the *B*'s are *A*'s and that they are accidentally (*kata sumbebēkos*) *A*. Similarly, '*A* no *B*' would entail that no *B*'s are *A*'s and that *A* is related accidentally to all *B*'s.

Now this would mean that in surveying a subject's essence, its genera, differentiae, propria, and accidents, with an eye to constructing syllogisms about it, the only source of assertoric premises should be the accidents. But this is clearly counter to what Aristotle says in chapters 27 and 28 of *Pr. An.* A, where *all* these relations provide assertoric premises.

That necessary relations provide assertoric premises is manifest also in Aristotle's examples of assertoric premises in chapters 4–7 of the *Prior Analytics.* Specifically, the free mixing of such genus-species examples as Animal/Human with accidentally linked pairs such as White/Cloak throughout the assertoric syllogistic of those early chapters – and in fact, the *predominant* use there of necessary relations as the source of assertoric propositions[66] – supports the implication of A.27–28 that necessary relations entail both necessity and, *a fortiori,* assertoric propositions.

At the same time, Wieland is right that applying (or failing to apply) in the manner of an accident, on the one hand, and necessarily applying (whether as essence, genus, differentia, or proprium), on the other, are mutually exclusive. This gives no support to his view, however, but is quite in keeping both with the *cop* reading of these modalities and with the view that Aristotle's assertoric statements are intended to cover all cases in which a predicate belongs to a subject, whether necessarily or accidentally.

Finally, perhaps I may be permitted to mention that although Wieland tries to avoid taking a definite position on the internal structure of modal propositions, he is, in effect, forced to do just that at more than one critical point in his discussion. And when he does so, the reading he favors is that which modalizes the copula. For example, he wishes to argue that any preeminence of ''apodeictic'' propositions rests not on their being logically stronger than assertoric (and one-way possibility) ones but on their role in theoretical science. ''Between such sentences [assertoric and necessary] there obtains no relation of implication. They indicate not different degrees of certainty, *but rather different types of relation between subject and predicate.*''[67] With this last clause, the truth, as Aristotle would put it, has forced itself upon the investigator. Only one must insist, for reasons given earlier, that a predicate ''applying to a subject'' is used by Aristotle in a broad sense covering cases in which the predicate belongs as genus, essence, or proprium, as well as cases in which it belongs, but only as an accident. And where Animal, say, applies to some horse as its genus, the non-modal statement that that horse is in fact an animal will follow.

With Wieland's remark about different types of relations between subject and predicate, we return full circle to the modal copula reading. In the meantime, we have seen that the primary task facing the interpreter is not to devise a reading, whether formal or informal, that recognizes only one type of modal proposition. Rather, one must first uncover the sources of the modal system in those modal facts and relations one wants to express and about which one wishes to reason. This means exploring the connections between Aristotle's modal logic and his essentialist metaphysics, on the one hand, and working out the details of the logical system, on the other. Before returning to the latter task, we must examine two recent attempts to produce a consistent formal modal of the system as Aristotle left it *and* provide an Aristotelian semantics based on his essentialism.

### *4.3.7. Thom and Johnson: essentialist semantics at last?*

Fred Johnson[68] builds on the work of McCall, adopting as (''unstarred'') axioms McCall's axioms for assertoric and necessity propositions. These include, besides 'all a are a' (labeled **A1**) and assertoric Barbara and Darii (**A2, 3**), axioms for Barbara, Celarent, Darii, and Ferio *NAN* (**A5–8**) and Baroco and Bocardo *NNN* (**A9, 10**), conversion of the necessary *I* proposition (**A11**), the assumption that for each property there is at least one subject to which it applies essentially (**A4**), and three axioms stating that the necessity versions of *A, I,* and *O* propositions entail their assertoric counterparts (**A12–14**).[69]

He then introduces an elegant set-theoretic model for the system that provides a semantics based on key notions of Aristotelian essentialism. Thus, he has not just given us *a* semantics (thereby, as Johnson remarks, filling that lacuna in McCall's treatment[70]), but the kind of semantics that has been called for in this study. Paul Thom takes another stride forward, using the same sort of set-theoretic approach, but correcting for certain ways in which he believes that Johnson (whose interest is explicitly mathematical rather than historical[71]) has departed in significant ways from Aristotelian metaphysics. At the same time, neither author relies on the introduction of two readings of necessity propositions. (Both papers are restricted to assertoric and necessity propositions, and so do not discuss two-way premises or any mixtures of these with necessity or assertoric ones.[72]) All of this constitutes a very welcome development, but also a most remarkable one: If some of the principal theses argued here are correct, it simply is not possible to model Aristotle's system, retaining all the elements at the base of the system, while also providing a genuinely Aristotelian semantics that does not introduce two readings of his modal propositions.

So let us see whether this has in fact been accomplished. Johnson's formal presentation of his semantics[73] has been usefully described more informally by Thom[74]:

> To each term variable *a* there are assigned four sets, two of which exclude each other and exhaust the universe, the other two being respectively included in the first pair. These sets are to be thought of as (i) the a's, (ii) the non-a's, (iii) the essential a's, and (iv) the essential non-a's. The set of a's may or may not coincide with the set of essential a's. Johnson assigns truth-conditions to assertoric propositions as follows:
>
> $ab^a$ is true iff the a's include the b's;
> $ab^e$ is true iff the non-a's include the b's;
> $ab^i$ is true iff the a's overlap the b's;
> $ab^o$ is true iff the non-a's overlap the b's.
>
> "Apodeictic" (necessity) propositions are interpreted as follows:
>
> $Lab^a$ is true iff the essential a's include the b's;
> $Lab^e$ is true iff the essential non-a's include the b's;
> $Lab^i$ is true iff the essential a's overlap the essential b's;
> $Lab^o$ is true iff the essential non-a's overlap the essential b's.

Thom criticizes Johnson on several counts, some of which are telling, but others of which seem at least in large part answerable.[75] I would like to focus here, however, on what seems to me the critical problem in Johnson's treatment of Aristotelian semantics – one not criticized by Thom, but in fact the one that allows Johnson to believe that he has preserved all of Aristotle's basic principles (including the validity of Barbara *NAN and* of $I_n$ conversion) without introducing any alternation between two readings of necessity.

The problem lies in Johnson's interpretation of necessity propositions. Two of these are perfectly in order:

> $Lab^a$ is true iff the essential a's include the b's.

For example, let $a$ = Horse and $b$ = White Thing in the Agora, in a possible situation in which all white things in the Agora are horses. Because the $b$'s are all in fact horses, and all horses are essentially horses, the $b$'s will be included in the essential $a$'s. Thus, in this situation, 'All White Things in the Agora are Essential Horses' will be true. Here we are on familiar ground, for this captures what is traditionally regarded as the *de re* reading of the proposition. (Johnson appears to intend a modal predicate formulation, but let us put aside any objections to that for now.)

> $Lab^e$ is true iff the essential non-a's include the b's.

For example, let $a$ = Horse and $b$ = White Thing in the Agora, where all the latter are in fact cats. Then 'All White Things in the Agora are essentially non-Horses' will be true. Again we have a version of *de re* modality.

For the *I* proposition, we would expect, given the interpretations of *A* and *E* just given,

> $Lab^i$ is true iff the essential a's overlap the b's (i.e., if at least one b is an essential a).

For example, let $a$ = Horse and $b$ = White Thing in the Agora, where at least one white thing in the Agora is a horse. But instead we get

> $Lab^i$ is true iff the essential a's overlap the essential b's.

Why, now, the *essential* $b$'s, rather than just the actual $b$'s, as in the other two definitions? This is extremely odd, not only because it departs from the style of definition given for $A_n$ and $O_n$, but even more so because it seems that now we cannot even *say* 'Some *accidental* White Things in the Agora (e.g., Pegasus, Silver) are Essential Horses'. In Johnson's system, the only affirmative particular necessity proposition we could make (or, the only available way to read such a proposition) would be 'Some *Essential* White Things in the Agora are Essential Horses'. And that is a very different claim. Moreover, given the existential import of Aristotle's universal affirmatives, $Lab^a$ (as defined by Johnson) *entails* 'the essential *A*'s overlap the *B*'s. This is standard fare, but in Johnson's system we can no longer even express this entailment, for we cannot express the entailed statement, 'Some *B* is an essential *A*'. Again, the only necessary universal particular in the system is 'Some essential *B* is an essential *A*'.

My objection is not that the Johnsonian $I_n$ proposition is odd, for in his use of "Essential" to modify *both* subject and predicate terms, Johnson has hit upon an interesting propositional form in which some important Aristotelian statements might well be expressed. (We shall return to this in Chapter 6.) The problem is rather that in departing from the style of definition he gave for $A_n$ and $E_n$, Johnson has deprived us of the ability to make a kind of $I_n$ statement that Aristotle can and does make within his system and that manifestly follows from statements that Johnson does include. Similar objections hold, of course, for Johnson's interpretation of $O_n$.

Why this difference between the treatment of $I_n$ and $O_n$ and that of $A_n$ and $E_n$? Johnson does not offer any explanation or comment on the matter. (Nor does he provide anywhere in the article any concrete Aristotelian substitution instance of any formula.) I conjecture that his treatment of $I_n$ and $O_n$ has to do with the fact that the only necessity conversion presupposed at the base of the system is that of $I_n$ (axiom **A11**). But $I_n$, if interpreted in the style in which Johnson interprets $A_n$ and $E_n$, would *not* convert: The fact that some *b* is among the essential *a*'s would not entail that some *a* is among the essential *b*'s. (Again, let $a$ = Horse and $b$ = White Thing in the Agora, where some of the latter are in fact horses.) Interpreted as "the essential *a*'s overlap the essential *b*'s," however, $I_n$ obviously does convert. So we have some good news and some bad news: $I_n$ does, on Johnson's definition, convert; but Johnson's semantics cannot be considered an adequate representation of Aristotle's semantics.

The converse problem arises with Johnson's versions of $A_n$ and $E_n$: Although they do capture one sort of necessity statement (i.e., *de re* ones) precluded by his $I_n$ and $O_n$, they cannot assert the kind of *per se* connections between predicate and subject that one will need to assert in certain contexts (above all, that of scientific demonstration).

Finally, a problem that is now familiar: $E_n$, according to Aristotle, does convert, but on Johnson's interpretation the conversion is liable to counterexamples: Let $A$ = Human and $B$ = White Thing in the Agora, where the *B*'s are all swans (or cloaks). Then all the *B*'s are among the essential non-*A*'s, so that Johnson's '$Lab^e$' is true; but his '$Lba^e$' ('all Humans are among the Essential non-White Things in the Agora') will be false. Similarly, in Aristotle's system, $A_n$ converts "to a particular," but on Johnson's interpretation it does not: The terms and situation described earlier constitute a counterexample. This implies, in addition, that none of Aristotle's proofs using conversion of $E_n$ or $A_n$ will go through in Johnson's system.

I think it will be agreed that these are severe problems. The remedy will not be difficult to find, however (assuming, for the sake of the argument, that Johnson's interpretations with two occurrences of "essential" are the Aristotelian readings one would need for scientific purposes). We must recognize *two* $I_n$ propositions, $Lab^i1$ and $Lab^i2$, one interpreted as 'Some *b*'s are among the essential *a*'s', the other as 'Some essential *b*'s are among the essential *a*'s'. (The second is stronger, because it entails the first, but not vice versa.) Then among the axioms, we include the conversion of $Lab^i2$, but not that of $Lab^i1$. And similarly for the rest of Aristotle's propositions of necessity. This results in a variation on the two-reading approach advocated here.

After developing a number of his own criticisms of Johnson, Paul Thom describes how he will nonetheless build on one "very attractive idea" behind Johnson's interpretation:

> This is the idea of distinguishing some classes (what he calls the essential a's, etc.) and then basing the semantics on standard (non-modal) set-theoretic relations among these and other classes. This is elegant. It supposes no special kinds of predication, and makes no appeal to the complexities of 20th-century modal logic.
>
> In what follows I shall suppose that among the terms which may be substituted for the variables of Aristotle's modal syllogistic there are some which may be distinguished as having a special character. These may be thought of as *kath' hauto* terms. I shall call them essences. My initial aim will be, à la Johnson, but (if possible) consistently with Aristotelian logic and metaphysics, to specify truth-conditions for apodeictic categorical propositions purely in terms of set-theoretic relations among the classes picked out by these distinguished or undistinguished terms.[76]

(The main difference between Thom and Johnson on the nature of essences is that whereas Johnson allows that for the same property *P*, one thing might be essentially *P* and another accidentally *P*, Thom's essential terms apply essentially to everything to which they apply. Thus, for example, White cannot apply essentially to one thing, and accidentally to another.)

There are many points of interest in Thom's paper, some of which I have already quoted with approval, and some of which (e.g., his discussion of modal *ekthesis*) deserve fuller consideration on another occasion. But I think it best that we focus here, as with Johnson's paper, on the critical question whether or not Thom has succeeded in giving a coherent, "single-reading" interpretation while providing an adequate Aristotelian semantics. His main criticisms of my approach are that (a) it would "be better to have a single reading which validates first figure apodeictic syllogisms *and* the principles of modal conversion" and (b) it would "be better to require of any interpretation that it validate all the elements in the base of the system being interpreted.''[77]

That second remark seems a trifle strong in any case: to *require* of any interpretation that it validate the basic elements of the system is to presuppose that there could not be any mistakes at that stage. But even ''the master of those who know'' should not be presumed infallible at any stage. Perhaps Thom would agree with the more modest requirement that an interpretation validate the elements at the base of Aristotle's system or give good reason why not. In any event, the main issue is joined in Thom's first criticism: Has he really succeeded in showing that there is no need for a two-reading approach – not only for the purpose of constructing a consistent formal model of Aristotle's system that ''validates first figure apodeictic syllogisms *and* the principles of modal conversion'' but also for providing a plausible Aristotelian semantics?

To answer this question, we need to look closely at how Thom has used his concept of an essence to interpret Aristotle's necessity propositions:

> $Lab^a$ is true iff b is included in an essence which is necessarily included in a.

Thom correctly remarks that this is essentially my ''weak'' *cop* necessity.[78]

> $Lab^e$ is true iff a and b are included in a pair of essences which necessarily exclude each other.

Thom correctly remarks that this is ''distinct from both of Patterson's senses. Like his strong sense, this sense is convertible; but unlike Patterson's strong sense, this one does not require either term in the proposition to be *kath' hauto.*''[79]

This leaves the two particular propositions, which he defines in terms of universal ones:

> $Lab^i$ iff $\exists d\, (Lad^a \,\&\, bd^a) \vee \exists e\, (Lbe^a \,\&\, ae^a)$.

> $Lab^o$ iff $\exists f\, [Laf^e \,\&\, bf^a]$.

Of these four, the first and last have the look of old friends, while the second and third wear an unfamiliar aspect. Recall that it is the second and third ($E_n$and $I_n$) that must convert ''simply'' or ''without qualification'' (*haplōs*); the first is supposed to convert ''to a particular.'' In a nutshell, my response to these interpretations of $E_n$ and $I_n$ is that (a) they differ from what one would expect given the interpretations of $A_n$ and $O_n$, (b) they depart from the style of $A_n$ and $O_n$ because they must do so if they are to convert, and (c) in departing from the style of $A_n$ and $O_n$ in order to maintain convertibility they cease to interpret accurately Aristotle's $E_n$ and $I_n$ propositions. (To avoid a possible misunderstanding: I would contend that Thom's versions of $E_n$ and $I_n$ are not Aristotelian, even putting aside the fact that they are defined quite differently from his $A_n$ and $O_n$.)

First, given Thom's interpretations of $A_n$ and $O_n$, one would expect the following for $I_n$:

> Lab$^i$ iff some b is included in an essence which is necessarily included in a.

That is,

> Lab$^i$ iff $\exists$d (Lad$^a$ & bd$^a$).

This would give a natural parallel to Thom's interpretation of $A_n$: Instead of (all) *b* being included in an essence that is necessarily included in *a*, it would simply say that *some b* is included in such an essence. It would also – as Thom remarked about his own version of $A_n$ – amount essentially to weak *cop* $I_n$.

Why not interpret $I_n$ that way? Thom does not pose any such alternative interpretation nor, therefore, explain why it would not be appropriate. But the fact is that on such an interpretation, $I_n$ obviously would *not* convert. Once again, let $a$ = Animal and $b$ = White, in a possible situation in which all the *b*'s are cats. Then it holds that all the *b*'s are included in an essence (Cat) that is necessarily included in *a* (Animal). So $I_n$ would be true. But it does not hold, conversely, that all *a*'s (animals) are included in an essence that is necessarily included in *b* (White), for none of the essences in which Animal is included will be necessarily included in White.

How, then, does Thom's actual interpretation of $I_n$ – which, as he says, does convert – differ from what we might have expected? It differs in interpreting '*A N i B*' (or, in his notation, Lab$^i$) as '*A N i B* or *B N i A*' (or, 'Lab$^i$' is in effect interpreted as 'Lab$^i$' or Lba$^i$'). This can be seen at a glance from the full interpretation of Lab$^i$ quoted earlier. Put another way, a Thomistic $I_n$ proposition converts only because it has been interpreted as the disjunction of itself with its own converse! So I would submit that although he has used only a single reading of necessity, Thom has not succeeded in accurately interpreting what Aristotle would mean by an $I_n$ proposition.

Meanwhile, notice that the conversion of Thom's $A_n$ proposition will be liable to counterexamples. (Let $A$ = Animal and $B$ = In the Agora, where all the *B*'s are horses.) Moreover, his $A_n$ proposition will not be adequate for scientific demonstration. Let *A* = Animal, *B* = Human, and *C* = In the Agora, where all the *C*'s are humans:

$$\frac{\begin{array}{c}\text{Lab}^{\text{a}}\\ \text{Lbc}^{\text{a}}\end{array}}{\text{Lac}^{\text{a}}}$$

This is the valid and complete syllogism Barbara *LLL* (or *NNN*). But there is no *per se* connection between Animal and In the Agora, nor does the conclusion assert that there is. Indeed, the problem is that no $A_n$ proposition *can*, on Thom's interpretation, actually assert a universal affirmative *per se* connection. All it can assert is that all its logical subjects (the *B*'s) are included in an essence that is necessarily included in *A;* but that does not entail any *per se* connection between *A* and *B*. Thom's Lab$^{\text{a}}$, like weak *cop* $A_n$, does presuppose a *per se* link between *A* and some appropriate essence *E;* but it does not assert any such connection between its own terms. But the premises and conclusion of a scientific demonstration would have to assert such a connection between their own terms. Otherwise (as we saw in discussing *cop* necessity syllogisms in Chapter 3), the syllogism cannot provide a principled explanation of why the predicate of the conclusion must apply *kath' hauto* to its subject.

This last objection would apply equally to Thom's interpretation of $E_n$, because, as he himself points out, it does "not require either term in the proposition to be *kath' hauto.*"[80]

This brings us to another problem with Thom's $E_n$ proposition ("*a* and *b* are included in a pair of essences that necessarily exclude each other"). This is convertible and, like his version of $A_n$ ("*b* is included in an essence which is necessarily included in *a*"), restricts itself to a single reading of necessity that correlates naturally with my weak *cop* necessity. And yet it seems still more distant than his version of $I_n$ from an accurate interpretation of Aristotle. This can be brought out by considering a concrete example. Let *a* = Sleeper and *b* = In the Agora, in a situation in which all sleeping things are horses and everything in the Agora is human. Then the *a*'s and the *b*'s are included in a pair of essences (Horse, Human) that necessarily exclude one another. So this satisfies the conditions under which Thom's version of $E_n$ is supposed to be true. But the $E_n$ statement in question (in effect, 'All sleepers are necessarily not in the Agora') is manifestly false. Indeed, it is not true either that the sleepers (i.e., certain horses) are necessarily not in the Agora or that the things in the Agora (i.e., certain humans) are necessarily not sleepers. Put another way, on

Thom's interpretation of '$A\ N\ e\ B$', it can still be true that $A$ possibly applies to every $B$ (Sleeper possibly applies to every Thing in the Agora) and $B$ possibly applies to every $A$! Thus it does not seem to me to express *any* plausible reading of Aristotelian necessity.

Here I conjecture that two things have gone wrong. First, Thom has perhaps not tried out his interpretation on appropriate concrete sets of terms, and so has not noticed that his definition will not fit many standard Aristotelian examples. Second, I believe that Thom has again departed from what one would expect in the way of $E_n$ propositions (given his interpretations of $A_n$ and $O_n$) in order to produce something that converts.[81] I would suggest that the natural parallel to his interpretation of $A_n$ and $O_n$ would have been

The $b$'s are included in some essence that is incompatible with $a$.

For example, let $a$ = Horse and $b$ = In the Agora, where all things in the Agora are humans. Let "some essence" mentioned in the definition be Human. In this situation our $E_n$ proposition ['The things in the Agora (i.e., certain humans) are necessarily not horses.'] comes out true. This would parallel Thom's interpretations of $A_n$ and $O_n$ and would, like them, correspond to weak *cop* necessity. But this proposition does not convert, as the same terms show.

In sum, however Thom may have arrived at his readings of $E_n$ and $I_n$, I conclude that he has not succeeded in constructing a consistent model of Aristotle's system using a single reading of necessity *and* giving an adequate Aristotelian semantics. Let us return, then, to the investigation of the *cop* approach, turning now from necessity propositions to the even more problematic area of two-way possibility.

# *Chapter 5*

# Two-way possibility: some basic preliminaries

Although he briefly considers the notion of contingency (or two-way possibility) in chapter 3, Aristotle does not formulate his "official" definition of this modality until chapter 13: "I mean, by being contingent, and by that which is contingent, whatever is not necessary but, being assumed to obtain, entails no impossible consequences" (32a18–20). "Impossible" in the definiens must refer here to "one-way" possibility, defined as "not necessarily not," with contingency (two-way possibility) defined in terms of necessity and/or one-way possibility.

We shall consider, however, Wieland's claim that two-way possibility is a modality *sui generis*.[1] And from a larger perspective, one is confronted with difficult issues concerning the relation of contingency to "belonging by nature," or "applying always or for the most part," and, in turn, of the importance of Aristotle's logic of two-way possibility (developed in chapters 14–22 of *Pr. An.* A) for Aristotelian science. Aristotle touches on this last problem without resolving it in A.3 (25b14–31) and again in A.13 (32b4–22), where he asserts that there are scientific demonstrations concerning things associated "by nature or for the most part," but not concerning chance associations.

Prominent among the more specific issues are, once again, those having to do with conversion. Here we encounter the usual questions about conversion of terms within *A, E, I,* and *O* propositions, only now in a way that requires resolution of a related issue about the "quality" (affirmative or negative) of two-way possibility propositions. Because, as Aristotle argues, *all* such assertions are in fact affirmative, the end results concerning conversion depart in striking ways from those for the corresponding *A, E, I,* and *O* plain or necessity propositions. In addition, there arises now a new form of conversion wherein *A* and *E* propositions, on the one hand, and *I* and *O*, on the other, are shown by Aristotle to be equivalent to one another. For example, '*A* two-way possibly applies to all *B*' converts to '*A* two-way possibly fails to apply to all *B*', and vice versa. As remarked in the Introduction, Ross calls this "complementary conversion"; Aristotle uses the same term for it as he does for term conversion, *antistrophē*. I shall call it "qualitative conversion," because such conversion is always a matter of converting from *A* to *E*, *I* to *O*, or vice versa – that is, of simply switching the apparent quality of the given proposition. This provides Aristotle with an additional important means, beyond those of term conversion, *reductio*, and *ekthesis*, of validating incomplete syllogisms by reduction to complete ones.

Aristotle introduces yet another important complication in chapter 13 with the observation that a statement of the sort '*A* two-way possibly applies to all *B*' (e.g., 'Being White two-way possibly applies to every Cloak') may be read either as '*A* two-way possibly applies to everything to which *B* applies' or '*A* two-way possibly applies to everything to which *B* (one-way? two-way?) *possibly* applies'. This "ampliation" of the logical subject term has far-reaching implications (which Aristotle does not himself systematically follow up) for the validity both of term conversions and of syllogisms with two-way possibility premises.

Finally, there is the overarching issue, with repercussions for all these questions, of the basic structure of two-way possibility propositions. Again I shall defend the modal *copula* approach textually and on philosophical grounds. And here, too, we shall find two possible *cop* readings – one at the level of relations between the natures signified by subject and predicate terms, and one pertaining to the relation of a given term to the essence of all individuals introduced by a second term. In Chapter 6 we shall trace the implications of these two readings (along with ampliation and qualitative conversion) for Aristotle's modal syllogistic and for his views on scientific demonstration.

## 5.1. THE STRUCTURE OF TWO-WAY POSSIBILITY PROPOSITIONS

First, the issue of placing the modal operator: Is contingency a property of assertions, as in 'It is contingent [or 'problematic' or 'two-way possible'] that *A* applies to all *B*'? Or have we to do with some sort of modal predicate statement, as in 'Being-contingently-white applies to all cloaks', or with a modal copula, as in 'Being white is contingently related to all cloaks'? As in the case of necessity propositions, the purely textual evidence favors a *cop* reading. Recall the opening of chapter 3: "belonging is one thing, necessarily belonging another, and possibly belonging another'' (29b29–30). Granted, there is, in one respect, a syntactical asymmetry among these modalities: Wolfgang Wieland observes that on the linguistic level Aristotle can express the predicational structure of (two-way) possibility assertions through use of *endechesthai* as a verb (as at 33a2, *to men A endechetai mēdeni tōi B*), whereas the necessity factor is regularly expressed adverbially (*ex anangkēs huparchein*)[2]. This is true enough, but is not of any consequence for the present point, because the independent verbal use of *endechesthai* is merely a variant on a common adverbial use of *endechesthai* to modify the copula. Thus Aristotle adverts to the adverbial modification of the linking expression at 29b30, *endechesthai huparchein* (so also 29b32, 32b25–26, 27, 31–32, 33a1, a26, *passim*). We shall see that there *are* peculiarities about these propositions, but they do not affect the question of whether Aristotle attaches modality to whole sentences, predicates, or copulae. Nor does Wieland's point undermine the implication of 29b29–30 that Aristotle treated all modalities [necessity, two-way (and one-way) possibility] in the same way, by adverbially modifying the copula.

Still, the *cop* reading must here face a major challenge precisely because metaphysical considerations might seem to favor a modal predicate reading of two-way possibility assertions. Specifically, one might argue that these assertions are simply attributing (natural) *potentialities* to subjects and that such ontological predications are best represented linguistically by a modalized predicate term. Thus 'All *B*'s have the potential for being *A*' would be read in more regimented form as 'Being potentially *A* applies to all *B*', or '*ppA* all *B*'.[3]

The case is strongest with natural potentialities that are realized ''always or for the most part.'' To say that the acorn may grow into an oak or that a male human will grow chin whiskers (the latter is Aristotle's example at 32b4–7) is not just to say that such occurrences are neither necessary nor impossible: It is to say that in the normal course of events the subjects in question are such that they will undergo these changes, that these things will happen. (Or, these things will happen unless something interferes with the natural course of events.) And this attribution of a natural potentiality might seem to be best expressed by use of a special modal predicate whose job is simply to signify the disposition itself (e.g., 'being by nature potentially an oak'), as opposed to a term like 'white' signifying a merely accidental property of cloaks or humans.

But closer examination shows this apparent advantage of the modal predicate reading to be illusory, and in fact discloses a number of significant problems. First, Aristotle's two-way possibility must cover two types of cases: not only what happens ''by nature'' or ''for the most part'' but also the "indefinite" (*to aoriston*) or what comes about by chance (*apo tuchēs*). If the point is left open in chapter 3 (where only the "by nature" reading of two-way possibility is mentioned), Aristotle makes this quite clear in chapter 13 when he prepares to actually introduce such propositions into the syllogistic system:

> The possible (*to endechesthai*) is said in two ways, one as 'happening for the most part and intermitting the necessity' (*epi to polu ginesthai kai dialeipein to anangkaion*), for example a man's growing grey or growing or declining, or in general what applies by nature (*to pephukos huparchein*) . . . another as the indefinite, which is possible both thus and not thus, for example an animal's walking, or there being an earthquake while someone is walking, or in general what comes about by chance (*to apo tuchēs gignomenon*). For this is by nature no more thus-and-so than the opposite. (32b4–13)

Aristotle does not distinguish these two cases of *endechesthai* in order to single one out as the strict or proper sense of *endechesthai* (a kind of procedure common enough elsewhere in Aristotle), much less to notify us that the ensuing discussion applies only to one reading rather than the other. On the contrary, he immediately asserts that *both* are subject "to (qualitative) conversion," and each for its own reason: "The natural because of its not necessarily applying (for thus it is possible that a man not grow grey), the indefinite because of its being no more this than that" (32b14–18).[4] And although he goes on to point out that there is understanding and scientific demonstration (*epistēmē de kai sullogismos apodeiktikos*) in the case of what is by nature (32b19) but not in the case of the indefinite or the chance (32b19), he does not suggest in any way that the discussion of modal syllogisms to follow has in view only or even primarily "the natural." Rather, he explicitly indicates that there *can* be syllogisms about the indefinite (even if not scientific demonstration), only we do not ordinarily seek them out (b21–22). In other words, from the logical point of view expressly defined here, the two Aristotelian cases of "two-way possibility" are entirely on a par, and propositions of both types are equally well covered by the definition of the modality itself, by the special rule of qualitative conversion, and by the discussion to follow of syllogisms containing two-way possibility propositions. So it is not possible to argue for a modal predicate reading of two-way possibility on grounds (a) that the modality is meant to capture or express what is distinctive about natural potencies and (b) that that would be best accomplished by use of a modal predicate,[5] for (a) is false. Finally, one could, in fact, just as easily express such potencies by an appropriately modified copula, as in Aristotle's own phrase "applies by nature and for the most part." So (b) is also false.

Furthermore, as noted earlier, in Chapter 2, Aristotle asserts a logical equivalence between the two universal *endechesthai* propositions ($A_{pp}$ and $E_{pp}$) and between the particular ones ($I_{pp}$, $O_{pp}$): If *A* might and might not apply to all *B*, then *A* might and might not fail to apply to all *B*, and vice versa. This again tells against thinking of the modality exclusively or even primarily in terms of attribution of natural potentialities or dispositions, for the attribution of a (positive) potency does not equally entail the attribution of the corresponding (negative) potency, where the latter may be thought of as the absence or privation (*sterēsis*) of the former, or in general simply as the non-realization of the former. Put another way, to say that *A* "naturally" or "always or for the most part" applies to *B* does not entail that *A* naturally or for the most part fails to belong to *B*.[6] Again, the only way to preserve equivalence between $A_{pp}$ and $E_{pp}$ is precisely to eliminate from the definition of *endechesthai* any stipulation of what is peculiar to the "by nature" variety of two-way possibility – to capture what is common to the "by nature" cases and the "indefinite" cases. That is what Aristotle does in his definition of the modality as, in effect, "neither necessary nor impossible."

This entire line of argument is reinforced by the fact that Aristotle does *not* introduce natural potentialities or things that obtain "for the most part" as examples of two-way possibility in the following chapters. On the contrary, his examples of two-way possibility are regularly ones of "accidental" subject-predicate connections.[7]

But just as in the case of necessity, so here we find an ambiguity in all two-way possibility propositions. One reading has to do simply with relations between the natures signified by the terms *A* and *B*, whereas the other takes account of the identity of the actual *B*'s. The former group of four propositions (*A, E, I, O*) would go as follows:

(I) *A PP a B:* Nothing entailed by the definition of *B* signifies anything either contrary to, or entailed by, anything entailed by the definition of *A*, and vice versa.[8]

A stock example would be White applying to all Cloaks/Humans: Nothing about being human or being a cloak either necessitates or excludes any human's or cloak's being white. Nor would its being white entail or exclude any white thing's being a cloak or a human. This first reading simply asserts an accidental relation between the natures *A* and *B* themselves.

The remaining "term relation" *cop* readings would follow suit:

| | |
|---|---|
| *A PP i B* | For some *C*, *B* $N_s$ all *C*, and *A PP* all *C*. |
| *A PP eB* | Nothing entailed by the definition of *B* either entails or precludes anything entailed by the definition of *A*. |
| *A PP oB* | For some *C*, *B* $N_s$ all *C*, and *A PP e C*. |

The *i* and *o* propositions assert that there is some proper subspecies *C* of *B* (not just any group of *B*'s) to which *A* itself bears an accidental relation.

By contrast the second reading of the two-way possibility universal affirmative asserts a relation between *A* and the essence of the actual *B*'s, rather than between the natures *A* and *B* themselves. For this I shall use lowercase *pp:*

(II) *A pp a B:* Nothing in the essence of any *B* either precludes or entails (its being) *A*.

The example 'White (two-way possibly) applies to all Human' would fulfill both definitions (I) and (II), so that it would be true on either reading. But 'Human *pp* all White' brings out the difference between two readings: Although we do know that at the level of what-it-is-to-be-a-human and what-it-is-to-be-white there is no relation either of entailment or of exclusion [which meets condition (I) and would explain the obvious truth of a *de dicto* reading of this example], we do not know, unless we know *which* actual things are in fact white, whether or not Human could two-way possibly apply to all of the things that are in fact white. As Aristotle himself remarks, some white things may be such that Human is necessarily inapplicable to them (37a7–9): Possibly there exist some white cloaks or horses; if so, then 'Human *pp* all White' is false, because '*A* $N_w$ some *B*' is true and is inconsistent with '*A pp* all *B*'. (If both were true, there would be at least one *B* to which *A* would both necessarily belong and not necessarily belong.)

A statement of the type 'White *pp* all White', in which the logical predicate signifies something that might belong either accidentally or essentially to different sorts of subjects, illustrates the point even more starkly. If it so happens that all the things that are white are in fact accidentally white, then the statement is true. [Recall that the lowercase *pp* signifies reading (II).] But on reading (I) (or on a *de dicto* reading), the statement 'White *PP* all White' (uppercase *PP*) is false, because being white *entails* being white. If, on the other hand, some of the actual things that are white are necessarily white (swans, snow, as in some of Aristotle's examples), then 'White *pp* all White' is false on reading (II) also. But even in this situation, where both readings come out false, reading (II) is false for a different reason than reading (I), namely, that for some actual white thing, the essence of that white thing entails being white.

The remaining three definitions in group (II) are

| | |
|---|---|
| *A pp i B* | For some *B*, (its being) *A* is neither precluded nor entailed by the essence of that *B*. |
| *A pp e B* | Nothing in the essence of any *B* either precludes or entails its not being *A*. |
| *A pp o B* | For some *B*, its not being *A* is neither entailed nor precluded by the essence of that *B*. |

The key feature of these last definitions is not that *A* itself is related accidentally to *B* itself (as with the uppercase *PP* readings, or a *de dicto* reading), but that *A* is related two-way possibly to *things* that are in fact *B*'s (suggesting the traditional *de re* label). But as with *cop* necessity, further analysis shows that matters are not quite as tidy as that. Ascertaining the truth or falsity of propositions of type (II) requires knowledge of (i) what items in particular are *B*'s and (ii) what properties are essential to those items. At this level, however, it becomes a matter of the relation between two *terms*, namely, between the originally given logical predicate *A* and the *essence* of (some, all) objects that are in fact *B*'s. One could say that reading (I) makes a direct, and reading (II) an indirect, assertion of a term-term relation – the former relating *A* and *B*, the latter relating *A* and the essence of each *B*. In this respect, the case is parallel to that of weak *cop* necessity.

So although reading (II), being framed in terms of what is essential to all/some *B*'s, does make two-way possibility depend on certain nonnecessary facts about the world (i.e., what things happen actually to be *B*'s), it does not ultimately depend on what might accidentally be true of those actual *B*'s at a given time. That is, the question of whether a given subject is or is not two-way possibly *F* depends on the *essence* of that subject: If it is essentially *F* and only accidentally *H*, then the test for a contingent relation between it and some property *G* would take account of what-it-is-to-be-an-*F*, but not what-it-is-to-be-an-*H*. For example, 'Sitting *pp* all Human' says that sitting is contingently related to all humans even if all humans happen to be standing at some time or even at all times. It asserts, in effect, that given the essential nature of humans, it is true of all of them that they might or might not sit. Likewise, 'Standing *pp i* Sitting' will come out true where there is something actually sitting whose essence (human) is compatible with standing.

This is to be contrasted with the temporally relativized test for possibility explored by Sarah Waterlow Broadie with particular reference to *De Int.* 9 (on "future contingencies") and *De Caelo* I.12 (on the relation between necessity and omnitemporality, possibility and being at some time). That test for possibility is designed to take account of even the accidental properties of a given subject at a given time. For example, given that Socrates is sitting now (time *t*), it is not possible that he stand *now;* but if nothing impossible follows (given, again, the facts that now obtain) from the assumption that he stand at $t + 1$, then it is possible at *t* that Socrates stand at $t + 1$.[9] In *Pr. An.* A.1–22, Aristotle (with the possible exception of one highly controversial passage of chapter 15 discussed later in connection with two-way possibility syllogisms) makes no reference to such temporally relativized modalities at all.

These approaches are not incompatible, but in fact complementary. More precisely, the notions underlying the readings defined here are presupposed by Broadie's temporally relativized modalities. The fact that Socrates' sitting at time *t* leaves open the possibility that he stand at $t + 1$ obtains at least in part *because* Socrates' essence (Rational Animal) is compatible with standing – which is just the test for type (II) contingency. Meanwhile, Socrates' sitting at *t* rules out the possibility that he stand at *t because* the essences (the what-it-is, the *ti esti*) of the postures Sitting and Standing are incompatible – which is just a case of non-contingency between two terms. (Obviously, none of this constitutes an objection to Broadie's definitions or applications of temporally relativized modalities to *De Caelo* or *De Int.*)

Readings (I) and (II), then, capture distinct but closely related Aristotelian concepts. But what is the logical relation between (I) and (II)? If they are independent, is there any philosophical motive for introducing a third reading, composed of conjunctions of corresponding members of (I) and (II)? Does Aristotle employ predominantly any one of these readings in his chapters 14–22?

To begin with, note that the two sorts of conditions are in fact logically independent of one another. To show, for example, that '*A PP* all *B*' does not entail '*A pp* all *B*', let *A* = Human and *B* = White, where some white things are cloaks. To show that '*A pp* all *B*' does not entail '*A PP* all *B*', let *A* = Standing and *B* = Sitting, where all things that are sitting may or may not stand.

As for the idea of defining a "strong" two-way possibility as the conjunction of (I) and (II), there simply seems to be no call for such a modality. Based on Aristotle's own usage, there certainly are cases in which he seems clearly to have in mind condition (II) – as when he denies

'Human *pp e* White' on grounds that Human might be necessarily inapplicable to certain white things (37a7–9). There are no such *direct* indications that he ever has securely in mind either the condition (I) of "direct term-term contingency" alone or a combination of the two conditions.

Nonetheless, the assumption of the direct term-term condition *alone* [as opposed to reading (II) or the conjunction of (I) and (II)] would at least explain certain claims he makes about term conversion of *PP* propositions – in particular, that '*A PP* all *B*' and '*A PP* some *B*' both convert to '*B PP* some *A*'. These conversions are obviously invalid given *just* condition (II). They are also invalid, though less obviously so, with *PP* defined by reference to both conditions. Consider 'White *PP* all Awake': If all waking things are humans, then both conditions are met. But 'Awake *PP* some White' does not follow, because it is not guaranteed that both conditions will still be met: Suppose that the same situation as before obtains and that all actual white things are cloaks; because cloaks cannot possibly be awake, 'Awake *PP* some White' – if read as asserting *both* conditions – is false. But even this evidence (from Aristotle's claims about two-way term conversion) for a straight condition (I) reading is rather weak, not only because it is indirect but also because, as Section 5.4 on term conversion will show, there is another plausible explanation for why Aristotle would have thought these conversions valid. Meanwhile, we may use our definitions to investigate a number of other Aristotelian claims involving two-way possibility propositions.

## 5.2. THE AFFIRMATIVE FORM OF TWO-WAY POSSIBILITY PROPOSITIONS

Having discussed in chapters 9–12 syllogisms with either two necessity premises or one necessity and one assertoric premise, Aristotle returns in chapter 13 to those with at least one two-way possibility premise. After setting out (at 32a18–20) the official definition of this modality, he takes up three important issues directly concerned with two-way possibility. The first of these concerns qualitative conversion (simply called conversion, *antistrophē*, by Aristotle):

> [i] For since what is ⟨two-way⟩ possible is not necessary, and what is not necessary might not apply, it is clear that if *A* ⟨two-way⟩ possibly applies to *B*, it also ⟨two-way⟩ possibly fails to apply to *B*, and if *A* ⟨two-way⟩ possibly applies to all *B*, it ⟨two-way⟩ possibly fails to apply to all *B*. Similarly for the particular affirmatives, for the proof is the same. (32a36–40)

The second has to do with the quality of contingency propositions:

> [ii] Such premises are affirmative and not negative. For the '*endechesthai*' is situated similarly to the '*einai*', just as was said before. (32b1–3)

Both passages are of considerable interest. The proof labeled [i] first reminds us of the definition of two-way possibility, according to which this modality excludes necessity, then invokes the principle that what does not of necessity apply can (one-way) possibly not apply. Thus, what two-way possibly applies also one-way possibly fails to apply. It is taken as obvious that two-way possibly applying entails one-way possibly applying. Thus, whatever two-way possibly applies also one-way possibly fails to apply (as the text explicitly says) and one-way possibly applies (hence, two-way possibly fails to apply). The implication in the other direction (from $E_{pp}$ to $A_{pp}$) is equally obvious and is also taken for granted, along with the proofs for $I_{pp}$ and $O_{pp}$.

His additional remark [ii] that these statements are *all*, appearances to the contrary, affirmative (rather than two affirmative, two negative) is not crucial to the validation of qualitative conversion, but it does remove one possible source of doubt: How *could* an affirmation be mutually convertible with its own negation? The answer is that $A_{pp}$ and $E_{pp}$, $I_{pp}$ and $O_{pp}$ are not in fact related as affirmation to negation. The negation of '*A pp* all *B*' would be '*A* does *not* two-way possibly apply to all *B*', rather than '*A* two-way possibly fails to apply to all *B*'. And that genuine negation is in turn equivalent to '*A* necessarily applies to some *B* or necesarily fails to apply to some *B*'. Aristotle does not here spell out this negation of '*A pp* all *B*' (nor, of course, tell us how to express this categorically!), but he has the idea well in hand, as shown by its skillful deployment in a passage we shall consider in the next section.

W. Wieland reads more into this passage, saying that these statements are all positive because they are all, in effect, attributions of a natural disposition to some subject.[10] Thus the pairs $A_{pp}$ and $E_{pp}$, $I_{pp}$ and $O_{pp}$ respectively, actually attribute the same disposition to the same subjects. But in the first place this is not the reason Aristotle gives for their all being affirmative. Second, this cannot be right in any case, for reasons stated earlier (Section 5.1): First, not all these two-way possibility propositions do attribute natural dispositions to subjects, and second, even when they do, the attribution of a disposition toward realization of a "natural" or "for the most part" potentiality would not be equivalent to the attribution of a disposition toward non-realization of that potentiality.

Aristotle's own explanation for the affirmative quality of $E_{pp}$ and $O_{pp}$ is that " '*endechesthai*' is situated similarly to the '*einai*' " (32b1–3). In this context the remark is slightly cryptic: One would expect some comment on the placement or function of the '*mē*' in '*A endechetai mē huparchei B*', because that is the item creating the appearance of negation. Aristotle's point is still discernible, although it is made somewhat more clearly in the earlier passage to which he here alludes ("just as was said before," 32b3), namely, chapter 3, 25b20–25:

> To possibly belong to none or not to some is affirmative in form, for the 'possibly' is situated similarly to the 'is', which always makes all statements to which it is added in predication (*hois an proskatēgorētai*) affirmative. For example, 'is not good' or 'is not white' or in general 'is not *this*'. This will be shown in what follows. And these convert like other affirmatives.

The text we possess of the *Analytics* contains no fuller demonstration "in what follows," but Aristotle's point is simply that just as

*X* is not-good (*X esti mē agathon*)

is not the negation of

*X* is good

and should rather be regarded as the affirmation of a (negative) predicate of a subject, so also

*X* (is) possibly not good (*X endechesthai mē einai agathon*)

is not the negation of

*X* is possibly good

but rather the affirmation of a certain relation between a (negative) term and a subject (as in "*Not-good* might-and-might-not-apply to all *X*"). The negation of '*X* is possibly good' would be '*X* is not possibly good', just as the negation of '*X* is good' is '*X* is not good'.

Aristotle's idea that the negation of a given sentence affirming some predicate of a subject is just the sentence that denies that predicate of that subject is easily generalized on the *cop* reading of modality: Where some sentence asserts that *A* applies in a certain way to *B* (e.g., '*A* possibly applies to *B*'), its negation will simply deny that *A* applies in that way to *B* ('*A* does not possibly apply to *B*'). It will assert neither that some negative predicate applies in the designated way ('not-*A* is possibly applicable to *B*') nor that the original predicate fails to apply in the designated way ('*A* ⟨two-way⟩ possibly fails to apply to all *B*', which is equivalent to, rather than the negation of, '*A pp* all *B*').

In the canonical form in which these propositions appear in Aristotle's logic there are no negative predicates, either in the plain or in the modal syllogistic. Nonetheless, his informal remarks are sufficient to allay one misapprehension about the claim that $A_{pp}$ and $E_{pp}$, $I_{pp}$ and $O_{pp}$ – contrary to the situation with regard to assertoric and necessity propositions, and to what one would expect judging by surface grammar – are equivalent pairs. Aristotle's "discovery" of the affirmative form of all two-way possibility propositions would, however, lead him seriously astray in the evaluation of certain second-figure syllogisms (see Chapter 6).

## 5.3. QUALITATIVE CONVERSION ON THE *COP* READING

We saw in Chapter 2 how the modal predicate and *de dicto* readings fail already at the syntactic level to accommodate Aristotle's qualitative conversion of contingency propositions. By contrast, a *cop* reading of type (II) is both philosophically plausible *and* in agreement with Aristotle's claims about conversion. On this reading, the universal affirmative is taken to affirm, of everything that is a *B*, that *A* might and also might not apply to it. From this it follows trivially that for every *B, A* might not and also might apply to it. Or, tracing qualitative conversion through Aristotle's official definition of two-way possibility plus his own conversion proof at 32a36–40, we have, given '*A pp a B*': Of every *B*, it is true that *A* does not necessarily apply to it, and true also that the assumption that *A* applies to it entails nothing impossible. But if it is true of every *B* that *A* does not necessarily apply to it, then it is true of every *B* that *A* (one-way) possibly *fails* to apply to it. Moreover, because *A* also one-way possibly applies to every *B* (its application gives rise to nothing impossible), it one-way possibly does not fail to apply to each *B*. This gives the desired conclusion that *A* might and might not fail to apply to each and every *B* (i.e., '*A pp e B*'). Thus Aristotle's conclusion that $A_{pp}$ and $E_{pp}$ are equivalent (as well as his brief chain of reasoning) makes perfectly good sense on a *cop* reading of type (II).

However, version (I) (the direct term-term reading) also yields a valid qualitative conversion: The qualitative conversion of '*A PP* all *B*' would on this reading give, 'Nothing in the definition of *B* either precludes or entails the non-application of *A* to any *B*'. And this will follow trivially from '*A PP* all *B*': 'Nothing in the definition of *B* either precludes or entails the application of *A* to any *B*'. Moreover, this version can also make sense of Aristotle's proof. That proof would use, as the definition of this modality, that nothing about being *B* or being *A* as such necessitates or makes it impossible that *A* apply to any *B*. Indeed, it almost looks as if

Aristotle first asserts the conclusion in terms of (I) when he says, without use of quantification, "if *A* may belong to *B*, then it may also not belong" (32a37–38), but then moves away from a simple term-compatibility version to a quantified, type (II) conception in the next line: "and if it may apply to all, it also may be inapplicable to all, and similarly for the particular affirmatives" (32a38–40). But this is very far from showing that Aristotle had in mind a distinction between the two readings, let alone a statement deliberately covering both options. Moreover, insofar as one does find in this wording any such distinction, it looks rather as though Aristotle takes it for granted that the two formulations assert the same thing – as if the second clause were epexegetical – or, at the very least, that the first entails the second.

If this is so, the passage is of particular interest not only for its introduction of the operation of qualitative conversion but also for its demonstration of how easy it can be to blur the distinction between (or at least obscure the logical independence of) the two readings of '*A PP* all *B*'. That is, not only does each entail its qualitative converse, but also one's proof of this may, by essentially the same form of words, establish both conversions. Here again the *cop* reading helps one appreciate how Aristotle might have failed to distinguish two readings of his modal propositions.

## 5.4. TERM CONVERSION

Aristotle deals much more briefly with the term conversions of two-way possibility propositions than with their qualitative conversions. He does not, in fact, take up the matter directly at all in chapter 13, apparently on grounds that his brief remarks in chapter 3 to the effect that these are affirmative propositions even in their *E* and *O* versions, and so convert in the usual manner of affirmatives rather than negatives (25b15–25), will suffice until some particular problem may demand further comment. Such a problem does arise in chapter 17 (36b35–37a31), where Aristotle explicitly denies the term convertibility of $E_{pp}$ and feels compelled to convince the reader of this. But there he does not fall back solely on the fact that $E_{pp}$ is affirmative in form, but gives three other arguments. And it is this passage that provides our primary source of information on term conversion of two-way possibility assertions in general. The first argument runs as follows:

> First it must be shown that the negative ⟨two-way⟩ possibility premise does not convert; that is, that if *A* ⟨two-way⟩ possibly fails to apply to all *B*, it

is not necessary that *B* ⟨two-way⟩ possibly fails to apply to all *A*. For let this be assumed, and let *B* ⟨two-way⟩ possibly fail to apply to every *A*. For since the positive ⟨two-way⟩ possibility propositions convert to the negatives, both the opposites and the contradictories [i.e., *A* to *E*, and *I* to *O*], then if *B* ⟨two-way⟩ possibly fails to apply to all *A*, clearly *B* would ⟨two-way⟩ possibly apply to all *A*. But this is wrong. For it is not necessary that if this ⟨two-way⟩ possibly applies to that, then that ⟨two-way⟩ possibly applies to this. Thus the negative does not convert. (36b35–37a3)

In brief, because $A_{pp}$ and $E_{pp}$ are equivalent (*via* qualitative conversion) to one another, if $E_{pp}$ were term-convertible, then $A_{pp}$ would be term-convertible to an $A_{pp}$ proposition (rather than just convertible to an $I_{pp}$ proposition). But $A_{pp}$ is not convertible in that way. Therefore, $E_{pp}$ is not term-convertible.

The basic structure of this *reductio* proof has been widely recognized. The critical question, however, is whether and why $A_{pp}$ itself converts or fails to convert. Aristotle says nothing further by way of explanation. He might have had in mind a term-thing reading and an appropriate counterexample: The one we shall cite in just a moment as his second argument would do. But as we shall also see, a little reflection would reveal that that very sort of counterexample could show that $A_{pp}$ does not convert even "to a particular." I suspect that he did not give the matter any new thought here, but simply assumed that $A_{pp}$ would behave like other affirmatives.[11]

But as usual we need to disambiguate the proposition, then assess the validity of conversion for each reading. On reading (I), the direct term-compatibility version of two-way possibility, $A_{pp}$ *will* simply convert: Just as nothing in the essence of *B* precludes or entails anything in the essence of *A*, so nothing in the essence of *A* precludes or entails *B*. $E_{pp}$, too, will simply convert on the same reading: Where there is neither conflict nor entailment between the terms *B* and *A*, there will be none between the terms *A* and *B*.

But on reading (II), both $A_{pp}$ and $E_{pp}$ fail to convert – as the counterexample supplied by Aristotle's second argument shows:

Furthermore nothing prevents *A* from ⟨two-way⟩ possibly failing to apply to *B*, but *B* being necessarily inapplicable to some *A*. For example white ⟨two-way⟩ possibly fails to apply to all humans (for it ⟨two-way⟩ possibly applies), but human is not rightly said to ⟨two-way⟩ possibly fail to apply to all white. For it necessarily does not apply to many ⟨white things⟩, and the necessary was not ⟨two-way⟩ possible. (37a4–9)

Aristotle will have in mind such white things as swans or white cloaks, to which Human is necessarily, therefore not two-way possibly, inapplicable. The main interest of this argument lies in its clear use of a *cop* (II) rather than (I) conception. Despite the compatibility of humanity and whiteness (or non-whiteness) in themselves, being human is nonetheless incompatible with something in the essence of at least some of the actual things that are white. Aristotle is perfectly correct in his use of this example to invalidate term conversion of $E_{pp}$ [on a *cop* reading (II)]. But the important truth he overlooks is that with this same sort of example one may show that $A_{pp}$ does not convert *even to an* $I_{pp}$ *proposition*, as would be normal for a universal affirmative. Let $A$ = Awake and $B$ = Human, in a situation in which *all* wakeful things are horses. Then Human will not two-way possibly apply (or fail to apply) to *any* wakeful thing. Thus, $A_{pp}$ does not, on a *cop* (II) reading – the reading Aristotle apparently has in mind in this passage on $E_{pp}$ – convert to $I_{pp}$.

The third argument is the most complex:

> [i] Nor can it [$E_{pp}$] be shown to convert by a *reductio ad impossibile (ek tou adunatou)*, as if someone should suppose that since it is false that $B$ possibly applies to no $A$ – true that it is not possible for $B$ to apply to none (assertion and negation) – that if this is so, it is true that $B$ must of necessity apply to some $A$. And thus also $A$ must ⟨of necessity apply⟩ to some $B$. But this is impossible.
>
> [ii] But it is not the case, if $B$ does not ⟨two-way⟩ possibly apply to no $A$, that $B$ must apply to some $A$. For 'not ⟨two-way⟩ possibly ⟨applying⟩ to no' is said in two ways: one, if [$B$] necessarily applies to some [$A$], the other, if [$B$] necessarily fails to apply to some [$A$.] For if something necessarily fails to apply to some $A$, then it is not true to say that it ⟨two-way⟩ possibly fails to apply to all $A$, just as it is also false to say of that which applies of necessity to some [$A$] that it ⟨two-way⟩ possibly applies to all [$A$].
>
> [iii] If, then, someone should suppose that since $C$ does not ⟨two-way⟩ possibly apply to all $D$, it is necessarily inapplicable to some, he would assume something false. For if it should belong to all, but belong of necessity to some, for this reason we say that it is not ⟨two-way⟩ possible to all. Thus to ⟨two-way⟩ possibly applying to all are opposed both necessarily applying to some and necessarily not applying to some. And similarly for ⟨two-way⟩ possibly applying to none.
>
> [iv] It is clear, then, that with respect to what possibly applies or fails to apply in this way, in the way defined at the outset [i.e., two-way possibly], that not [only] 'belonging of necessity to some', but 'necessarily not belonging to some' must be taken. But if this [latter] is taken, nothing impossible follows, so there will be no syllogism.
>
> [v] It is clear then from what has been said that the negative does not convert. (37a9–31)

With one reservation, the passage is a model of lucidity and locates precisely the flaw in the argument set forth in paragraph [i]. Here, in slightly more formal dress, is the argument that Aristotle envisions, then refutes:

To be proved: If *A pp* no B, then *B pp* no A.
Suppose the negation of *B pp* no *A*, namely,
(1) not: (*B pp* no *A*).
If (1), i.e., if *B* does not two-way possibly fail to apply to all *A*, then
(2) *B N* some *A*.
But if (2), then by term conversion of $I_n$,
(3) *A N* some *B*.

But this contradicts our original antecedent,

*A pp* no *B*.

Therefore, if '*A pp* no *B*' is true, (1) must be false, which is to say that '*B pp* no *A*' is true. Thus if '*A pp* no *B*', then '*B pp* no *A*'.

Aristotle's objection (justified in paragraph [iii]) is that (1) does not entail (2). The transition from (1) to (2) is facilitated by the fact that the same term *endechesthai* regularly covers both one- and two-way possibility, and the step *is* valid if read with one-way possibility. Thus a second possible response to the fallacious argument would be not (as suggested earlier) that the concept of two-way possibility has been mishandled (in the misidentification of its negation) but that there is an equivocation on *endechesthai*, with one-way possibility substituted for two-way in order to get from (1) to (2). Either way one analyzes the mistake, Aristotle is right to object that the contradictory of (1) would be, in effect, the disjunction '*B N i A* or *B N o A*'. So for the proposed argument to go through, one must show not only that if '*B N* some *A*' is true one can derive a contradiction to the initial antecendent (*this* much is what the imagined argument did accomplish) but also that if '*B N o A*' is true we can derive such a contradiction. Thus Aristotle says, "*B N o A* must be taken" (37a28–29). But as Aristotle also points out, taking '*B N o A*' will not yield the required contradiction ("nothing impossible follows").

The small reservation is that in view of our results in earlier chapters, both the *reductio* argument that Aristotle attacks and Aristotle's objection to it accept the convertibility of $I_n$. The *reductio* argument uses it explicitly to get from (2) to (3); Aristotle himself recognizes the conversion at 25a32–34 and never questions it thereafter. So it could be objected that the proposed *reductio* breaks down here also, at least on the weak *cop* reading of necessity, because on that reading neither $I_n$ nor $O_n$ converts.

To evaluate this suggestion, one would have to know which reading of two-way possibility, (I) or (II), Aristotle uses in this proof, and thence what reading of necessity will appear in the appropriate contradictory. The context, and especially Aristotle's concrete example (Human/White), definitely indicates a reading of type (II), whose contradictory would use weak *cop* necessity. But again, weak *cop* $I_n$ does not convert. So whereas one may approve Aristotle's own objection to the argument (that the *reductio* proof *for* conversion fails, and only appears to succeed because it gets the contradictory of '*A pp e B*' wrong), one must add that he ought also to have objected to his antagonist's use of $I_n$ conversion to get from step (2) to (3).

Regarding the remaining conversions of $A_{pp}$, $I_{pp}$, and $O_{pp}$, we may concentrate on $I_{pp}$, for given that subalternation holds (so that '*A pp* all *B*' entails '*A pp* some *B*'), *if* '*A pp* some *B*' converts to '*B pp* some *A*', then '*A pp* all *B*' entails '*B pp* some *A*'. Thus, $A_{pp}$ and $I_{pp}$ would convert in the usual manner of affirmative propositions. On the other hand, because $I_{pp}$ is equivalent (by qualitative conversion) to $O_{pp}$, if $I_{pp}$ converts then $O_{pp}$ will also. Thus we would have the overall situation Aristotle foretold back in chapter 3 (25b16–18), that although the two-way possibility affirmatives behave in the same manner as other affirmatives, the negatives do not convert as other negatives: The universal negative does not convert, but the particular negative does.

What, then, about the validity of $I_{pp}$? Read in version (II) (there is some actual *B* such that nothing in its essence either precludes or entails the possession of *A*), it does not convert. Adapting Aristotle's own example for $E_{pp}$, let *A* = White and *B* = Human, where all actual white things are cloaks. In this possible situation, 'White *pp* some Human' true, but 'Human *pp* some White' false, because Human is necessarily inapplicable to *all* the actual white things, hence does not two-way possibly apply to some white. The same terms will show that $A_{pp}$ and $O_{pp}$ also fail to convert on this reading. Thus, in chapter 17, the only place where Aristotle stops to examine any of these conversions, he has in mind a type (II) conception; he realizes the consequences for $E_{pp}$, but does not apply to $A_{pp}$, $I_{pp}$, or $O_{pp}$ the principle underlying his own counterexample to $E_{pp}$.

On reading (I), the term-compatibility version (there is some *C* such that '*B* $N_s$ all *C*', and nothing in the definition of *C* either precludes or entails anything in the essence of *A*), $I_{pp}$ does convert. (If so, $O_{pp}$, being equivalent to $I_{pp}$, will also convert.)

To prove this, we may observe quickly that $E_{pp}$ [on reading (I)] converts: If nothing about *C* itself either entails or precludes *A* itself, then obviously nothing about *A* itself either entails or precludes *C* itself. And since '$B\ N_s a\ C$' holds, we know that *B* consists of or entails only such properties as are entailed by *C*. (For example, let *A* = White, *B* = Animal, *C* = Horse.) Therefore, if *A* neither excludes nor entails anything essential to *C, A* will *a fortiori* neither entail nor exclude *B* itself. (A similar argument can be made in which *A, B*, and *C* are any appropriate triple of genus, species, differentia, or *idion.*) So it will follow that if '*A PP* some *B*', then '*B PP* some *A*'. In fact, on this reading we could equally well have concluded that '*B PP all A*'. We, in effect, demonstrated that '*C PP* all *A*' by showing that nothing about being *A* precludes or entails being *C*, nor, then, the less specific *B*. So from '*A PP* some *B*' we can derive, *at the level of term relations alone*, both '*C PP* all *A*' (for some '*C*') and '*B PP* all *A*'. This is not so surprising, however. All it says is that if *A* is accidentally related to some nature that essentially entails *B*, then both that nature and *B* itself are accidentally related to *A*. [Notice also that if one term (*A* = White) is related accidentally to some subspecies of the other (*B* = Animal, *C* = Human), then even if it is related necessarily to another subspecies (*D* = Swan), that initial term (*A*) will be related accidentally to all *genera* of those lower species (Animal, Living Thing, etc.). Again, the genus will contain only what is essential and *common* to all its species; if at least one of *B*'s species contains nothing incompatible with *A*, it is impossible that *B* or any of its *genera* contain any such item.]

## 5.5. AMPLIATION

The final section of chapter 13 introduces the concept dubbed "ampliation" – perhaps quite misleadingly – in later tradition. Becker deletes some parts of the passage in order to produce a more orderly chain of reasoning, but the new concept emerges clearly enough in any case. First, the entire text (with Becker's useful subdivisions[12]) as it has come down to us:

a) These things will be determined further in what follows.
b) But for now let us say when and how and what will be a syllogism from possible premises
$c_1$) since "this possibly belonging to that" can be taken in two ways,
$c_2$) for either that to which [*B*] belongs, or to which [*B*] possibly belongs ⟨can be the subject⟩ –
$c_3$) for "*A* possibly applies to that to which *B* applies" signifies one of two things, either that of which *B* is said or that of which *B* is possibly said.

$c_4$) "*A* possibly applies to everything of which *B* is said" and "*A* can apply to all *B*" do not differ.

$c_5$) Obviously one might say in two different ways that *A* possibly applies to all *B*.

$d_1$) First, then, let us say, if *B* possibly applies to that of which *C* is said, and *A* ⟨possibly⟩ of what *B* is said, which and what sort of syllogism there will be.

$d_2$) For in this way both premises are taken in terms of possibility.

$e_1$) But when *A* possibly applies to that of which *B* is said, one is belonging, the other possibly belonging.

$e_2$) So that one should begin from ⟨premises⟩ of the same form, as in the other cases. . . .

f) Whenever, then, *A* possibly applies to *B* and *B* ⟨possibly applies⟩ to *C*. . . . (32b23–37)

Certain small obscurities notwithstanding, it is sufficiently clear that Aristotle's main concern in a–c is to distinguish two readings of any given two-way possibility premise, differing in that the first two-way possibly attributes the predicate to all things of which the subject is said (e.g., '*A* two-way possibly applies to all *B*'), whereas the second two-way possibly attributes the predicate to all things of which the subject is possibly said ('*A* two-way possibly applies to everything to which *B* possibly applies'). These are the only doubly modalized propositions recognized by Aristotle. In our terms, they will employ a second modalized copula, as the wording of the preceding sentence shows.

Those minor obscurities create no significant problems. It is possible that, as Becker maintains, $c_1 - c_5$ are a later addition, probably by Aristotle himself, with $c_3$ and $c_4$ perhaps constituting an explanatory gloss on $c_1$, $c_2$, and $c_5$; $d_2$ *may* be a "later" explanation of (the consequences of) $d_1$, which would then lead naturally to ($e_2$ and) f; $e_1$ *may* be a confused and fragmentary attempted repetition or summation of $c_1$–$c_5$.[13] If we do accept these proposals, what emerges as "original" is a smooth and appropriate introduction to chapter 14 – the first treatment of two-way possibility syllogisms – consisting of the sequence a, b, $d_1$, $d_2$, $e_2$, f. It is also now more evident that only $c_1$–$c_5$ actually introduce that new (ampliated) way of treating the subject term. Their wording is highly compressed, but intelligible. To be fair to $e_1$, although it could attach confusedly to c (says Becker), it could also – as I believe makes better sense – attach to $d_1$ and $d_2$. In that case it refers, albeit in a compressed and awkward manner, to syllogisms with one two-way possibility and one assertoric premise, by way of contrast to those containing only the former sort of premise. ('That of which *B* is said' will then refer in $e_1$ to whatever subject *B* applies to in some minor premise.[14]) Thus $e_2$ would not be so silly (*alberne*, says Becker) after all: It simply says that, as before (i.e., when syllogisms with plain and necessity premises were examined), one should start with syllogisms having both premises of the same form (*homoioschēmonōn arkteon*). Accordingly, chapter 14 considers first-figure syllogisms with two contingent premises, chapter 15 "mixed" syllogisms of the first figure with one assertoric and one contingent premise, chapter 16 those with one necessity and one contingent premise, and so forth.

The main interest of these textual questions, and their implication that "ampliation" was a kind of afterthought, lies in their connection to the role of that operation in Aristotle's system. Aristotle does not, as he works his way through two-way possibility syllogisms, make explicit the kind of systematic distinction between ampliated and non-ampliated propositions that this passage would lead us to expect, and which purely logical considerations would require. Combined with the fact that a, b, $d_1$, $d_2$, ($e_1$), and f form such a smooth and natural introduction to chapter 14 without ever mentioning ampliation, this lends support to Becker's conclusion that Aristotle only later realized the possibility and the potential importance of ampliation and that, although he himself then inserted notice of this concept into chapter 13, he never returned to chapters 14–22 to work out systematically its effects. On the other hand, it is just possible that he simply intended ampliation to apply where needed and did not consider it necessary to point out such places. As it happens, there are only a few syllogisms whose validity requires ampliation, and these can be treated in Aristotelian fashion by showing (1) that certain syllogisms in the first figure are complete, given ampliation, and (2) that appropriate conversions are valid, given ampliation, so that the relevant incomplete moods can be validated by reduction to the first figure. Both points will be established later, in Chapter 6 (Sections 6.1 and 6.4, respectively).

In following up these issues, however, we shall uncover one further ambiguity of logical significance. Aristotle apparently thinks of ampliation as involving two-way rather than one-way possibility. If so, the ampliated version of '*A pp a B*' is '*A pp a B pp*', not '*A pp a B p*'. But because, in accordance with his customary practice in the *Prior Analytics*, he uses *endechesthai* to cover whichever sort of possibility he intends at the moment, it is not crystal clear which sort is involved in the ampliation of two-way possibility propositions. With two-way ampliation (as in '*A pp a B pp*'), one does not really ampliate (amplify) anything, because *B* might apply (necessarily) to many items, but *two*-way possibly to none. (Let *B* = Human.) By contrast, *B* will *one*-way possibly apply to everything to which $B$ actually applies – whether necessarily or accidentally – *and* to everything to which $B$ does not actually apply but is two-way possibly applicable. The last clause will not always "amplify" the application of $B$, but it will guarantee that the set of things to which $B$ one-way possibly applies at least includes the set of things to which $B$ applies. So here the traditional label "ampliation" makes more sense. But again, this in itself is no evidence for how *Aristotle* thought of the operation, for he used no term corresponding to "ampliation." (For convenience, I shall nonetheless use "ampliation" to cover both one- and two-way operations on the subject term.) Still, one would think that if Aristotle meant one-way possibility, within a larger context in which he was speaking of two-way possibility syllogisms, he surely would have said so, because otherwise the potential for confusion would be not only great but also obvious. This seems to me to tip the scale in favor of ampliation *via* two-way possibility. But this is not decisive, because Aristotle did not always say what one expects that he surely would have said. Moreover, one needs to consider what might have been his *motive* in using one or the other sort of ampliation. This, too, supports two-way ampliation, in a way to be explained in Chapter 6.[15]

Meanwhile, from the logical point of view, there arises another small surprise, one that gives some urgency to this question. I say "surprise" because I have not found notice of the fact that among syllogisms valid only with ampliation, some are valid only when ampliated *via* one-way possibility, and others only with two-way possibility. Let us turn now to these issues and to the notorious problem of relating two-way possibility to Aristotelian science.[16]

# Chapter 6

# Two-way possibility syllogisms

In chapters 14–22, Aristotle methodically considers all the various combinations of premise pairs involving at least one two-way possibility ("problematic" or "contingent") premise. Chapters 14–16 take up first-figure moods having, respectively, two problematic premises, one problematic and one assertoric premise, and one problematic and one necessity premise. Chapters 17–19 take up the same combinations, now in the second figure, and 20–22 carry the plan through the third figure. (For a chart of the ground plan of chapters 8–11 and 14–22, see the first page of Chapter 3 herein.) As with the necessity syllogisms of 8–11 and the assertoric ones of 4–6, Aristotle singles out the "complete" or "perfect" (*teleios*) moods, those whose validity is obvious on the basis of the premises precisely as given, and then validates other moods by reducing them to perfect moods by use of term or qualitative conversion or *reductio ad impossibile*, or by validating them through *ekthesis*. Some portions of these chapters are fairly routine and so will be presented here in summary fashion. This will leave us free to focus on a number of logical curiosities and on some significant philosophical issues, including that of the relation of these syllogisms to Aristotelian science.

## 6.1. TWO PROBLEMATIC PREMISES: FIRST FIGURE

All the perfect moods with this combination of premises fall into the first figure and correspond exactly to the four perfect plain moods of *Pr. An.* A.4. Thus, chapter 14 consists in a discussion of Barbara, Celarent, Darii, and Ferio *pp, pp/pp*, and of several invalid moods. The terminology of perfection is the same as in earlier chapters: Barbara is called "perfect" (*teleios*, 32b39), and its validity "obvious" (*phaneron*, 32b40; cf. *pha-*

*neron* in connection with Darii, 33a24, and the denial, at 33a31, that a certain argument type yields a *phaneros* syllogism).

Unfortunately, nothing can make the validity of these moods obvious, for both are, on at least one natural reading, invalid. More specifically, without ampliation, and on the type (II) reading, one has

$$\frac{\begin{array}{c}A\ pp \text{ all } B\\ B\ pp \text{ all } C\end{array}}{A\ pp \text{ all } C}$$

The second premise brings all the *C*'s under the things which are two-way possibly *B*'s. But the first brings only the *actual B*'s, not all the things that are two-way possibly *B*'s, under the two-way possibly *A*'s. Because it is entirely possible that some two-way possibly *B* things are not actual *B*'s, the premises do not guarantee that all the *C*'s come under the two-way possibly *A*'s. [For a counterexample, let *A* = White, *B* = Walking, and *C* = Raven, where ravens are (as in Aristotle's own use of the example) necessarily black, and where all things actually walking (humans, horses) are two-way possibly white.]

Both moods are, however, perfect syllogisms if their subject terms are two-way possibly ampliated:

$$\frac{\begin{array}{c}A\ pp \text{ all } B\ \ pp\\ B\ pp \text{ all } C\ (pp)\end{array}}{A\ pp \text{ all } C\ (pp)}$$

For then all the things to which *C* two-way possibly applies are in fact things to which *B* two-way possibly applies (by the second premise), and all of the latter are (by the first premise) among the things to which *A* two-way possibly applies. Notice that although we could have ampliated both the middle and minor terms (*B* and *C*), only the middle need be ampliated to give a complete syllogism. If the minor term is not ampliated, the conclusion must, of course, be '*A pp* all *C*', rather than '*A pp* all *C pp*'. Although he has just introduced ampliation, Aristotle's brief remarks on Barbara *pp*, *pp/pp* do not indicate clearly whether or not he wishes to apply that operation (using either one- or two-way possibility) here:

> When, then, *A* possibly applies to all *B* and *B* to all *C*, there will be a complete syllogism that *A* possibly applies to all *C*. This is manifest from the definition. For that is what we meant by possibly applies to all. (32b28–33a1)

With one-way ampliation of the middle term, we would have

$$\frac{\begin{array}{l} A\ pp\ a\ B\ p \\ B\ pp\ a\ C \end{array}}{A\ pp\ a\ C}$$

(With modal predicates, notice that we would have another ill-formed argument with four terms.) This is also valid, and is arguably complete, because the things to which *B* two-way possibly applies, including the *C*'s, are plainly included in the things to which *B* one-way possibly applies. But if the motive is simply to produce a perfect syllogism, two-way possibility fits more neatly. And again, Aristotle has just spoken unambiguously of these premises as involving two-way possibility in the main copula, so that a gratuitous and unannounced shift to one-way possibility ampliation seems unlikely.[1] But this is hardly decisive.

It is striking that neither Aristotle's brief description of this mood nor his treatment of any other moods of this chapter shows unequivocally that he realized the need for an ampliated middle term. On the contrary, one might well read the chapter as *not* calling for ampliation, since that operation receives no *clear* mention at any of the several places where one would expect it (33a5–6, 33a11, a14–15, a23–24, *et al.*).

But there are some considerations on the other side, and I strongly suspect a deliberate allusion to ampliation in Aristotle's treatment of Celarent immediately following the lines (32b38–33a1) just quoted on Barbara:

> Similarly also if *A* possibly applies to no *B* and *B* to no *C*, ⟨there is a complete syllogism concluding⟩ that *A* possibly applies to no *C*. For *A*'s possibly not applying to that to which *B* possibly applies just was to leave out none of the things that are possibly under *B*. (33a1–5)

The key question is whether the second sentence refers to the major premise alone or, telescopically, to the two premises together. If the former, then the major premise is definitely ampliated: '*A pp a B pp* (or *B p*)'. Here, 'that to which *B* possibly applies' refers to all the things possibly falling under *B*, with no reference yet to the *C*'s. If the latter, then, 'that to which *B* possibly applies' will allude to a minor premise of the form '*B pp a C*' and will say nothing either way about ampliation.

One minor point in favor of the first reading is that the phrase 'that to which *B* possibly applies' (*to gar kath' hou to B endechesthai*) recalls the wording of the preceding passage, which introduced ampliation:

> '*A* possibly applies to that to which *B* applies' signifies either 'that of which *B* is said' or '*that of which B is possibly said*' (*to gar kath' hou to B to A endechesthai toutōn sēmainei thateron, ē kath' hou legetai to B ē kath' hou endechetai legetai*). (32b27–30).

This last phrase is surprisingly rare in *Pr. An.* A.1–22. (There are several places in which Aristotle says that a given letter stands for a particular term, but there he regularly uses '*eph' hōi to A*': 31b5, 30a30, 34b33, 37b4, 38a31, 38b20.) So there is some evidence that he had in mind at 33a1–5 the earlier passage on ampliation and hence has in mind ampliation of Celarent.

More important, the phrase invoking the definition of 'applies to all' (33a3–5: "not leaving out any of the things possibly falling under *B*") has several parallels, all in passages that, like this one, purport to say why some complete syllogism is complete. The issue of completeness of syllogisms will be more fully discussed later, in Chapter 7. For present purposes, let it be noted that these phrases *always refer to the major premise of a complete syllogism.* In the assertoric case, he would say, in effect, that because (by the major premise) *A* applies to all *B* – which means that "none of the *B*'s will be left out," that all of them will be included among the things to which *A* applies, for that is what we mean by "applies to all". And because (by the minor premise) some (or all) of the *C*'s are *B*, it follows that some (or all) of the *C*'s are *A*'s. In the present case we would have the following: Because *A* possibly fails to apply to *all* the things to which *B* possibly applies, and because the *C*'s are among the things to which *B* possibly applies, it follows that the *C*'s are among the things to which *A* possibly fails to apply, for that is how we defined "possibly applies to all/none." So here he will allude to a major premise of the form '*A pp e B pp*'. We shall "not leave out any of the things possibly falling under *B*": They are all included among the things to which *A* possibly fails to apply; hence, if the *C*'s fall under the things that are possibly *B*, then *A* will possibly apply to them.

For these reasons – plus the fact that the need for ampliation is rather obvious – I am inclined to think that Aristotle did have ampliation in mind here, and fails to mention it elsewhere because he takes it for granted. Whether or not Patzig is correct in saying that the only reason Aristotle introduces ampliation at all is to produce a valid syllogism[2] remains to be seen.

With two-way ampliation Barbara and Celarent are complete: then too, $A_{pp}$, $E_{pp}/A_{pp}$ of the first figure will, by qualitative conversion of the second premise, reduce to Barbara *pp, pp/pp* (33a5–12). Moreover, if both premises are universal negatives, the same operation on both will give Barbara again (33a12–17). Similarly, Darii and Ferio are perfect and equivalent to one another – as well as to the heretofore unnamed Bambino, Bamboni, Feroco, and all their siblings by qualitative conversion.[3]

## 6.2. PROBLEMATIC BARBARA AND SCIENTIFIC DEMONSTRATION

Our complete syllogism in Barbara, especially in its ampliated version, raises a substantial issue that was postponed in the last chapter: How are Aristotle's two-way possibility syllogisms related to his view that certain connections obtaining "for the most part" and "by nature," but not universally or necessarily in a strict sense, are nonetheless included in the realm of scientific demonstration? There would seem to be a serious problem if scientific demonstration concerns only what is necessary and universal (*Post. An.* A.4, 6) rather than contingent. Of course, we are not talking about "chance" contingencies. But even so, we have seen that the definition of two-way possibility does not go beyond "neither necessary nor impossible"; and since "by nature and for the most part" connections are classed under that definition, it is hard to see how Aristotle thought he could bring them into the scientific fold.

I am happy to record my sympathy for, if not complete agreement with, Gisela Striker's proposal to treat general statements of this type (e.g., 'Adult male humans grow grey hair') as, in effect, a kind of necessity statement – one asserting an "intermittent" necessity or one with "gaps" – while conceding that in an individual instance the instantiation of a "natural" property is yet contingent.[4] Given that Aristotle nowhere settles upon (see note 14) nor, within the *Prior Analytics*, ever pursues the alternative of introducing a new operator for $A$'s "applying by nature" to $B$, Striker's may be the most plausible way to preserve Aristotle's brief comments linking such propositions to scientific demonstration, on the one hand, and to contingency, on the other.

But as Striker says, given the importance for science of syllogisms in Barbara, one must figure out how to express "natural" connections in the form of universal affirmative statements.[5] If connections holding by nature are those that would hold in *every* case *unless something prevented it*, then one might define a modal operator $\phi$:

$A \phi B$ (i.e., $A$ applies by nature to every $B$) $= N(S \rightarrow A \, a \, B)$

where $N$ stands for 'necessarily' and $S$ stands for 'if nothing hinders'.[6] Then one could construct a scientific demonstration in Barbara:

$$\frac{\begin{array}{c} N(S \rightarrow A\,a\,B) \\ N(S \rightarrow B\,a\,C) \end{array}}{N(S \rightarrow A\,a\,C)}$$

If Aristotle could adopt this analysis, he would not have to say, paradoxically, that one can know that all sheep by nature have four legs even though that is contingent; rather, he could say that all sheep normally have four legs, and one can know this because it is necessary: Only the matter of whether or not a particular sheep has four legs will be contingent.[7] But, Striker remarks, this analysis – which seems to provide what Aristotle needs – cannot be expressed in Aristotle's system, which lacks an expression for implication.[8] Within Aristotle's system, any statement whatever must be formulated using such terms as *A* and *B*, the traditional symbols *a, e, i*, and *o*, and one of his modal operators *N, PP*, and *P*. But no such formulae can express the content of the syllogism just given.

Without adopting the foregoing analysis, what can Aristotle do? Striker suggests that, much to his credit,[9] he does not simply allow himself to be straitjacketed by his modal system, but tries to modify his problematic propositions so as to make them fit to serve the cause of science.

This modification appears in chapter 13, with the introduction of ampliation. Striker suggests that the assumption that the predicate term (possibly) applies to everything (possibly) falling under the subject term might be connected with Aristotle's having in mind cases in which the one term applies to the other by nature, even if not, strictly speaking, always and of necessity. For if all the *B*'s are by nature *A*'s, then the assumption lies close to hand that anything that could be a *B* could, at least, be an *A*.[10] Now if "could" is read the same way in both its occurrences, we are talking about two-way ampliation, as in '*A pp a B pp*'. (We shall consider one-way ampliation in a moment.)

There may be some internal difficulties with this proposal, but before looking into these, let us remind ourselves that there is a more direct explanation of why Aristotle introduces two-way ampliation at this point. As we just saw, without ampliation Barbara *pp, pp/pp* is simply not valid; with ampliation it is not only valid but also complete by Aristotle's usual criterion.[11] So Patzig is right that this would suffice to explain the introduction of ampliation.

Turning more directly to the proposed interpretation, notice first that ampliation does not express specifically, but covers only in a generic way, statements about natural connections – just as the *unampliated* '*A pp* all *B*' covers only generically all cases of *A* two-way possibly applying to all *B*, including those of the "for the most part and by nature" variety. For example, the *unampliated* '*A pp* all *B*' would cover 'White *pp* all Adult Male Human', along with the "natural" fact that Grows Grey Hair *pp* all Adult Male Human. Similarly, the *ampliated* '*A pp* all *B pp*' would cover such accidental connections as 'White *pp* all Walking *pp*' along with 'Physically Declining *pp* all Growing Grey Hair *pp*'. In other words, there is nothing intrinsically scientific about these two-way ampliated statements.

For scientific use, one needs the general principle that on Aristotelian grounds '*A pp a B pp*' will be true just in case there is some natural connection between *A* and *B*. Given that, 'Standing *pp* all Sitting *pp*' might be true because those anatomic features that make sitting two-way possible for something also, in the normal course of events, make standing two-way possible. On the other hand, 'White *pp* all Sitting *pp*' would turn out to be false, since there is no natural connection between its terms. But without introducing a new, stricter reading of two-way possibility, it would seem just a mistake to rule that second proposition false in advance, as it were – i.e., without looking to see what things could be sitting. Nor do I see how one could begin to demonstrate that general principle. (On the contrary, it would seem to be false.) Thus if we retain Aristotle's concept of two-way possibility, and if that general principle cannot be demonstrated, our (two-way) ampliated propositions will be no more scientific than unampliated ones, and for essentially the same reasons.

Second, if contingency is the two-way possibility defined in chapter 13, and if one says that instantiation (e.g., of "growing chin whiskers") in an individual case is contingent, then one is making a very weak claim about those individual cases. Again, nothing more will be *explicitly claimed* than in the "chance" cases, namely, non-impossibility and non-necessity. But this seems problematic in two ways. First, the individual cases (Plato grows chin whiskers and Ockham doesn't) are not supposed to be *just* contingent. Aristotle's position seems to be rather that in the *individual* case the natural course of events will come about unless something intervenes. Indeed, our ground for asserting a "by nature and for the most part" general statement, and the justification for thinking of it as *necessary* in the sense discussed by Striker, would seem to be that all relevant individuals naturally have a certain property. Similarly, the general statement "adult male humans all grow chin whiskers" falls short of strict universality *because* every now and then something happens to individual men to prevent their growing chin whiskers as they would in the normal course of events. But this means that we do not want to say of *any* individual cases, even those in which some normal property fails to apply, merely that they are contingent. For the individuals who fail to possess the natural property ψ are just as much ψ ''by nature'' as those that do possess ψ. It is just that in their case, something intervenes to block the normal course of events. In any event, making the application in individual cases merely two-way possible would break the desired connection between individual cases and the general statement: If individual *B*'s were merely two-way possibly *A*, there would be no ground for asserting a natural or necessary connection between the natures *A* and *B* themselves.

Third, the ''by nature or for the most part'' statements that Aristotle has in mind in *Pr. An.* A.13 are such propositions as 'Grows Chin Whiskers *pp* all Male Humans' or 'Declines Physically *pp* all Humans'. But these will *not* have any use for ampliation of the subject term, because, for example, Human will apply necessarily, not two-way possibly, to all humans. Ampliation will be relevant, even on the proposal in question, only where *both* terms apply two-way possibly to their subjects. Certainly Aristotle may want to formulate such propositions, and some of these may correspond to propositions of natural science; but these are not the kinds of examples he actually gives in *Pr. An.* A. Again, this throws doubt on the suggestion that ampliation was consciously introduced with scientific demonstration involving ''by nature and for the most part'' propositions in mind.

Finally, there is the testimony of Aristotle's own examples of two-way possibility propositions in the chapters (14–22) treating of actual two-way possibility syllogisms. These are *all* of the humdrum ''accidental connection'' variety, rather than of the ''by nature'' sort. This is somewhat odd if he is consciously concerned throughout these chapters with the question of scientific demonstration. The best reply is perhaps that that is due to Aristotle's having only later on had the (not very good) idea to use two-way ampliation for scientific purposes, then adding the remark on ampliation in chapter 13 – without, however, indicating the purpose of the operation, or adding any scientific examples.

But perhaps one-way ampliation offers more hope? This would give '*A pp a B p*' or '*A* two-way possibly applies to everything to which *B* one-way possibly applies'. As noted earlier, in Chapter 5, the things to which *B* one-way possibly applies will cover all things to which *B* necessarily applies and all things to which *B* relates as an accident, whether it actually applies or not. But this is no help, for unless we explicitly represent ''by nature'' connections with some kind of necessity operator, we are left with two classes of things to which *B* one-way possibly applies, neither of which can be identified with the things to which *B* applies by nature rather than in a strictly necessary way.

## 6.2 *Problematic Barbara and scientific demonstration*

There is, however, a way of using one-way ampliation to produce a necssity proposition. In chapter 15, Aristotle (or perhaps some other peripatetic) calls for omnitemporally true universal assertoric propositions.[12] Now if modality is given a temporal interpretation (on which 'necessary' = 'true at all times'), and the one-way possibly *B*'s are taken to be all *B*'s, past, present, and future, we could even arrive at 'nec: (*A pp a B*)'. This would be an even more forceful expression of Striker's idea that if *A* could belong to anything to which *B could* belong, there must be some underlying natural connection between the natures *A* and *B* themselves. But it seems to me there is a major problem for this or any other attempt to modify Aristotle's '*A pp a B*' in the direction of science by adding two-way or one-way ampliation, or a condition of omnitemporality or a temporal interpretation of necessity. For as long as our primary connective (the first occurrence of '*pp*' in '*A pp a B pp*' or '*A pp a B p*') is not itself upgraded, all the modifications one likes will only produce a sentence that no more asserts the natural application of *A* to the *B*'s than its natural *non-application.* Even 'Nec: *A pp a B*' gets us no closer to '*A N a B*' than to '*A N e B*'. ['*N(A p a B p)*' *would*, but it is not a two-way proposition.] I see no way around this problem short of introducing a new modal operator.

As for that alternative, I am more optimistic than Striker about fitting a new operator into Aristotle's system. In fact, '*A* $\phi$ *a B*' – with $\phi$ standing for 'applies by nature' or 'applies unless something intervenes' – would do. This, in fact, has the advantage over '$N(S \rightarrow A\ a\ B)$' of stating directly the basic fact that *each B* is such that it will be *A* unless something intervenes. The resulting syllogisms in Barbara

$$\begin{array}{c} A\ \phi\ a\ B \\ B\ a\ C \\ \hline A\ \phi\ a\ C \end{array} \qquad \begin{array}{c} A\ \phi\ a\ B \\ B\ N_w\ a\ C \\ \hline A\ \phi\ a\ C \end{array}$$

would be both valid and complete. But neither could be scientific, because the middle term might be only accidentally related to the extremes. (Thus we have the same problem as with Barbara *NAN* and *NNN* with weak *cop* necessity.) But just as with other modalities, we could define two versions of $\phi$ propositions, one of which would require not only that *A* naturally apply to each *B* but also that *A* bear some natural relation to *B* itself. (Here *A* and *B* might be types of events that were causally related. Such propositions would then, on one reading of *Post. An.* A.6, fall under *per se* predications of Aristotle's fourth type – an idea defended recently by Michael Ferejohn.[13]) Then on this "*per se*" reading

$$\text{(i)}\ \frac{A\ \phi\ a\ B \quad B\ \phi\ a\ C}{A\ \phi\ a\ C} \qquad \text{(ii)}\ \frac{A\ \phi\ a\ B \quad B\ N_s\ a\ C}{A\ \phi\ a\ C} \qquad \text{(iii)}\ \frac{A\ N_s\ a\ B \quad B\ \phi\ a\ C}{A\ N_s\ a\ C}$$

would seem to be strong candidates for scientific demonstrations.[14] But to investigate their credentials in light of the often highly controversial details of Aristotle's theory of science would take us far afield. And in any case, as far as the *Prior Analytics* goes, there is no evidence that Aristotle meant to introduce such an operator.

## 6.3. TWO INVALIDITY PROOFS

One further feature of chapter 14 calls for special comment. Having discussed four allegedly complete moods, Aristotle goes on to propose an efficient proof for the invalidity of arguments with any of four sorts of premise pairs with particular major and universal minor premises: $I_{pp}$, $A_{pp}$; $O_{pp}$, $A_{pp}$; $O_{pp}$, $E_{pp}$; $I_{pp}$, $E_{pp}$. All are equivalent, by qualitative conversion, to

$$\frac{A\ pp\ i\ B\ (pp) \quad B\ pp\ a\ C}{A\ pp\ i\ C}$$

And all are invalid, because

> nothing prevents *B* from extending beyond *A* and not being predicated to an equal extent. But then let *C* be that by which *B* exceeds *A*. For ⟨it does not follow that⟩ *A* ⟨two-way⟩ possibly belongs to all or none or some or not to some of this, if ⟨two-way⟩ possibility premises convert and *B* ⟨two-way⟩ possibly belongs to more than *A*. (33a38–b3)

The point is that the *C*'s might all be among the things to which *B* two-way possibly applies, but fall outside the "some" *B* to which *A* two-way possibly applies. Thus, so far as our premises go, *A* might be either necessarily applicable or necessarily inapplicable to each and every *C*.[15] Because the premises are compatible both with '*A N a C*' and with '*A N e C*', they cannot entail any two-way possibility relation of *A* to *C*, either affirmative or negative, universal or particular. For as Aristotle will add a few lines later on, at the conclusion of his second proof of invalidity, "that which is necessary is not ⟨two-way⟩ possibly applicable" (33b17).[16] In short, everything goes smoothly so long as one realizes that Aristotle (in the first line quoted earlier) has in mind that the things to which *B* two-way possibly applies may extend beyond the things to which *A* two-way possibly applies. This has been missed by some commentators (understandably so!), but is more clear in the closing line: "*B* two-way possibly belongs to more than *A*", where "two-way possibly belongs to" seems clearly understood.

## 6.4. ONE PROBLEMATIC, ONE ASSERTORIC PREMISE

### 6.4.1. *Outline of chapter 15*

The general plan of chapter 15 is basically that of chapter 14, but it becomes in the details of its working out substantially more complex because of the introduction of (1) a section necessary for construction of a single *reductio* proof and (2) a passage on the temporal qualification of syllogistic premises. Both have recently been objects of intense study: The first *may* show Aristotle's recognition of some very important principles of modal propositional logic; the second may be a later, bungling intrusion or may contain an important insight utilizing a temporal version of Kripke-type possible-worlds semantics.

First, an overview of the chapter, on syllogisms with one assertoric and one problematic premise in the first figure.

- I. Moods with two universal premises
  - A. Perfect syllogisms, 33a25–40
  - B. Imperfect syllogisms
    1. RAA proof for *A*, $A_{pp}/A_p$
       - (a) Modal principle: If *S* is a valid plain syllogism, and if the premises of *S* are both possible, then the conclusion of *S* is also possible, 34a5–24
       - (b) Modal principle: If a proposition *p* entails *q*, then if *p* is at worst false and not impossible, then *q* is at worst false and not impossible, 34a25–33
       - (c) Application of (a) and (b) to the RAA proof, 34a36–b2
       - (d) Alternative proof "through the first figure," 34b2–6
       - (e) Remarks on temporal qualification of premises, 34b7–18
    2. Proof for *E*, $A_{pp}/E_p$
       - (a) RAA gives one-way possibility conclusion, b19–31
       - (b) Proof that that mood cannot give two-way possibility conclusion, b31–a2

3. Proofs by qualitative conversion, 35a3–20
   C. Invalid configurations, a20–30

II. Moods with one universal and one particular premise
   A. Perfect moods
   B. Imperfect moods
   C. Invalid configurations

### 6.4.2. *The perfect moods*

The perfect moods are just the usual suspects: those in Barbara, Celarent, Darii, and Ferio with two-way possibility major premise and assertoric minor. Their perfection derives from the usual source: The minor premise serves to bring (some or all of) the minor term "under" (*hupo*) the middle, and the major premise brings all of the middle under the things to which the major term two-way possibly applies (33b34; cf. 35a35). Therein lies the importance of the minor premise being the assertoric one, rather than the major. Because we shall have, for example, 'All *C*'s are *B*'s', then if *A* is related to all the *B*'s in some particular way, *A* will thereby be related to all the *C*'s in that same way. Thus Barbara *pp, A*/*pp* is obviously valid:

*A pp* all *B*
*B* all *C*
―――――
*A pp* all *C*

It is for this reason, too, that ampliation is *not* needed here – although Aristotle makes no mention of the fact. More interesting is that with *two-way* ampliation of the major premise, the mood is *invalid* (because one would not know that all the actual *B*'s, including the *C*'s, are among the two-way possibly *B*'s), whereas with *one-way* ampliation it is valid (because the actual *B*'s are all included among the one-way possibly *B*'s). Possibly Aristotle did not bother himself about any of this simply because with an assertoric minor the mood is so obviously valid as it stands. Still, it is once again clear that Aristotle's distinction between ampliated and unampliated (two-way) propositions, and that between one- and two-way ampliation, point to a large area for potential theorizing that he himself did not explore.

Making the major premise assertoric and the minor problematic gives quite a different picture:

## *6.4 One problematic, one assertoric premise*

$$\frac{\begin{array}{c} A \text{ all } B \\ B\ pp \text{ all } C \end{array}}{A\ pp \text{ all } C}$$

If we know only that all the *C*'s might and might not be *B*'s, it does *not* follow, from the added fact that *A* is related in a certain way to all the actual *B*'s, that *A* is related in that same way to all the *C*'s. Aristotle realizes that this argument is invalid but maintains that such a premise pair will still yield a *one*-way possibility conclusion (*A p* all *C*). His proof, the preparatory introduction of two new modal principles, the alternative proof "through the first figure," and an appendix on temporal qualification of premises are all of great interest and all highly problematic. These points will occupy us for the next several sections.

### *6.4.3. Principles of propositional modal logic in chapter 15?*

First of all, what precisely are those modal principles, and how are they related to one another? At first sight it may appear that Aristotle has in mind a *de dicto* reading and that he has discovered the important principles of propositional modal logic (PML) that if *p* entails *q*, then poss: *p* entails poss: *q*, and if *p* entails *q*, then nec: *p* entails nec: *q*. The complete passage reads as follows:

> First, let it be said that if, when *A* is, *B* must be (*tou A ontos anangkē to B einai*), then likewise if *A* is possible, *B* must be possible (*dunaton ontos tou A dunaton estai kai to B ex anangkēs*). For i) let such be the case [i.e., that *A* entails *B*] and ii) let that for which *A* stands be possible, that for which *B* stands impossible. iii) If, then, that which is possible, when it is possible, should come about, but the impossible, when it is impossible, would not come about, and iv) *A* is possible, *B* impossible at the same time, then v) *A* could come about without *B*. vi) But if it comes about, then it also is. For what has come about, when it has come about, is. But it is necessary to take the impossible and the possible not only with regard to their coming about, but also their being true and their obtaining, and however else the possible is spoken of. For in every case the same point will hold. ⟨vii) But this is impossible for it was assumed that if *A* is, *B* must be also.⟩
>
> viii) Further, 'if *A* is, *B* is' must not be taken as 'if one thing, *A*, is, then *B* will be'. For nothing necessarily obtains on grounds of one thing being the case, but two at least – as when the premises are such as was described with regard to the syllogism. ix) For if *C* ⟨is predicated⟩ of *D*, and *D* of *Z*, then *C* will necessarily be predicated of *Z*. x) And if each ⟨premise⟩ is

> possible, the conclusion will also be possible. xi) Just as, then, if one should put *A* as the premises, *B* the conclusion, xii) it would follow not only that if *A* is necessary, *B* will also at the same time be necessary, xiii) so also ⟨if *A* is⟩ possible, ⟨it will follow that *B* is⟩ possible. (34a5–33; textual divisions added)

The critical portion for Aristotle's possible discovery of those rules of PML are viii–xii, 34a19–24. The *de dicto* reading would be that *A* is here supposed to stand for the *conjunction* of the premises, and *B* for the conclusion, so that Aristotle is correctly observing that, given a valid plain syllogism, if the conjunction of the premises is necessarily (or possibly) true, then so is the conclusion. But proposition ix suggests that what the passage is saying is that if *each* (*hekateron*, 34a21) premise (as opposed to their conjunction) is necessary (or, by implication, possible), then so is the conclusion.

There are two related issues here, one having to do with Aristotle's ability to express a conjunction, and the other with the validity of his argument. As for the former, Aristotle neither devises any way of expressing a conjunction within his categorical syntax nor recognizes explicitly that he might adopt a *de dicto* syntax. Still, he speaks early in the passage of *A* being possible and *B* impossible "at the same time," and presumably he would use this language to capture the idea of a conjunction of the two propositions for which *A* stands (in his example, '*C a D*' and '*D a Z*'). This is still slightly charitable, because a clear expression of the idea would require recognition of the distinction between having both '*C a D* at time *t*' and '*D a Z* at time *t*', on the one hand, and '*C a D* & *D a Z* at time *t*', on the other. In some contexts this distinction is critical: With plain or necessity propositions throughout, the two alternatives would be equivalent, but not with possibility propositions.

This is directly relevant to the second issue. There does seem to be an error here, or at least a looseness in the handling of the modal principle sometimes attributed to him. Specifically, proposition x) represents either an outright error (inferring in effect that the conjunction of two possibility propositions is true from the fact that each of them is true) or a potentially misleading manner of speech. It looks as though he has in mind a set of plain premises in Barbara, then uses 'possible' (and 'necessary') metalogically to specify the modality of the propositions in question,[17] and then says that if each of the premises is possible (i.e., if they are '*C p a D*' and '*D p a Z*' rather than plain '*C a D*' and '*D a Z*'), then the conclusion will be '*C p a Z*' (rather than just '*C a Z*'). But of course the possibility conclusion will follow (given the validity of *plain* Barbara) if and only if one has '*C p a D* & *D p a Z*' true, not just '*C p a D*' true and '*D p a Z*' true.

Aristotle makes a parallel claim about the pure necessity versions of his valid plain syllogisms. He is not, however, introducing a new principle (if *A* entails *B*, then nec: *A* entails nec: *B*) by which one could validate at a stroke all the pure necessity syllogisms of chapters 9–11. Rather, he merely recalls the fact, already demonstrated in those chapters by a variety of proofs, that if one starts with a valid plain syllogism and then substitutes necessity propositions throughout for their plain counterparts, one obtains a valid necessity syllogism. What is new, and what he now adds because he will need it in the specific *reductio* proof for Barbara *A*, *pp/p* to follow, is that in the same way, if a given plain syllogism is valid, then so is its (one-way) possibility counterpart. And that, as suggested earlier, is where error creeps in.

### *6.4.4. A second modal principle*

There follows one more paragraph preparatory to the validation of Barbara *A*, *pp/p:*

> This having been demonstrated, it is clear that if something is assumed that is ⟨at worst⟩ false and not impossible, then that which follows on account of the assumption will be ⟨at worst⟩ false and not impossible (*pseudos estai kai ouk adunaton*). For example, if *A* is false but not impossible, and if when *A* is, *B* is, then *B* also will be ⟨at worst⟩ false and not impossible. For since it was shown that if *A* is, *B* is, and if *A* is possible, *B* is possible – and ⟨it was⟩ also assumed that *A* is possible – then *B* will be possible, too. For if ⟨*B*⟩ were impossible, then the same thing would at the same time be possible and impossible. (34a25–33)

Aristotle here puts to work what had just been shown, but with "at worst false but not impossible" substituted for "possible." The change in terminology is not entirely superfluous, for it will facilitate the application of the principle in question (if *A* entails *B* and *A* is at worst false, then *B* is at worst false) to Barbara *A*, *pp/p* – to which application we now turn.

## 6.5. FIRST PROOF OF BARBARA *A*, *pp/p*

First, the proof itself:

> These things having been determined, i) let *A* ⟨apply⟩ to all *B*, *B* possibly ⟨apply⟩ to all *C*. ii) It is necessary, then, that *A* possibly apply to all *C*. iii) For let it not possibly apply, iv) but assume *B* to apply to all *C*. v) This

> may be false, but it is not impossible. vi) If, then, *A* does not possibly apply to every *C*, but *B* applies to every *C*, *A* will not possibly apply to every *B*. vii) For there comes about a syllogism through the third figure. viii) But it was assumed that ⟨*A*⟩ possibly applied to every *B*. ix) Therefore it is necessary that *A* possibly apply to every *C*. x) For something ⟨at worst⟩ false and not impossible was postulated, but the consequence is impossible. (34a34–b2)

The syllogism to be proved (with disambiguation of Aristotle's 'possibly applies') is

(a) *A a B*
(b) *B pp a C*
———
(c) *A p a C*

The proof begins with the standard first step of a *reductio ad impossibile*: Assume as true the negation (d) of the desired conclusion (c):

(d) *A N o C* (there is some *C* to which *A* does not possibly apply)

Then comes an unusual step: Combine (d) with

(e) *B* all *C*

Given the initial premise (b), '*B pp* all *C*', (e) might well, for all we know, be false, but it is not impossible. This step is clearly correct, for '*B pp* all *C*' entails '*B p* all *C*', which simply says that *B* (one-way) possibly applies to every *C*. So while *B* might not actually apply to every *C*, there is nothing impossible about its applying to every *C*.

Meanwhile, because (d) is assumed *true* for purposes of the *reductio*, it obviously follows that it is assumed possible. (This obvious point is left implicit in the text.) Then we have the two premises

(d) *A N o C*

and

(e) *B* all *C*

each of which is at worst false, and not impossible. These two constitute in turn the premises of Bocardo *NAN*. Therefore, if each of (d) and (e) is at worst false, then anything they entail should be at worst false. But these premises entail, *via* Bocardo *NAN*,

(f) *A N o B*

So (f) should be at worst false, not impossible. But given (a), '*A* all *B*', (f) is not just false, but impossible. So *if* (d) is true, then we can derive something impossible from something that is at worst false. But because such a derivation is impossible, (d) must be false. Therefore the negation of (d) – that is, (c), '*A p* all *C*' – is true.

Alas, this elegant reasoning is invalid. The problem, as noted earlier, is that the possible truth of *each* premise of a given valid syllogism does not guarantee the possible truth of their conjunction, hence not the possible truth of that syllogism's conclusion. On the other hand, this analysis of Aristotle's argument is liable to more than one possible objection.

First, Mario Mignucci, in a lengthy and careful discussion of the entire passage, contends that Aristotle appeals to a principle that if *p* is possible, then *p* will at some time be true.[18] Mignucci says that Aristotle "constantly" asserts this,[19] and cites *De Caelo* A.12, 281b15ff. Whether or not, and in what sense, Aristotle held this view is highly controversial. (And there are, in any case, only a few passages in which he even *seems* to assert it.) But putting this aside, I do not find that the principle is asserted in the present passage of the *Prior Analytics.* What Aristotle says (in Mignucci's own, quite accurate translation) is, "If then that which is possible, when it is possible for it to be, might happen, and if at the same time *A* is possible and *B* impossible, it would be possible for *A* to happen without *B*, and if to happen, then to be . . ." (34a8–12). Aristotle's argument is that if *A* is possible at *t* and *B* is impossible, then it is possible that at *t, A* will be the case and *B* not. (Or, nothing impossible would result from the assumption that at *t A* obtains but *B* does not.) But this will conflict with the assumption that "*B*'s being follows necessarily from *A*'s being" (34a5–6).

There is no ground here for saying, first, that Aristotle believes that anything that is possible *will* at some time be actual, nor, further, that the argument proceeds by assuming that at "that" time it will actually be the case that *A* obtains and *B* does not. All Aristotle says, and all his argument requires, is that if it is possible for some statement to obtain at a certain time, then nothing impossible should follow from assuming that it actually does obtain at that time. If, meanwhile, some second statement is impossible, then it will fail to obtain at all times, including any time at which that first statement might be hypothesized to obtain. But if the first statement entails the second, then something impossible will follow (i.e., the second statement's being simultaneously true and false) from supposing the first statement true at a certain time. Nothing in this reasoning requires that either statement ever *actually* be true, but only that a (hypothetical, even counterfactual) supposition of their truth entails an impossibility.

A second, independent issue is raised by both Mignucci and Becker, who contend that the *reductio* syllogism is not Bocardo *NAN*

$$\frac{\begin{array}{l} A\,N\,o\,C \\ B\,a\quad C \end{array}}{A\,N\,o\,B} \quad \text{(note the necessity conclusion)}$$

but only Bocardo *NAA* (with plain conclusion, '*A o B*').[20] The text, however, calls for a necessity conclusion [*ei to A mē endechetai panti tōi C* (if *A* is not ⟨one-way⟩ possibly applicable to every *C*) . . . *to A ou panti tōi B endechetai* (*A* is not possibly applicable to every *B*), 34a38–39], not just for a conclusion that necessarily follows from the premises. The parallel between the two clauses just quoted leaves no room for assuming a merely assertoric meaning for the second. In fact, the Greek of the second clause could not mean that, even taken in isolation.[21]

The motive for introducing such an unlikely reading of the text is that a passage in chapter A.11 (31b37–39) seems to deny that Bocardo *NAN* is valid:

> But if one of the premises is affirmative, the other negative [among mixed assertoric/necessity syllogisms of the third figure], whenever the universal should be negative and necessary, the conclusion will also be necessary [Ferison *NAN*]. . . . But when the affirmative is necessary, whether universal or particular, or the negative is particular, the conclusion will not be necessary [i.e., Bocardo *ANN*, Ferison *ANN*, Bocardo *NAN* are invalid].

Ross's response, that Aristotle, "forgetting the rule laid down in 31b37–39, draws the conclusion [in the proposition vi cited earlier] Some *B* cannot be *A*,"[22] would be more reasonable than the imposition of an impossible reading on the Greek of 34a38 – or the emendation of that line (by adding *huparchein* against all external evidence) *plus* the imposition of a still unlikely, even if barely possible, reading. Ross's response might lead one to suppose that had Aristotle remembered his own rule (I would not call it a "rule" so much as a summary of results), he would have used Bocardo *NAA* in chapter 15 (in the complex proof under consideration) rather than Bocardo *NAN*. Then he could also have erased the *endechesthai* in 34a41 (*all' hupekeito panti endechesthai huparchein*) as superfluous, so using precisely the *reductio* syllogism Becker and Mignucci want him to use.

However, this misses two interesting points: First, the "rule" stated in chapter 11 is *wrong*: Bocardo *NAN* is valid (read with weak *cop*, as with the mixed assertoric/necessity moods in general) and can easily be proved valid by a standard *ekthesis* proof. (See Chapter 4, Section 4.2. It could also be shown valid by *reductio*, using Barbara *p,A/p*; but Aristotle does not discuss such syllogisms.) Thus his summary statement of results back in chapter 11 could, even without further evidence in the same direction, be regarded quite reasonably as a minor oversight.

Second, Aristotle *needs* a necessity conclusion here. In a standard *reductio* proof it will be sufficient to show that the negation of the desired conclusion, combined with one of the initial premises, entails the contradictory of the other initial premise. From this it follows that if the premises are true, then the negation of the desired conclusion is false, the desired conclusion true. But his argument here in chapter 15 is not a standard *reductio*: To produce the reducing syllogism, Aristotle here combines the negation of the conclusion *not* with one premise, or with something entailed by one premise, but with something whose *possibility* is thus entailed (i.e., something that is "at worst false" given the premise). Specifically, given the initial premise '*B pp a C*', it will be "at worst false and not impossible" that '*B a C*'. And it is this latter proposition that combines with '*A N o C*' (the negation of the initial conclusion to be derived) to entail '*A N o B*'. Thus, as Aristotle says, the latter, because it follows from premises that are at worst false, should itself be at worst false. At this point the entire proof hinges on the question of whether or not '*A N o B*' is – given the premise '*A a B*' – at worst false, and not impossible.

Well, given '*A a B*', '*A N o B*' is indeed *impossible* and not just false. Contrariwise, given '*A a B*', it is *not impossible* that '*A o B*' (the latter being the conclusion that Becker and Mignucci wish to see Aristotle using here). Of course, it is impossible that *at any given time* '*A a B* & *A o B*' hold; Aristotle would not question that. What he would assert, however, is that given just '*A a B*', *A* might be accidentally related to each of the *B*'s and so just happen to apply to all of them. If that is so, then *A* might possibly fail to apply to some (or all) of the *B*'s.

The point is clear at *De Caelo* 281b9ff. (cf. *Soph. Elen.*, 166a23–31, 177b23–26), where Aristotle distinguishes falsehood from impossibility:

> That you are standing when you are not standing is false but not impossible (*pseudos men, ouk adunaton de*). Similarly to say that a man who is playing the lyre, but not singing, is singing is false but not impossible (*pseudos all' ouk adunaton*). To say, however, that you are at the same time standing and sitting, or that the diagonal is commensurable, is to say what is not only false but also impossible . . . a person has, it is true, the capacity at the same time of sitting and standing, because when he possesses the one he also possesses the other; but not in such a way that he can at the same time sit and stand, but at different times.

The basic test of possibility here involves the natures or essences involved – that of being human, and of the "postures" (as the *Cat.* has it) standing and sitting. Relative to the information that Socrates is sitting, the supposition that Socrates is standing would be false but not impossible. Similarly, to return to the *Prior Analytics*, given merely that '*A a B*', we know the supposition '*A o B*' to be false, but we do not know it to be impossible. By contrast, given '*A a B*', we know that '*A N o B*' *could not* be true. For example, if all humans are standing, it would be merely false to say that some human was sitting; but it would be impossible to say (correctly) that some human *could* not stand. (Conversely, if it were true that some human *could* not fly, it would be not only false but also impossible to suppose that all humans were flying.) And that is why Aristotle not only *does* (*pace* Becker and Mignucci and, as we shall see, Nortmann) use Bocardo *NAN* (rather than NAA), as the text itself asserts, but must do so.

Put another way, the phrase '*B* is impossible, given *A*' is ambiguous in much the same way as '*B* must be false, given *A*'. It might mean simply that *A* and *B* are incompatible (either as contraries or contradictories), so that if *A* is true *B* must be false. This does not assert that *B*'s falsity is itself necessary (necessary *haplōs*), but rather that its falsity necessarily follows from *A*'s being true. But the phrase could also mean that *B* itself is impossible, and that this follows from the truth of *A*. The Greek text and the requirements of Aristotle's argument show that it is in this latter sense that "something impossible" has been derived, where the impossible statement is '*A N o B*'.

## 6.6. SECOND PROOF FOR BARBARA *A*, *pp/p*

Aristotle's second *reductio* proof for this syllogism has been denounced and excised as the work of a confused commentator,[23] as so full of logical blunders that it "cannot receive a meaningful interpretation."[24] Even its claim to be a *reductio* is rejected with contempt.[25] The text reads as follows:

> It is possible to produce the impossibility through the first figure, positing that *B* applies to *C*. For if *B* applies to all *C*, and *A* possibly applies to all *B*, then *A* would also possibly apply to all *C*. But it was assumed that it did not possibly apply to all. (34b2–6)

To begin with, Ross is right that the argument is not the usual sort of *reductio*, in which the negation of the original conclusion combines with one of the premises to yield something inconsistent with the other premise. Ross says that "here the original conclusion (For all *C*, *A* is possible) is proved by a manipulation of the original premises, and from its truth the falsity of its contradictory is inferred.''[26] But I believe a close look at this ''manipulation'' shows that the argument is coherent, that it is a *reductio*, and that it, like its immediate predecessor, uses the modal principle that if *p* entails *q*, then if *p* is at worst false, *q* is at worst false, but that it is simply more complex than one might suspect from its terse five-line presentation (and perhaps more complex than its author suspected). The strategy is to prove valid

(A) (1) *A a B*
(2) *B pp a C*
(3) *A p a C*

by supposing it invalid (i.e., by supposing that the two premises could be true and the conclusion false). Then we would have '*A a B*', '*B pp a C*', and (4) '*A N o C*' all true. But if '*A a B*' is true, then '*A p a B*' is at worst false, not impossible. And if '*B pp a C*' is true, then '*B a C*' is also at worst false. If that is so, then anything those two entail must be at worst false (by principle I.B.1.b). But those two statements clearly entail, by the following perfect syllogism, '*A p a C*':

(B) (5) *A p a B*
(6) *B a C*
(3) *A p a C*

Therefore '*A p a C*' must itself be at worst false. But if (4), '*A N o C*', then (3), '*A p a C*' is not just false, but impossible. Therefore we must reject '*A N o C*', hence accept '*A p a C*'. (Notice, by the way, Aristotle's recognition of a mixed syllogism involving a one-way possibility premise.)

This gives a coherent reading of the text, and one that does follow the general *reductio* procedure of showing that something impossible follows from assuming both the initial premises and the negation of the desired conclusion to be true. Again, it departs from Aristotle's usual implementation of that procedure insofar as it does not immediately pair the negation of the desired conclusion with one of the given premises in order to derive something inconsistent with the other premise. Rather, it derives from the two premises something that is inconsistent with the negation of the desired conclusion.

But at this point one might wonder if this argument doesn't go around behind the barn to shoe the horse. If the ''at worst false . . .'' modal principle is to be used, why not just prove from premises (1) and (2) that (5) and (6) are "at worst false" (which the argument does do), then derive directly from them the desired '*A p a C*', as in the foregoing syllogism (B). Well, this isn't quite what one needs, for this syllogism would only show that '*A p a C*' *is at worst false*, rather than that '*A p a C*' *is true.* That is, the premises do not entail plain '*A a C*'; if they did, they would entail the truth of '*A p a C*'. Rather, they entail only the "being at worst false" of '*A p a C*'. (One could remedy this situation if the system contained an appropriately constructed counterpart to the principle 'poss: poss: $p \rightarrow$ poss: $p$'; but it doesn't.) So in order to clinch the point, Aristotle's argument appeals in effect to the idea that if *A* necessarily fails to apply to some *C*, then it is not possible that *A* possibly apply to all *C* [which would be a kind of hybrid counterpart to 'Nec: $\neg p \rightarrow \neg$poss: (poss: $p$)']. Thus, given '*A N o C*', it is impossible that '*A p a C*'. But given premises (1) and (2), '*A p a C*' is not impossible, but at worst false. Thus if we assume (1) and (2), we must reject '*A N o C*', hence accept '*A p a C*'.

One final point of interest is the fact that here, as elsewhere, Aristotle theorizes in a way that cannot be fully represented within his formal system. For it seems that here he is implicitly prefixing a *de dicto* operator ('it is at worst false, and not impossible, that') to a ground-level modal *cop* proposition to produce statements of the form 'possibly: *A pp* (or *N*) *a B*'. There are various ways in which one might use two modal operators within the *cop* framework, as in '*A N a B pp*' or '*A N a B N*' or of course '*A pp a B pp*'. But it does not appear that Aristotle means to express his concept of "at worst false and not impossible" in that way. Rather, he here shows an intuitive grasp of principles that would today be expressed by use of iterated or nested modalities.

## 6.7. OMNITEMPORAL PREMISES?

The passage 34b7–18 is a kind of appendix to the treatment of Barbara *A, pp/p* but *may* claim to draw a moral applicable if not to the whole of the system then to the large portion of it concerned with contingency syllogisms. Above all, the passage has received a great deal of attention from antiquity to the present for its introduction of omnitemporal premises – and, some would say, for its discovery of (a temporal version of) possible-worlds modal semantics.

(i) One must take belonging to all not as temporally defined, e.g., as now, or at this time, but without qualification (*haplōs*).
(ii) For it is through such premises that we produce syllogisms,

(iii) since if the premise is taken as ⟨holding⟩ now, there will not be a syllogism.
(iv) For perhaps nothing precludes Human from belonging some time to all moving things, e.g., if nothing else should be moving ⟨at that time⟩.
(v) And Moving possibly applies to every horse.
(vi) But Human does not possibly apply to any horse.
(vii) Further, let the first term be Animal, the middle Moving, the last, Human. The premises will then be similar (*homoiōs hexousi*) but the conclusion will be necessary, not possible (*anangkaion, ouk endechomenon*).
(viii) Clearly then the universal ⟨premise⟩ must be taken without qualification, not temporally defined (*haplōs, kai ou chronōi dioridzontas*). (34b7–18)

More than one commentator has suggested that the insistence on universal assertoric premises holding true at all times indicates that Aristotle has in mind scientific demonstration (cf. the demand for premises true at all times at *Post. An.* 73a28–34), that in fact this passage is best explained on such an assumption.[27] It is also possible to view omnitemporality as a condition intended simply to preserve validity, with no special connection to science. In either case, opinion would then divide as to whether that condition is supposed to apply to all syllogisms in the system or only to the type of case at hand.[28] The passage is in any case a prickly one, and we shall have to take a very close look before deciding about its possible implications for science or for temporality – or about its authenticity, which there is serious reason to doubt.

The first problem is that the passage is moderately obscure. Becker proposes to clarify matters by excising what I have labeled (ii), (iii), and (vii).[29] This leaves the general statement (i) about taking premises without temporal restriction, then the syllogism contained in (iv)–(vi):

Human all Mover
Mover *pp* all Horse
―――――――――
Human *p* all Horse

(Recall that Aristotle is in this context considering a syllogism with one two-way premise and a one-way conclusion.) Item (vii) supplies a second counterexample, independent of the first, (i)–(vi), to the practice of allowing temporally indexed syllogistic propositions. Although we shall see why Becker wished to deal with it surgically, we shall also find that Ulrich Nortmann has proposed a coherent reading of the entire passage with rather far-reaching implications. But to this we shall have to return after a look at the first argument, (i)–(vi).

There appears to be some reason to eliminate (ii) and (iii), because (ii), if taken to mean that one must use only premises that are in fact always true, is flatly contradictory to Aristotle's frequent practice (especially with regard to examples serving to invalidate a given mood) of using premises that are not true at all times, but only in fact true for a limited time. Indeed, 'Human all Mover' would be a good representative of such Aristotelian examples. (See, e.g., 'Moving *a* Animal', 30a28; cf. 30b5–6, b33, 31a17, b27, b31, 32a1.) Still, it would be better to find an interpretation that makes sense of the text as it stands, and one might try to answer the objection by limiting the scope of the remark in an appropriate way. Nortmann's suggestion that it is not meant to apply to the pure assertoric, pure necessity, or mixed assertoric/necessity syllogisms of chapters 4–12, but only to the chapters (14–22) on two-way possibility syllogisms, is certainly reasonable. It is far from trouble-free, however, as we shall see in a moment.

Meanwhile, notice that item (iii) is also problematic, because it is patently false – at least, if taken as a completely general statement. For if the premises of a given assertoric, necessity, or mixed assertoric/necessity syllogism validly entail a certain conclusion, the mood will remain valid if all three propositions are temporally indexed to the same time (e.g., the present). Taken in a more limited way, however, it is true, as the example in the text shows. Here again one could retain the given text and restrict it (pending further developments) in the same way we did (ii) – that is, to contexts involving at least one two-way premise.

Returning, then, to the concrete example in (iv)–(vi), we can at least say that it does show Barbara *A*, *pp/p*, without omnitemporality, to be invalid. As a matter of fact, the concrete example offered fits exactly the pattern Aristotle follows time and again when invalidating proposed syllogisms – that of specifying a *possible* situation (not necessarily an actual situation) in which the premises are both true and the conclusion false. Moreover, the particular major premise offered here of a possible, even if clearly not actual, situation is of exactly the same type as several of Aristotle's unquestionably authentic examples. (Some of these were just cited.)

But even this moderate reading, on which omnitemporality is meant to apply merely to the universal assertoric premises of syllogisms involving two-way possibility premises (i.e., those of ch. 14–22), will not appease those who doubt its authenticity, for there remain two important reasons to question the whole of 34b7–18, or at least to regard it as, at best, a later addition by Aristotle, and one whose implications he did not work out.

First, consider Aristotle's own examples of universal assertoric propositions in chapters 14–22. Outside the passage in question, chapter 15 itself contains few occurrences of such statements:

(a) Raven *e* Thinking (34b4–5)
(b) Moving *e* Knower[30] (34b38)
(c) Animal *e* Snow, Animal *e* Pitch (35a24)

All of these examples occur as premises in counterexamples to proposed syllogisms. Of these, the ones under (a) and (c) would always be true. But (b) is clearly another example of a premise that would be true at most only at selected times. So only thirty lines after the condition of omnitemporality has been laid down (34b7–8), we find it violated.

Beyond chapter 15, I find only two universal assertoric propositions, chapter 18's 'Healthy *a* Animal' and 'Healthy *a* Human' (37b36–37); Aristotle invokes these "same terms" twice more in the chapter, but without listing them again (38a2, a12). So here we find once more the banished true-only-at-some-time universal affirmative propositions – and none of the supposedly mandatory omnitemporal ones. Recall that these propositions, like (b), are used in constructing counterexamples that ought to be ruled out on the same grounds as the Human/Mover/Horse case of chapter 15. And yet Aristotle employs them as if he had never heard of requiring omnitemporal premises. Thus, what evidence we have runs counter even to the limited claim that Aristotle requires omnitemporal universal plain premises in A. 14–22. This strongly suggests either that the passage on omnitemporality is inauthentic or that if it is genuine, it was tacked on later (and its implications not worked out) and so cannot be taken as basic for the understanding of chapters 14–22 in their original intent.

Second, there remains the curious fact that apparently nothing in Aristotle's argument for this mood (34a5–34b6, discussed at some length earlier) even hints at or logically depends on any condition of omnitemporality. This suggests once again that the omnitemporality passage is not by Aristotle, or was added by him later on. If the latter, he ought to have gone back to that proof to see if in fact it required, for reasons he had not noticed before, any omnitemporal premises. In that case, his remarks (i)–(vi) could simply be taken retroactively. Or, if the proof worked just as well without such a requirement, he ought to have smelled something rotten in the state of Denmark. (I argued earlier that the latter alternative is correct.) But, supposing (i)–(vi) authentic, Aristotle never raises the question of how it is related to his previous proof for Barbara *A*, *pp/p*. Still, this would not be a decisive argument for excising (i)–(vi): We have seen that Aristotle's own discussion of the conversion of $E_{pp}$ ought to have caused him to rethink his earlier remarks on other conversions; but it did not.

Supposing the passage authentic, however, how does the counterexample work? We shall examine in the next section a reading that presupposes temporal definitions of necessity and possibility, along with a temporal version of possible-worlds semantics. But there is a more innocent and, to my mind, more plausible reading. Now, without the condition of omnitemporality, '*A a B*' (Human all Mover) will be true if we suppose merely that all the things that are *B*'s at some time *t* are also *A*'s at that time. [As the text says, "perhaps nothing precludes the application of Human to all movers at some time (*pote*)," 34b11–12.] As for the second premise, if we concentrate on the positive side of our two-sided possibility, we have '*B p a C*' (Mover *p* all Horse), and this will be true just in case nothing impossible follows from the assumption that in fact '*B a C*' holds. It is obvious that 'Human *a* Mover' and 'Mover *a* Horse' cannot be true at the same time. But again, without the condition of omnitemporality we may coherently assume the truth of '*B a C*' by supposing that at some time *other than t* everything that is a *C* at that time is also then a *B*. Then we have both premises true and the conclusion (Human *p* all Horse) false. (Notice that this analysis does not make the original premises true at different times. Both may be true at *t* because '*B p a C*' will be true at *t* if the essence of the things that are *C* at *t* does not preclude their being *B* at *some* time or other, whether the same or different from *t*.)

The condition of omnitemporality blocks this sort of counterexample by making '*A a B*' true at all times, so that we cannot make the minor ('*B p* all *C*') true by supposing the *C*'s to be *B*'s at some time other than that at which all the *B*'s are *A*'s. If that is so, it is no longer possible to make both premises true, and the counterexample collapses. Thus, probably, reasoned the author of the counterexample, and the reply to it, at 34b7–18.

But blocking a certain class of counterexamples does not establish the validity of the mood in question. And the sad fact is that Barbara *A, pp/p*, even with omnitemporal major (or major and minor) premise, is of dubious validity. Here is a counterexample:

| | |
|---|---|
| Human all Drinks Hemlock | (taken omnitemporally) |
| Drinks Hemlock *pp* all Horse | (taken omnitemporally) |
| Human *p* all Horse | |

It certainly seems possible that all things past, present, and future that drink hemlock be human beings; and the minor premise is always true.

Yet the conclusion can never be true. [Notice that the same counterexample works with a (weak *cop*) necessity major.] One could fend off such cases by defining 'possible' as 'true at some time'. Then, because it is possible that some horse drink hemlock, it would actually happen at some time that some horse drank hemlock, with the result that our major premise would no longer be true. But Aristotle does not define his modalities temporally, as we shall see in a moment. A more plausible course would be to use a strong *cop* necessity premise (along with the appropriate [Type I] reading of the other premise). This would give Barbara $N_s$, *PP/P*:

$$\frac{\begin{array}{c} A\, N_s\; a\, B \\ B\, PP\, a\, C \end{array}}{A\, P\; a\;\; C}$$

which is easily validated by *reductio*. But clearly the author of our passage does not have in mind a strong *cop* proposition, either.

But if Barbara *A*, *pp/p* is invalid even with omnitemporal premises, what is wrong with Aristotle's argument for it? Again, the mistake is, in effect, to suppose that if each of two propositions is possible, then their conjunction is possible. And the Human / Drinks Hemlock / Horse example shows that the conjunction of the premises of Barbara *A*, *pp/p* need not be possible even if each conjunct is.

Finally, concerning the *motivation* of the passage, it seems dubious that omnitemporality was supposed to make assertoric propositions scientific, for even if we allowed only universal assertoric propositions that were always true (or, more reasonably, if one read such propositions as *asserting* that their predicates always applied to their subjects), that still would not be enough to make them fit for scientific service. A scientific premise would have to assert not just that, say, *A* always applied to *B*, but *that A was related per se to B* itself. But there is not the slightest evidence anywhere in the *Prior* or *Posterior Analytics* that Aristotle intended *any* of his assertoric premises to be read *that* strongly. On the contrary, they are consistently contrasted with all propositions of necessity. Moreover, the *Posterior Analytics* expressly denies that statements asserting the application of predicates incidental to their subjects can figure in scientific demonstrations, even if they should be always true: "For what is incidental is not necessary, so that you do not necessarily know why the conclusion holds – not even if it should always be the case (*oud' ei aei eiē*) and not in itself (*kath' hauto*)" (75a28ff., Barnes translation). *Perhaps* (even this is *highly* dubious) there are more global and indirect Aristotelian arguments linking "always applies" to *per se* predication; but even if so, that is hardly to say that a universal affirmative plain proposition (taken omnitemporally) *asserts* some kind of *per se* connection between its predicate and subject.[31]

This would indicate that (a) if the requirement of omnitemporal assertoric universal statements is being made in the name of science, then the author of the passage is not Aristotle, and (b) if the passage is authentic, its motivation must have been the rescue of Barbara *A*, *pp/p* from an otherwise fatal counterexample.

Those who remain unconvinced of the passage's authenticity will want to add that *if* its task is to preserve the validity of Barbara *A*, *pp/p*, requiring omnitemporality is an extremely odd way for Aristotle to have gone about it, even aside from the fact that this device won't work. He had just (ch. 14) had to do something about the validity of Barbara *pp*, *pp/pp*:

$$\frac{\begin{array}{l} A\ pp\ a\ B \\ B\ pp\ a\ C \end{array}}{A\ pp\ a\ C}$$

We noted earlier that as it stands, the minor premise brings the *C*'s under the two-way possibly *B*'s; but in order to bring them under the two-way possibly *A*'s (as asserted in the conclusion), the minor would have to bring the *C*'s under the actual *B*'s. The problem was remedied by ampliation: Given a major premise

$$A\ pp\ a\ B\ pp$$

the minor now can bring the *C*'s under the two-way possibly *A*'s by bringing them under the two-way possibly *B*'s. And this is precisely what it does.

In chapter 15 we have a similar situation:

$$\frac{\begin{array}{l} A\ a\ \ B \\ B\ pp\ a\ C \end{array}}{A\ p\ \ a\ C}$$

The minor brings the *C*'s under the two-way possibly *B*'s, rather than the actual *B*'s, whereas the major brings the actual *B*'s under the *A*'s. But just as before with the Barbara of chapter 14, so would ampliation now save the Barbara of chapter 15:

$$\frac{\begin{array}{l} A\ a\ \ B\ pp \\ B\ pp\ a\ \ C \end{array}}{A\ a\ \ C}$$

Notice that we now can derive an assertoric rather than just a one-way possibility conclusion. I am, of course, not saying that Aristotle *did* apply ampliation here, for if he had he surely would have seen that an assertoric conclusion would follow. The point is rather that *if* Aristotle had seen a problem with Barbara *A*, *pp/p*, it seems to me much more likely that he would have repaired it by ampliation rather than by omnitemporality. By using ampliation, Aristotle would have solved a problem in chapter 15 in the same way he had just solved an essentially identical problem in chapter 14. Notice, too, that this allows a reading of both Barbaras in the usual way as (implicitly) indexed to the present time (*or* as both indexed to some other time, or as true at all times). For example, everything that is a *C* now is two-way possibly *B* now. Because all the things that are two-way possibly *B* now are now two-way possibly *A*, it follows that everything now a *C* is now two-way possibly *A*. By contrast, requiring omnitemporality of an assertoric premise is not only unparalleled but actually in conflict, as we saw earlier, with his practice both before and after chapter 15.

A final problem pertains to all readings based on a temporal definition of modality (as on Nortmann's interpretation, discussed in the next section), as well as the more moderate reading of the counterexample just given, which does not go so far as to take possibility as truth-at-sometime, necessity as truth-at-all-times. Either way, the reasoning that underlies the passage on omnitemporality differs radically from Aristotle's thinking about other syllogisms involving two-way possibility and will wreak havoc on most of A.14–22, including even Aristotle's perfect moods in Barbara *pp*, *A/pp* (and *pp*, *N/pp*):

$$\frac{\begin{array}{l} A\ pp\ a\ B \\ B\ a\quad C \end{array}}{A\ pp\ a\ C}$$

The author of the counterexample to Barbara *A*, *pp/p* ought to say that this syllogism, too, is invalid. We can verify that the (positive side of the) first premise is true by supposing that '*A a B*' holds, say, on Tuesday, and observing that nothing impossible follows from that; and we may suppose that the minor premise is true because '*B a C*' holds today (Friday), although perhaps not on Tuesday. But '*A a B* on Tuesday' and '*B a C* today' – both of which, let us suppose, are possible given the initial premises – do not jointly entail that '*A a C*' will be true at any time. (The point also holds taking the positive side of the major premise as, 'for each *B*, there is a time at which it is *A*'.) Hence if '*A p a C*' follows only if

'*A a C* at some time or other' follows, the original premises do not entail the possibility of *A*'s applying to all *C* at any time. [As before, I have kept these remarks at the same level of (im)precision as the text itself, which could be given a more precise reading either in terms of Broadie's temporally relativized modalities or in terms of Hintikka's or Nortmann's temporally defined modalities. On the former, the major premise could read, "It is possible now, given the present state of the world, including the accidental and essential properties of the *B*'s, that '*A a B*' hold on some future date." If the minor premise is not taken omnitemporally, and tells us only that all the present day *C*'s are also *B*'s, or merely that all the *B*'s existing at some time or other are *C*'s, then the argument is again invalid, and liable to the same sort of counterexample as given in the text.]

By contrast, on Aristotle's usual reading of his perfect moods, this one is in fact obviously valid, just as he says: If every actual *B* is possibly an *A*, and every actual *C* is an actual *B*, then every actual *C* is possibly an *A*. This reasoning clearly holds with three properly present-tensed statements. The fact that this simple, direct, and manifestly valid reasoning will have to be thrown out and the mood (read in the usual present-tense way) declared invalid shows how radically any "omnitemporal" approach that will support the text's counterexample to Barbara *A*, *pp/p* differs from Aristotle's thinking prior to and after our controversial passage in A.15, for exactly similar remarks will hold of the perfect moods of A.16 with necessity minor (with either weak *cop* or modal predicate) and contingent major. Moreover, the rescue of Barbara *pp*, *pp/pp* by ampliation in A.14 (32b38ff.) will have to be understood very differently than in the straightforward manner set forth earlier.

Do all these considerations prove the omnitemporality passage inauthentic? I suppose it could still be retained, although with serious reservations, if restricted in scope very severely, applying at most to moods with assertoric or necessity major and contingent minor [or, if it were (incorrectly) presupposed that all necessity statements would be omnitemporal anyway, to moods of the form *A*, *pp*/____]. My own conclusion is that the passage probably is not by Aristotle. He thought he had established, by an elaborate and highly ingenious argument, the validity of Barbara *A*, *pp/p*, and simply went on about his business. I am not in a position to identify an interpolator, but we do know that tense logic was of great interest to many later peripatetics, some of whom went so far as to read "necessary" and "possible" as "always true" and "true at some time."[32] So there is no lack of suspects. We shall see in the next section that this approach is liable to all the objections just raised against the more moderate temporal reading, and some additional ones as well.

## 6.7 *Omnitemporal premises?*

Meanwhile, the lines labeled (vii) in the earlier quotation, which introduce a second counterexample, also present difficult problems. They ask us to let $A$ = Animal, $B$ = Moving, and $C$ = Human, so that we have

Animal *a* Moving
Moving *pp* a Human
---
Animal *pp a* Human

They then appear to object that the example shows that the premises give a necessary conclusion 'Animal *N* all Human' ("the conclusion is necessary, not ⟨two-way⟩ possible, for man is of necessity an animal," 34b16–17). (Taken literally, this is wrong: The alleged necessity "conclusion" is merely, as Peter Geach observes, "a proposition that holds true with this special choice of terms."[33]) One might charitably take the objection as saying that the premises are at least *consistent* with a necessary conclusion (i.e., 'Animal *N* all Human') and that this necessary conclusion is inconsistent with the proposed two-way possibility conclusion.

But this calls forth the criticism that the conclusion of the mood under discussion, Barbara *A, pp/p*, was a *one*-way, not a two-way, possibility proposition, and as Aristotle well knew and explicitly declared, a universal affirmative necessity proposition is *not* inconsistent with a corresponding affirmative *one*-way possibility proposition, but rather with a *two*-way proposition. That is, although the truth of 'Animal *N* all Human' would show 'Animal *pp a* Human' false, it does not show that 'Animal *p* all Human' is false. Thus the second counterexample to Barbara *A, pp/p* is beside the point.[34] In fact, this blunder is so obvious that it would support Becker's excision of the passage.

But this criticism could be met by seeing the passage as a cogent counterexample to Barbara *A, pp/pp*, showing that without omnitemporality the mood will be invalid right along with Barbara *A, pp/p*, the target of the previous counterexample. This means that the *eti* at 34b14 must be read as adding a new observation rather than as giving a second counterexample to Barbara *A, pp/p*. This is natural enough, and even makes good sense: If we look back to the earlier discussion of Barbara with *A, pp/* as premises, we find that our text had never actually said that only a one-way conclusion was possible. Rather, it simply argued that the premises do entail a one-way conclusion, and then (in the first counterexample) that they cannot do this unless the assertoric universal major is taken omnitemporally. This is in contrast to the pure two-way moods of chapter 14, which did give a two-way possibility conclusion. Having gotten this far, it might have seemed worthwhile to prove quickly, by a separate counterexample, that those same premises cannot yield a two-way possibility conclusion, either, without omnitemporality. Logically speaking, this further counterexample is superfluous, because if two premises do not entail a one-way conclusion, they will not, *a fortiori*, entail a two-way conclusion. But this hardly shows that the passage is spurious; after all, Aristotle will sometimes give separate proofs of validity even for two moods that are manifestly equivalent *via* qualitative conversion.

The remaining objection to this positive reading of the second counterexample would be that for completeness our author ought really to have shown not that Barbara *A*, *pp/pp* – without omnitemporality – is invalid, but that Barbara *A*, *pp/pp* – *with* omnitemporality – is invalid. This would be a genuine and essential addition to the discussion, which apparently aims to show that Barbara *A*, *pp/*____, *with* omnitemporality, will give *only* a one-way possibility conclusion. As it stands, the text does not quite finish off this demonstration. This is not a fatal problem, however: With an *affirmative* major (and minor) premise it might have seemed obvious that no negative one-way conclusion, hence no two-way conclusion, would follow. So on this reading, we are left with no major blunders in 34b14–17 [(vii) in the earlier quotation], but rather a cogent, even if logically gratuitous, appendix to 34b7–14. So if that previous, larger body of the passage on omnitemporality is authentic, Dr. Becker's appendectomy of (vii) is not indicated.

## 6.8. NORTMANN ON A.15, AND POSSIBLE-WORLDS SEMANTICS

Ulrich Nortmann's treatment of Aristotle's modal syllogistic draws heavily on the resources of contemporary modalized predicate logic and on possible-worlds semantics. He proposes distinctive, doubly modalized renderings of Aristotle's modal propositions, on the basis of which he is able (a) to carry out formal proofs within now-familiar systems (especially S4 and S5) and (b) to assess the strength of Aristotle's modal logic relative to those systems.[35] *Pr. An.* A.15, and the section on Barbara *A*, *pp/p* in particular, are fundamental for Nortmann's approach (at least to chapters 14–22 on syllogisms involving two-way premises), which is worked out with close attention to the details of Aristotle's text. Nonetheless, I think there are some significant difficulties with his account.

First, Nortmann's formulations of various modal propositions, as they are affected, directly or indirectly, by the condition of omnitemporality announced in A.15:

| | |
|---|---|
| $A_n$ = nec: $(x)$ $(Bx \rightarrow$ nec: $Ax)$ | Necessarily: For all $x$, if $x$ is $B$, then it is necessary that $x$ is $A$. |
| $I_n$ = nec: $(\exists x)$ $(Bx$ & nec: $Ax)$ | Necessarily: For some $x$, $x$ is $B$ and it is necessary that $x$ is $A$. |
| $E_n$ = nec: $(x)$ $(Bx \rightarrow$ nec: $\neg Ax)$ | |
| $O_n$ = nec: $(\exists x)$ $(Bx$ & nec: $\neg Ax)$ | |

The variable $x$ ranges over all individuals at all times (or in all possible worlds).

| | |
|---|---|
| $A_{pp}$ = nec: $(x)$ $(Bx \rightarrow$ *PP:* $Ax)$ | Necessarily: For all $x$, if $x$ is $B$, then it is two-way possible that $x$ is $A$. |
| etc. | |
| $A_p$ = nec: $(x)$ $(Bx \rightarrow$ *P:* $Ax)$ | Necessarily: For all $x$, if $x$ is $B$, then it is possible that $x$ is $A$. |
| etc. | |

Assertoric statements (again, within the context of A. 15–22) are defined as

| | |
|---|---|
| *A a B* = nec: $(x)$ $(Bx \rightarrow Ax)$ | Necessarily: For all $x$, if $x$ is $B$, then $x$ is $A$. |
| *A e B* = nec: $(x)$ $(Bx \rightarrow \neg Ax)$ | |
| *A i B* = nec: $(\exists x)$ $(Bx$ & $Ax)$ | Necessarily: For some $x$, $x$ is $B$ and $x$ is $A$. |
| *A o B* = nec: $(\exists x)$ $(Bx$ & $\neg Ax)$ | |

The startling presence of necessity operators at the head of possibility and assertoric propositions arises from the demand for omnitemporal universal premises (A. 15), combined with a reading of necessity as truth-at-all-times, and one-way possibility as truth at some time.[36] So if '*A a B*' is always true, it is necessarily true; and if '*A p a B*' is always true, it is necessarily true; and so on.

Nortmann's validity proof for Barbara *A*, *pp/p* can then be carried out as follows:

(1) $(x)$ nec: $(Bx \rightarrow Ax)$

which entails

(2) $(x)$ (*P:* $Bx \rightarrow$ *P:* $Ax)$ [or, in S4, even (2'): $(x)$ nec: (poss: $Bx \rightarrow$ poss: $Ax$)]

The minor premise is

(3) $(x)$ nec: $(Cx \rightarrow PP{:}\ Bx)$

which entails

(4) $(x)$ nec: $(Cx \rightarrow P{:}\ Bx)$

From (2) or (2') and (4) we can conclude

(5) $(x)$ nec: $(Cx \rightarrow P{:}\ Ax)$

as desired.

On this reading the mood is valid. As for the counterexample, it does work against the mood with major premise

$(x)\ (Bx \rightarrow Ax)$

but not against the mood with major premise

$(x)$ nec: $(Bx \rightarrow Ax)$

In terms of Nortmann's temporal semantics, this account would say that what the minor premise (3) tells us (concentrating, for present purposes, on the *p* side of *pp*) is this: Every individual that is an instance of *C* in any possible world (including the real world) will also be an instance of *B* in some alternative to that world. Or, in temporal terms, "as Aristotle would prefer to express it": Everything that is now or at any time an instance of *C* will be an instance of *B* at some time when it is a *C*. If now our assertoric major '*A a B*' is restricted to the actual world (or to the present time), there is no chance to exploit the information contained in the minor. Expressed in the temporal idiom, the only thing one knows about any present instance of *C* is that it is *B* at a point in time possibly different from now. To derive the conclusion, it is not sufficient to have the additional information that anything that is *now B* is now *A* as well. What one needs is the information that everything that at *any* time is *B* is *A*, too, at that time. And given such temporally unrestricted information concerning *A* and *B*, one can conclude, given the minor premise that every *C* will at some time also be *B*, that any *C* will at some time be *A;* or, in other words: Everything that is now *C* is now *possibly A* (possibility being actuality-at-some-time in the temporal paraphrase of possible-worlds semantics).[37]

Some of my reservations about this sort of analysis are implicit in what has gone before. First, at a very general level, my own aim has been to analyze Aristotle's modal proofs in a way that allows us to think them through as he did. This is not so unusual. As Robin Smith remarks, John

Corcoran's formal model (and Smith's own, which essentially follows Corcoran's) of the assertoric syllogistic "stays very close to Aristotle's actual text, since it allows us to read formally precise natural deductions straight out of it." By contrast, Łukasiewicz's model incorporates the whole of the propositional calculus, and his proofs of the moods recognized by Aristotle are carried out using its resources, typically in ways that can hardly be read directly out of the text step-by-step.[38] The situation with regard to the modal syllogistic is analogous. This is not to deny any interest to the project of saying how Aristotle might have validated Barbara *A, pp/p* had he been Kripke. The point is simply that we would like, as far as possible, to express the relevant insights and proofs in the way Aristotle did. There was no formal model available to him, let alone one as powerful as quantified S4 or S5. What we find is the invention of a formal system step-by-step, with proofs typically carried out via complete syllogisms laid down along the way. (Accordingly, we are here and now, in effect, gathering materials for a formal model as we work through various conversion principles and allegedly complete syllogisms. Chapter 8 will indicate how these could be readily put together in a formal model of Aristotle's modal syllogistic.)

Second, as was demonstrated earlier, the key passage (if authentic) can be fully understood without reading necessity as truth-at-all-times or possibility as truth-at-some-time. But Nortmann would reply that even Aristotle's argument for Barbara *A, pp/p* actually requires such a condition. His reasoning is that in order to establish

(1) *A a B*
(2) *B pp a C*
―――――――
(3) *A p a C*

Aristotle supposes that

(4) *A N o C* [the contradictory of (3)]

which entails

(5) *A o C*

Then, in view of (2), it should be at worst false that

(6) *B a C*

But (5), '*A o C*', and (6), '*B a C*', entail, *via* plain Bocardo,

(7) *A o B*

Thus, it should be at worst false that (i.e., possible that) '*A o B*', or true that

poss: *A o B* (*to A ou panti tōi B endechetai*, 34a39)

But because this is supposed to contradict (1), '*A a B*', Aristotle must mean by (1) not merely '*A a B*', but 'nec: *A a B*'. And this, in turn, shows, however indirectly, that omnitemporality is equivalent to necessity and that Aristotle had this in mind when constructing his *reductio* proof of validity.[39]

The problem with this reply is that although Aristotle does begin the *reductio* by supposing '*A N o C*', he does not replace this with '*A o C*', but combines '*A N o C*' with the possibility (being at worst false) of '*B a C*' to obtain, via Bocardo *NAN*, '*A N o B*'. As we saw earlier, given '*A a B*', it will not be just false, but impossible, that '*A N o B*'. We also saw earlier that the major premise of the reducing syllogism is '*A N o C*', not merely '*A o C*', and that its conclusion is '*A N o B*'.[40] Given this, it is not necessary to upgrade the plain major premise '*A a B*' to a necessity statement in order to make the argument work.

Third, Nortmann's temporal reading seems to me intrinsically implausible. On his reading, a statement of the form '*B p a C*' will say that everything that is now or at any time a *C* will be a *B* at some time or other. But against this, one has Aristotle's express statement that some cloak that might possibly be cut up could nonetheless never in fact be cut up (*De Int.* 9, 19a12ff.). It was due in part to such considerations that Hintikka formulated a temporal reading in terms of *kinds* of situations or connections: Any *kind* of event that can occur will occur at some time.[41] Conversely, if some *C* is a *B* at some time, then being *B* is possible for *C*'s. This version is not without its own textual and philosophical problems.[42] The present point is just that Nortmann's own version seems implausible, and if he wishes to modify it in Hintikka's (or some other) direction, he needs to provide textual support and also show how it will justify his particular reconstruction of Aristotle's argument.

It is obvious by now that I do not believe that any temporal definition of Aristotelian modalities can be correct. The basic notion of necessity involved in the understanding of Aristotelian necessity statements of the *Prior Analytics* has to do with the relations of entailment and (in)compatibility among the natures and attributes introduced by subject and predicate terms. So, for example, cloaks can be cut up, because there is nothing about being a cloak that precludes being cut up. This holds for any particular cloak whether or not it is ever cut up, and even if it happens that no other cloak is ever cut up. Facts of this sort will explain the omnitemporality (or non-omnitemporality) of various statements, and not vice versa.

But Nortmann does not always stick to his temporal reading, speaking sometimes instead of possibility as truth-in-some-possible-world, and maintaining that Aristotle's temporal language is just a crutch, or a makeshift manner of speaking about possible situations or possible worlds. On this familiar approach, a particular cloak's possibly being cut up would entail not that it is cut up at some time in the actual world, but only that it is cut up at some time or other in some possible world accessible from ours. This escapes some of the manifest implausibility of his temporal reading. But it seems no longer to be supported by the text, whose call for an omnitemporal assertoric premise was the basis for Nortmann's temporal interpretation of modality and for the quite precise tense-logical understanding of that counterexample to Barbara *A pp/p*. But even putting this aside, the truth-in-a-world approach does not seem appropriate either. Unless one is going to say that Aristotle was a realist about possible worlds, I do not see how such formulations can be regarded as any thing more that a convenient way of describing the contents and consequences of modal statements, where these statements and consequences would hold only *because* certain relations obtained at the level of genus, species, differentia, proprium, and accident. It is true that Aristotle often describes possible situations, sometimes counterfactual ones, in constructing premises for counterexamples. But this does nothing to show that he thought that necessity statements linking genera with their species, species with their propria and so forth, were at bottom true because of facts about individuals in situations that were somehow real, but not part of our world.

And let us not forget that just as with the more moderate temporal reading, this one will upset not only present-tense Barbara *A, pp/p*, which is invalid in any event, but also present-tense Barbara *pp, A/pp*, which is, on a very direct and natural reading, obviously valid – as Aristotle rightly says. Indeed, defining possibility as truth-at-some-time (or true-in-some-world) and necesssity as truth-at-all-times will, as Nortmann points out, threaten all the present-tense syllogisms of A. 14–22, including all those that are in fact obviously valid on a natural, present-tense reading. This seems to me to argue once again either for a severe limitation on the scope of the omnitemporality requirement or for excision.[43]

## 6.9. A FEW REMAINING ASSERTORIC/PROBLEMATIC CURIOSITIES FROM THE FIRST FIGURE

One might suppose that after all this discussion of Barbara *A, pp/p*, the case of Celarent *A, pp/p* would be relatively straightforward. But the difficulties with this second text are such that Becker deletes the entire section of 25 lines (34b19–35a2),[44] whereas Wieland leaves all but four lines standing (b28–31),[45] and Ross prints the whole thing. I side with Ross, for the passage is, on Aristotelian grounds, perfectly sound in its own right and fits naturally into the larger plan of the chapter.

The main task of the passage is to prove, by *reductio ad impossibile*, the validity of Celarent *A, pp/p*:

(1) *A* no *B*
(2) *B pp* all C

(3) *A p*-not all *C* (or *A p e C*; *A* is one-way possibly inapplicable to every *C*)

The *reductio* syllogism will then go as follows:

(4) *A N* some *C*
(5) *B* all *C*

(6) *A N* some *B*

Proposition (4) is just the contradictory of (3), the initial conclusion to be proved; (5) is presented as being at worst false and not impossible, given premise (2), '*B pp* all *C*'. Because (4) and (5) entail (6) *via* Disamis *NAN* "of the third figure" (34b24–25), and because (4) and (5) are not impossible, then (6), '*A N i B*', should also be at worst false. But given the initial premise (1), '*A* e *B*', (6) is worse than false – it is impossible. So we must reject (4), the *reductio* assumption, which is to affirm the contradictory of (4), namely, '*A p e C*'.

As Aristotle then immediately points out, the conclusion just derived by this particular argument must pertain to one-way rather than to two-way possibility, for only the former gives the contradictory of our *reductio* hypothesis (4): "For it was supposed that *A* applied of necessity to some *C*, and proof by *reductio* pertains to the contradictory assertion" (34a28–31). The contradictory of '*A N i C*' is, of course, '*A p e C*', rather than '*A pp e C*'.

This much occupies the first part of the passage, lines 34b19–b31. The balance of the treatment of this syllogism, 34b31–35a2 (which Wieland omits), consists of a proof, *via* concrete substitution instance, that the premises can entail neither a two-way possibility conclusion nor a negative universal necessity conclusion. It is essential to show this, for the fact that the premises do prove '*A p e C*' leaves open the question whether they also entail '*A pp e C*' or '*A N e C*'. Aristotle wants to show that *only* the one-way possibility conclusion follows.

Regarding two-way possibility, then, let *A* = Raven, *B* = Thinking, and *C* = Human. Then we have

(1) *A* no *B*

"for nothing that is thinking is a raven" (34b34–35), and

(2) *B pp* all *C*

"for thinking [two-way possibly] applies to all humans" (b35–36). But *A* (Raven) is necessarily inapplicable to all *C* (Human). "Therefore the conclusion [of the original mood] is not two-way possible." The point is that because the premises given are consistent with '*A N e C*', and this is inconsistent with *any* two-way possibility relation between *A* and *C*, the premises cannot entail any such contingent relation between *A* and *C*.

Having established that the original premises do entail '*A p e C*' and that they do not entail any *pp* conclusion, Aristotle must now prove that they do not also entail '*A N e C*'. This addition is apparently intended to bring us to the true conclusion of the proof for Barbara *A*, *pp/p*: By ruling out the last alternative with which '*A p e C*' is consistent, namely, '*A N e C*', it is supposed to pinpoint the one conclusion these premises do entail. This is accomplished by producing a substitution instance demonstrating the consistency of the premises with '*A p a C*'. Let *A* = Moving, *B* = Science (*epistēmē*), and *C* = Human: "Then we have *A* no *B*, and *B pp a C*, but the conclusion will not be necessary. For it is not necessary that no human move [i.e., it is not the case that '*A N e C*'], but rather it is not necessary that any human [not move]. It is clear then that the conclusion is to the effect that [*A*] does not apply of necessity to any [*C*]" (34b39–35a2), or '*A p e C*'.

Because this second section (34b31–35a2) of the passage provides an essential step in the validation of Celarent *A*, *pp/p*, Wieland should not bracket it. Still, there are three problems with Aristotle's proof, two very minor, and one major. First, he does not quite rule out *all* alternative conclusions to '*A p* no *C*': The assertoric '*A* no *C*' is left unmentioned. Perhaps, having shown that the premises were consistent with '*A pp e C*', he simply took for granted (because '*A pp e C*' is equivalent to '*A pp a C*') the obvious implication that they were consistent with the *affirmative* assertoric proposition '*A* all *C*', for this *would* rule out the premises' entailing '*A* no *C*'. Or, equally plausibly, he may have relied on the principle enunciated in chapter 12 that a (positive) assertoric conclusion requires at least one (positive) assertoric (or stronger) premise.

Second, he remarks in the last line of this section that "terms must be chosen better" (35a2). Ross says this has to do with the ambiguity of *huparchein* (referring to ch. 34) and suggests, following Alexander (196.8–11), that the last set of terms be *A* = Walking, *B* = At Rest, and *C* = Animal. (Aristotle's were *A* = Moving, *B* = Science, and *C* = Man.[46]) Actually, Aristotle's own remarks in chapter 15 probably have to do not so much with any ambiguity in *huparchein* but with the need to tend carefully to whether or not one uses nominal or adjectival forms in setting out one's terms. 'Science' (*epistēmē*) is the awkward term: In 'Moving no Science', the nominal form seems quite appropriate, but in 'Science *pp* all Human', the nominal form, strictly speaking, makes the statement false. (No human is possibly a science.) If we change 'Science' to 'Knower', the example works (where Knowers are moving).

The third and, in this case, major problem is that the general proof technique is, as we saw in Chapter 3, Section 3.6, invalid. This is a shame, given that the Philosopher has so lucidly brought us through a complex and ingenious application. To see that the mood is invalid, however, let *A* = Bird, *B* = Walking, and *C* = Raven. It is possible that both premises be true ('nothing walking is a bird', 'all ravens are two-way possibly walking') and the conclusion false ('bird one-way possibly fails to apply to all ravens'). In other words, had he not convinced himself by his "at worst false" proof technique that the premises entailed at least a one-way possibility conclusion, he might well have seen that there are counterexamples to *that* sort of conclusion as well as to the ones he correctly eliminated.[47]

Aristotle next attempts to validate the first-figure moods $A$, $E_{pp}/A_p$ and $E$, $E_{pp}/E_p$ by qualitative conversion of the $E_{pp}$ premises to $A_{pp}$, thus reducing these arguments to the varieties of Barbara and Celarent incorrectly declared valid immediately before (35a3–20). The moods $A_{pp}$, $E/$__ and $E_{pp}$, $E/$__ are then shown to prove nothing, by the method of "contrasted instances." To show that no conclusion of any sort follows, he must provide substitution instances showing that the premises are consistent, on the one hand, with '*A N a C*' and, on the other, with the negative 'A N e C'. This he achieves by letting *A* = White, *B* = Animal, and *C* = Snow (here, as elsewhere, assuming that all snow is necessarily white and – perhaps counterfactually – that all animals are two-way possibly white and hence also two-way possibly not white). Given all that, the premises of both moods may be true, along with '*A N all C*'. He then lets

$A$ = White, $B$ = Animal, and $C$ = Pitch, so that the premises of both moods will be true, along with '$A$ $N$-not all $C$' (35a20–24).

There is a slight oddity here that has drawn some attention. Ravens are usually taken to be necessarily black, as swans are necessarily white in several examples. But in the present passage all animals are assumed, for purposes of the counterexample at hand, to be two-way possibly white. Evidently Aristotle is imagining a situation in which there are no ravens, or in which ravens are accidentally black. (On the possible significance of this, see Chapter 2, Section 2.8.)

Syllogisms with one universal and one particular premise now fall quickly into line. Darii and Ferio *pp*, *A/pp* are both valid and also complete (*teleios*, 35a34), for the same sort of reason as that which applied to Barbara and Celarent *pp*, *A/pp* (35a30–35). When the major premise is assertoric and the minor problematic, we have four moods, two "proved" by *reductio* (35a40), namely, those with positive *pp* premises ($A, I_{pp}/I_p$ and $E, I_{pp}/O_p$), and two "proved" *via* qualitative conversion to the first pair ($A, O_{pp}/I_p$ and $E, O_{pp}/I_p$). Ross rightly observes[48] that in this procedure, $A, I_{pp}/I_p$ and $E, I_{pp}/O_p$ correspond to the previous moods $A, A_{pp}/A_p$ and $E, A_{pp}/E_p$, all four proved by *reductio*. Meanwhile, the two particular moods that are converted to the two *reductio* moods correspond to the two universal moods converted before to their two *reductio* moods. But neither Ross nor Aristotle notices that all eight moods are invalid; that is, neither realizes (a) that the earlier syllogisms to which the current ones are reduced were themselves mistakenly declared valid and (b) that there are counterexamples of a routine Aristotelian kind to the syllogisms here in question.

## 6.10. ONE PROBLEMATIC, ONE NECESSITY PREMISE: FIRST FIGURE

Chapter 16 is not quite so colorful as 14 or 15, but the mix of necessity and problematic premises still produces some logically interesting results. Also, there are several errors calling for analysis – although it turns out that these are, for the most part, "carryovers" from earlier contexts. Aristotle lists the high points of the chapter in a brief introduction: (i) there will be four perfect moods in this figure, namely, those with a necessary minor premise (just as in ch. 15 the perfect moods were those with assertoric minors); (ii) if both premises are affirmative, one can derive a two-way possibility conclusion; but (iii) if the premises differ in quality, one gets such a conclusion only when the affirmative premise is the necessitated one; (iv) otherwise one may derive both an assertoric and a two-way possibility conclusion; (v) by no means can one derive in this figure a conclusion of the form '*A N o C*' (35b23–36). This rather dry summation does not really do justice to the interest of the chapter. Although Aristotle does not here (or later) dwell on the matter, it is intriguing that any combination of necessity and problematic premises should entail a plain conclusion, and worthwhile to consider in turn what this would mean on an Aristotelian (metaphysical) interpretation of the moods involved. Also, item (i) raises anew the fundamental question (pressed most vigorously by Wieland) about the relation between necessity and assertoric propositions.

Aristotle's treatment of individual moods proceeds relentlessly in the usual manner: (A) two universal premises: (1) both affirmative, (2) major negative, minor affirmative, (3) minor negative, major affirmative; (B) one universal and one particular premise; and so forth. The two perfect universal moods, then, are Barbara and Celarent *pp*, *N*/*p*:

> Let *A* ⟨two-way⟩ possibly belong to *B*, and *B* belong of necessity to *C*. There will be a syllogism to the effect that *A* ⟨two-way⟩ possibly belongs to all *C*, but not that it belongs, and a complete ⟨syllogism⟩ rather than an incomplete one. For it is completed immediately (*euthus gar epiteleitai*) through the initial premises. (36a3–7)

Similarly for the completeness of Celarent *pp*, *N*/*pp*. Aristotle's treatment of this mood is noteworthy for his claim – or rather for his supporting arguments – that the premises do not yield an assertoric conclusion. For this he gives two reasons. First, he says, the major premise is a two-way possibility proposition (36a21–22). Perhaps some readers will wish to eke out a *peiorem* rule from this remark, to the effect that *each* premise must be (at least) assertoric in order to entail an assertoric conclusion. But I think the simpler and more plausible explanation is that given the minor premise, which brings the minor term "under" the middle in familiar fashion, the *sort* of relation that the major premise asserts between the major term and all items falling under the middle term just is the sort of relation that will obtain between *A* and the *C*'s, and in the present case this is the two-way possibility relation. So Aristotle immediately explains, "for the premise was taken in this way [i.e., as pertaining to two-way possibility], the ⟨premise⟩ pertaining to the major term" (36a22–23). We shall return to this revealing remark in Chapter 7 when considering the perfection of modal syllogisms.

Aristotle also maintains that the conclusion will not be assertoric for a second reason, namely, that one cannot prove such a conclusion in this context by use of *reductio ad impossibile.* And this is itself evidence for the general view that Aristotle believed that any valid mood could be validated by *reductio* proof:

> One cannot prove ⟨an assertoric conclusion⟩ by *reductio.* For if one supposes that *A* applies to some *C* [the contradictory of the imagined assertoric conclusion, '*A* no *C*'] and that *A* ⟨two-way⟩ possibly fails to apply to all *B*, nothing impossible results from these. (36a22–25)

Although Aristotle does not pause to prove here the invalidity of this proposed reducing (as opposed to reduced) mood, he will, as Ross points out, establish at 37b19–22 that this second-figure combination of premises proves nothing, hence nothing contradictory to the initial premises of the mood currently in question.

The imperfect mood Barbara *N, pp/p* (35b38–36a2; notice that the *major* premise is now necessary) is proved by the "same (*reductio*) proof" as for *A, pp/p* at 34a34–b2. The additional first-figure mood "Barbera" *N, pp/p* is then proved by qualitative conversion of the minor premise (36a25–27) to obtain Barbara *N, pp/p* once again. The flaw here is that Aristotle's earlier proof for Barbara *A, pp/p* contains an error – one that we discussed at some length earlier. The present mood is invalid for the same reasons as the earlier one, as the same counterexample will show: Let *A* = Human, *B* = Mover, and *C* = Horse, where all moving things are humans. Then '*A N* all *B*' and '*B pp* all *C*' are both true, but '*A p* all *C*' is false. The same terms show the first-figure mood $A_n$, $E_{pp}/A_p$, the one "validated" by qualitative conversion to Barbara *N, pp/p*, invalid on any weak *cop* or *de re* reading. And with this we are back to the controversy about omnitemporal premises – only here we have to do with a weak *cop* necessity major premise that, as it happens, may be true only at a certain time. But here there is no mention of omnitemporality.

Similar remarks apply to the proof of Celarent *N, pp/A*, although the mood remains of interest for other reasons. The proof Aristotle has in mind involves the reducing syllogism:

(1) *A i C* (the contradictory of the desired conclusion, '*A e C*')
(2) *A N e B* (the initially given major premise)
―――――
(3) *B N o C* (which is incompatible with the original minor premise, '*B pp a C*')

Validation of this argument requires that (2) be converted to '*B N* no *A*'. Then we obtain, *via* Ferio *NAN*, the desired (3), which is inconsistent with, but not quite the contradictory of, the original minor premise, '*B pp* all *C*'. There is nothing wrong with Ferio *NAN*. Rather, the problem is in converting $E_n$ to get to the requisite version of Ferio. Again, on a weak *cop* reading of $E_n$, and the corresponding term-thing (type II) reading of two-way possibility, both that conversion and the mood Aristotle here wishes to validate are invalid. Let $A$ = White, $B$ = Walking, and $C$ = Swan, where all things walking are ravens, and ravens are, as in Aristotle's examples in *Pr. An.* A, necessarily black. Then '*A N e B*' and '*B pp a C*' are both true, but '*A e C*' false, because all swans are (necessarily) white.

Unfortunately (on the indicated reading of these syllogisms) this deprives us of the intuitively surprising result that a necessity premise combined with a problematic one can yield an assertoric conclusion. The inference was surprising because the problematic premise asserts neither the actual application of *B* to *C* nor the actual failure of *B* to apply. So it seems odd that one could conclude to an assertoric relation between *A* and *C*, no matter what relation the major premise may assert between *A* and *B*. Thus things are, after all, disappointingly predictable: We can establish no such relation between *A* and *C*.[49]

## 6.11. TWO CONTINGENT PREMISES IN THE SECOND FIGURE: DISCOVERY, BEFORE OUR VERY EYES, OF AN INGENIOUS "PROOF"

Chapter 17 is a bit peculiar. It begins routinely enough with the usual advance summary of results. Perhaps because the harvest of valid moods is so meager – there are *no* valid second-figure moods with two problematic premises – Aristotle gives us a quick peek at the more gratifying results (in this figure) for moods with one problematic and either one assertoric or one necessity premise (36b26–34), combinations of which occupy chapters 18 and 19, respectively. The chief peculiarity of chapter 17 lies not in the total absence of valid moods, nor in the inclusion of a long discussion of $E_{pp}$ term conversion preliminary to a decision on Cesare *pp*, *pp/p*, for that section is itself admirably worked out and fits perfectly into its surrounding context (see Section 2.3 herein), but rather in the fact that about two-thirds of the way through a quite involved discussion, Aristotle seems suddenly to hit upon a much simpler general strategy for showing all the moods under investigation to be invalid.

At the conclusion of the opening paragraph, he maintains that any possibility conclusions of valid second-figure moods involving one or more problematic premises will have to involve one-way rather than two-way possibility (36b33–34). Aristotle gives no reason for this claim, but Ross suggests that what he has in mind is that all moods in this figure must be proved by *reductio ad impossibile, ekthesis* being as obviously futile in this case as term conversion. This would imply that any possibility conclusion would have to involve *one*-way possibility, because that is the sort of proposition whose contradictory is the sort of necessity proposition that would be used in validating these moods by *reductio.* Logically speaking, this is not a very good reason for Aristotle to take that position, for he has not shown that a *reductio* using a disjunctive *reductio* assumption ('$A\ N\ i\ B \vee A\ N\ o\ B$', the contradictory of '*A pp a/e B*') cannot be valid, but has only, on this interpretation, conceded that such a proof could not be carried out in his system. Indeed, because, as Aristotle well knew (and shows himself aware in this very chapter; see 37a9–31, and see Section 5.4 herein), two-way possibility propositions do have uniquely determined, even if, in effect, disjunctive, contradictories, the present situation might have been seen as calling for some modification of the system to handle such propositions, either as disjunctions or in some categorical manner. So let us keep in mind the question whether or not Aristotle can provide any better reason for ruling out two-way possibility conclusions.

Turning then to individual syllogisms, he follows up his remark that any possibility conclusion would in this context pertain to one-way possibility by showing first, for Cesare *pp, pp/p*, that term conversion of the major premise does not work (for reasons set forth at length in lines 36b35–37a31), so that there will be no reduction by that means to the first figure. He then shows that assuming the contradictory of the one-way possibility conclusion does not lead to anything "false" (better, does not entail anything that *cannot* be true given the initial premises), thus showing that proof by *reductio*, starting from a presumed one-way possibility conclusion, cannot validate the mood (37a35–37).

But just there, instead of proceeding to the question of the possibility of a two-way conclusion for Cesare *pp, pp/*____ (or to the next syllogism), he apparently switches to a different and more general approach:

> In general (*holōs*) if there is a syllogism, it is clear that it will be taken in possibility because neither of the premises was taken in belonging (*mēdeteran . . . eilēphthai en tōi huparchein*), and this whether it is affirmative or negative. But neither way can there be a syllogism. (37a38–b1)

This general principle becomes somewhat clearer in the immediately following application:

> It being assumed to be affirmative, it can be shown by use of terms that it does not possibly apply, and if negative, that the conclusion is not possible but necessary. For let *A* [middle] = White, *B* = Human, *C* = Horse. *A* may possibly belong to all of one and possibly not belong to all of the other. But *B* neither possibly applies nor fails to apply to *C*. That it does not possibly apply is obvious, for no horse is ⟨even possibly⟩ a human. Nor does it possibly fail to apply. For it is necessary that no horse be a human, and the necessary was not possible. Therefore there will be no syllogism. (37b1–10)

Because Aristotle uses the term *endechesthai* throughout this chapter for "possibility," one might easily suppose, in light of his earlier remark that second-figure syllogisms can yield only one-way possibility conclusions (36b33–34), that in the text just quoted, when he says the moods of chapter 17 will yield, if anything, a possibility conclusion (*ei esti sullogismos, dēlon hoti tou endechesthai*, 37a38), he still means one-way possibility. But this reading cannot be maintained. For the crucial incompatibility of this "possibility" with necessity cited at 37b3 and b9–10 makes sense, as Aristotle well knows, only if *endechesthai* means two-way possibility.

Supposing, then, that the passage starting at 37a38 is concerned with two-way possibility, his strategy is to show the following: (a) *if* these moods are valid, they entail a two-way possibility conclusion; (b) they can entail no such conclusion; (c) therefore they are all invalid. The key question is why (a) should be true – as opposed, say, to his earlier claim that any valid conclusion would pertain to *one*-way possibility.

Aristotle's very briefly stated reason is that "neither premise was taken in belonging": *dia to mēdeteran tōn protaseōn eilēphthai en tōi huparchein* (37a39). Here one might suppose him to be saying that neither premise is assertoric (as opposed to necessary or one-way or two-way possible). But this will not hold up either, for Aristotle himself will maintain in a moment that the second-figure moods Cesare *N*, *pp/A* and Festino *N*, *pp/A*, neither of which contains an assertoric premise, will yield an assertoric conclusion (see 38a16–25 and 38b25–27, respectively; the same holds for the first-figure moods $E_n$, $A_{pp}/E$, 36a7–17, and $E_n$, $I_{pp}/O$, 36a34–39, and the third-figure moods $E_n$, $A_{pp}/O$, 40a25–32, $O_n$, $A_{pp}/O$, 40b3–8, and $E_n$, $I_{pp}/O$, 40b3–8). So he is probably not here invoking a rule to the effect that one needs at least one strictly assertoric premise to get something other than a (two-way) possibility conclusion.

I take him to be claiming, correctly, that neither premise asserts the actual application (i.e. the application, whether necessary or not) of the middle term to either the major or the minor. From this he correctly infers that there is no way one could derive the actual application (necessary or otherwise) of *A* to *C*. So there can be no necessary or assertoric conclusion. That leaves one- and two-way possibility propositions as possible conclusions. Why, then, does he say that any validly inferred conclusion would have to involve two-way possibility? Because, I would suggest, a one-way possibility proposition holds in three different cases: (i) where the appropriate necessity proposition holds, (ii) where the appropriate assertoric one holds, or (iii) where the appropriate two-way proposition holds. But if, as has just been established, one cannot infer either a necessity or an assertoric conclusion, Aristotle may have reasoned that one can infer a one-way possibility conclusion only if one can also infer a two-way possibility conclusion. Therefore it would suffice, to finish off the invalidation of the present moods, to show that no two-way possibility conclusion can be inferred. And this is what his counterexample at 37b1–10 is designed to show.

To return to the question left open a moment ago, Aristotle does not "prove" this by appeal to the claim that every valid syllogism can be proved by *reductio* and the fact that his system cannot handle the kind of *reductio* that would be needed in this case. Rather, he simply shows that the premises are consistent with '*B N e C*' (recall his terms Horse and Human). And given this, it is, as he says, obvious (*phaneron*, 37b7) that no positive (two-way) possibility conclusion can be inferred. But it is also obvious that no "negative" problematic one can be inferred either, because that, too, would be inconsistent with the *necessary* non-application of *B* to all *C*'s. So *no* sort of conclusion can be obtained in these cases.

As for the midstream swap of strategies, I would like to believe that Aristotle was unhappy about the dead end to which his initial approach was leading, looked for something better, and discovered a completely general and rigorous proof. There is no reason to suppose the first approach inauthentic, but only that neither Aristotle nor pious tradition ever bothered to "erase" it. Actually, it would not in any case have been erased by later commentators unless it had been seen that the following lines rendered it superfluous; but so far as I know, this has not been noticed.

The balance of the chapter is entirely routine, showing that the same triple of terms as before will rule out a two-way possibility conclusion (hence any conclusion) for *any* combination of two problematic premises, of *whatever* quality and quantity.

## 6.12. THE SPREAD OF A PROOF-THEORETIC INFECTION

The one feature of Aristotle's brief chapter 18 that is not routine is in fact quite striking. The first half of the chapter recognizes as valid a number of syllogisms that are invalid and that can be shown invalid by the same sort of counterexample Aristotle himself gives, in the second half of the chapter, for several other moods. Specifically, Aristotle concludes that one assertoric and one problematic premise can (in the second figure) sometimes yield a conclusion. He allows that with an $A_{pp}$ premise plus an assertoric negative universal, one gets a valid syllogism regardless of which is the major and which the minor. Thus $E, A_{pp}/$— and $A_{pp}, E/$— both give $E_p$ (see 37b24–28 and 37b29, respectively). $E, I_{pp}/O_p$ is also declared valid, but the premise pair $I_{pp}, E/$— is not mentioned, probably because Aristotle believed it would yield a conclusion of the form '*C P o B*', but not '*B P o C*'. (As remarked earlier, in such situations he often says "there is no syllogism.") A large number of other premise combinations are proved invalid by use of triples of terms showing the premises consistent both with '*B N a C*' and with '*B N e C*'. For example, $A, E_{pp}/$—:

*A all B*
*A pp e C*

Let *A* = Healthy, *B* = Animal, and *C* = Horse. This interpretation shows the original premises consistent with '*B N* all *C*'. Letting *A* = Healthy, *B* = Horse, and *C* = Human, one obtains premises consistent with '*B N e C*'.

There is nothing wrong with Aristotle's invalidation "by terms" of various arguments in this figure. The problem is that a similar selection of terms would also show those figures invalid that he declares and "proves" valid:

| | | |
|---|---|---|
| *A* no *B* | *A pp* no/all *B* | *A pp* no/all *B* |
| *A pp* no/all *C* | *A* no *C* | *A* not some *C* |
| *B p* no *C* | *B p* no *C* | *B p* no *C* |

For the first of these three, let *A* = Healthy, *B* = Animal, and *C* = Horse. It is quite possible that no animals be healthy and that all horses be two-way possibly healthy (or not healthy). But in this situation, '*B N* all *C*' also holds, because Animal is necessarily applicable to all horses. For the case in which '*B N e C*', let *A* = Healthy, *B* = Horse, and *C* = Human.

With obvious minor adjustments, the same terms will work for all three syllogisms – and these are the same terms Aristotle used to invalidate the other combinations in this figure (see 37b36–38, 38a12).

So why did Aristotle think these three syllogisms valid? Because he thought he had proved them valid. How had he proved that? By converting the plain universal negative (which is, of course, unobjectionable) to reduce these moods, respectively, to the first-figure moods $E, A_{pp}/E_p$, $E, A_{pp}/E_p$, and $E, I_{pp}/O_p$, respectively. And *these* were proved by *reductio ad impossibile* arguments using (a) the third-figure moods $I_n$, $A/I$ and $A_n$, $I/I_n$ and (b) the "at-worst-false-and-not-impossible" proof technique discussed earlier. Those third-figure moods are in fact valid (see Chapter 4), and Aristotle gives perfectly sound proofs for them. But his "at-worst-false . . ." proof technique is faulty, for reasons we have discussed. Once again, that proof-theoretic virus from *Pr. An.* A has infected an apparently healthy demonstration.

Let us recall that Aristotle (or some "later hand") had saved the first-figure moods Barbara *A, pp/p* and presumably Celarent *A, pp/p* from proposed counterexamples by insisting on an omnitemporal reading of the premises. The latter mood is the one to which he reduces two of the three syllogisms we are now considering. Meanwhile, the moods he here invalidates by counterexample are precisely those that do not reduce by conversion to a previously accepted first-figure mood. And yet the counterexamples he uses here could be used just as well not only against the moods here "proved" valid by reduction but also against the first-figure moods to which these are reduced:

Celarent *A, pp/p*

Human *e* Healthy<br>
Healthy *pp a* Horse<br>
———————————<br>
Human *p a* Horse

These terms work also against Barbara *A, pp/p* and follow exactly the same pattern, in terms of genus-species-accident relations, as the example proposed there. The only difference is that here we have Healthy instead of Moving.

This raises the question why he did not save his *second*-figure moods with such premise pairs as '*A a B*', '*A pp e C*' from counterexample by requiring omnitemporal premises. Why does he here let stand counterexamples using propositions that at best hold only at certain times? Having introduced the idea of omnitemporal premises, and having used them to save Barbara *A, pp/p* from counterexample, he ought to have at least tried a similar rescue of the moods here declared invalid, instead of simply letting these counterexamples stand. But he did not attempt this. On the contrary, he wrote as if the condition of omnitemporality had never occurred to him. This is further evidence that the passage on omnitemporality is not original – that it was added either by some later editor or by Aristotle himself. As for the question of validity with omnitemporal premises, the same sort of counterexample that worked in chapter 15 (see Section 6.7) will work here.

As a result, apparently there are *no* valid syllogisms in the second figure with one assertoric and one problematic premise. But one may hope not only to show this piecemeal but also to understand *why* this must be so on general grounds. This can be accomplished rather easily in terms of the underlying semantics of species, accident, and so forth. Now if we know only that the middle term, which in this figure will be the predicate term in both premises, *actually* applies to all/some of one of the "extremes" (major or minor terms) and that it *two-way possibly* applies to the other extreme, then, for all we know, the relation might in *both* cases be that of two-way possibility (i.e., that the *actually* applying predicate of the assertoric premise applies as an accident to its subjects). But if so, it is entirely possible that the middle term be only accidentally related to the items referred to by *both* the major and minor terms, *regardless of how the latter two terms are related* (necessarily, one- or two-way possibly) *to one another.* Thus, exactly as with pure two-way possibility syllogisms in this figure, the premises cannot guarantee any more than that the designata of the extreme terms are included within the range of things to which the middle two-way possibly applies – which is not enough to establish any specific relation between the two extremes themselves. [For example, *B* might be a genus to which *A* (the middle term) is related accidentally, and *C* a species of *B*. This was, in effect, what one had with Aristotle's counterexamples, letting *A* = Healthy, *B* = Animal, and *C* = Horse. Or *B* and *C* might be mutually exclusive subspecies of a common genus *D* to which *A* is related as accident: Let *A* = Healthy, *B* = Horse, and *C* = Human.] Such examples work equally well against all the moods Aristotle considers in this chapter.

## 6.13. AN IMPORTANT PRINCIPLE OVERLOOKED

Chapter 19 (second figure, one necessity, one problematic premise) extends our run of comparatively peculiar chapters. The principles employed in the (in)validation of various moods are all familiar, so there is nothing uniquely peculiar here. And yet it is surprising that Aristotle overlooks the *validity* of certain moods. The explanation is that he fails to recognize, for reasons that are quite understandable, an important logical principle. But first let us look briefly at two of Aristotle's specific proofs:

(1) *A N e B*
(2) *A pp a C*
_______________
(3) *B e C*

Suppose that '*B i C*'. This, combined with '*A N e B*', gives '*A N o C*', *via* the complete first-figure Ferio *N, A/N* (necessity here being, of course, weak *cop* necessity). But this conclusion is incompatible with (2), '*A pp a C*'. Hence the premises entail '*B e C*' (see 38a21–25). By contrast, consider the proof for

(1) *A N e B*
(2) *A pp i C*
_______________
(3) *B o C*

Here Aristotle converts the first premise to '*B N e A*', which, in combination with '*A pp i C*', gives, *via* Ferio *N, pp/A*, '*B o C*'. The alarm sounds, however, at the conversion of $E_n$, for it does not convert on a weak *cop* reading, but only on its strong *cop* version. Aristotle here takes no fresh thought about that conversion, for he is simply applying a principle he had long ago declared valid and used freely ever since. Fortunately, it is not necessary to establish the mood by use of that conversion, because a *reductio* proof of the type he had just given at 38a21–25 will suffice: Suppose '*B a C*'; then if '*A N e B*', it follows, by Celarent *NAN*, that '*A N e C*', which is incompatible with '*A pp i C*'.

Notice that all the combinations here declared valid involve a *negative* necessity premise and an *affirmative* two-way possibility premise (the latter being affirmative because, as Aristotle had earlier argued, *all* propositions of that modality are, at bottom, affirmative). This gives the kind of opposition between inclusion and exclusion of the extreme terms with respect to the middle term that Aristotle has looked for, in the second figure, from the plain syllogisms of chapter 5 on. For example, with plain Camestres, the middle term *A* would include all the *B*'s but none of the *C*'s. Hence none of the *C*'s can be *B*'s, or '*B e C*'. With two affirmative assertoric premises, one cannot prove any particular relation between *B* and *C*: All one then knows is that the middle includes all/some of the major and some/all of the minor, and this does not establish any particular relation between designata of the major and minor terms themselves.

Accordingly, it is not so odd that Aristotle should declare all combinations of an affirmative necessity premise with a two-way possibility premise (i.e., two *affirmative* premises in the second figure) invalid. It may then come as a surprise that, on the contrary, *all* those mixed *N/pp* premise pairs with two universal premises or with one universal and one particular premise – regardless of the quality of the necessity premise – do in fact yield an assertoric conclusion. (Some of the latter pairs yield only a '*C o B*', rather than a '*B o C*', conclusion, however.) In fact, exactly the same sort of *reductio* proofs Aristotle gives for some of the moods that he rightly recognizes as valid will work just as well for some of the cases he declares invalid. For example, $A_{pp}$, $A_n/E$:

(1) *A pp a B*
(2) *A N a C*
(3) *B e C*

Suppose '*B i C*'. This converts to '*C i B*', which, combined with '*A N a C*', yields '*A N i B*' *via* Darii *N*, *A/N* (rightly recognized as a complete mood at 30a37–b1). But '*A N i B*' is incompatible with (1), '*A pp a B*'. So the syllogism is valid. Similar *reductio* proofs will work for $A_n$, $A_{pp}/E$, and hence for the qualitative twins of these two moods, $E_{pp}$, $A_n/E$ and $A_n$, $E_{pp}/E$.

Moreover, the same holds for *all* combinations of a universal major necessity premise with a particular two-way possibility minor [$A_n$, $I_{pp}$ (or $O_{pp}$)/*O*; $E_n$, $I_{pp}$ (or $O_{pp}$)/*O*]. On the other hand, where the major premise is universal and problematic, and the minor particular and necessary, one must use RAA plus *ekthesis*:

(1) *A pp a B*
(2) *A N i C*
(3) *B o C*

Suppose '*B a C*'. Given this, plus '*A N i C*', it follows (*via ekthesis*) that '*A N i B*', which is incompatible with (1), '*A pp a B*'. Hence the mood (again, read weak *cop* throughout) is valid.

In light of all this, we shall, first, look again at Aristotle's stated reason for declaring invalid so many second-figure moods that he could have validated by exactly the same sort of *reductio* proof he used to validate other moods in this same chapter and, second, attempt to clarify a different method of proof (from RAA) – one growing out of an analysis of Aristotle's error in thinking these moods invalid – that will work for all the valid moods of this chapter.

Aristotle is right that any conclusion in this figure must be negative. And he is right that no negative two-way possibility conclusion can follow from any combination of an *affirmative* necessity with a two-way possibility premise, because such a combination will be consistent with a necessity proposition relating major and minor terms. [For the case of Camestres *N, pp/pp*, he gives the terms *A* (middle) = White, *B* = Swan, and *C* = Human; 38b18–20.[50]] But Aristotle reasons also that there can be no negative assertoric or necessity conclusion because both premises are affirmative (38b14–16). Consequently, he says, no conclusion at all can be drawn. He might simply have in mind that no genuinely negative conclusion can follow from two affirmative premises. Or he might (also?) have in mind reasoning parallel to that for the case of *plain* second-figure syllogisms: Two affirmative premises can at most place the designata of major and minor terms within the same domain, which is not enough to establish any of the four possible assertoric relations between either term and the designata of the other.

In any case Aristotle overlooks an important principle: In the second figure the key factor in deriving one's negative conclusion is not whether one premise is affirmative, and the other negative, but rather whether the relations between the middle term *A* and the *B*'s, on the one hand, and *A* and the *C*'s, on the other, are such that *A* excludes (whether necessarily or as a plain matter of fact) all or some *B*'s while including some or all *C*'s, or vice versa. When that obtains, some sort of negative assertoric *B–C* or *C–B* conclusion will follow. When will that combination of inclusion and exclusion obtain? Well, if we let $R_1$ be the relation between *A* and all or some of the *B*'s, and let $R_2$ be the relation between *A* and some or all of the *C*'s, we shall have the kind of situation we want – one in which an assertoric conclusion follows – if and only if *A*'s being related by $R_1$ to some subject is incompatible with *A*'s being simultaneously related by $R_2$ to that same subject. (For convenience, let us say, "just in case $R_1$ is incompatible with $R_2$.") Within the assertoric system, $R_1$ will be "applies to," and $R_2$ will be "does not apply to." Obviously, applying to a thing is incompatible with not applying to it, and it is because of this incompatibility that we know that if *A* applies to all the *B*'s and does not apply to any *C*, then the *B*'s and *C*'s must not overlap at all – that '*B* no *C*' and '*C* no *B*' follow. If they did, then there would be something to which *A* both applied and did not apply. Similarly, if one of the premises is universal and one particular, as with Baroco and Festino, there will be at least some failure of overlap such that '*B o C*' holds. This is the whole ball of wax as far as the assertoric second figure is concerned: It explains, in light of the arrangement of the terms, precisely why we get only negative conclusions, and how we get them.

Returning to chapter 19, we have another pair of relations, "belonging necessarily to," and "two-way possibly (not) applying to." Now it may be that both are, at bottom, as Aristotle claims, positive in form. But these two relations are, nonetheless (as Aristotle elsewhere recognizes), mutually exclusive. Therefore, if '*A pp* all *B*' and '*A N* all *C*', none of the *C*'s can be *B*'s, or vice versa; otherwise there would be something to which *A* both necessarily and two-way possibly applied (or, to which *A* did and did not necessarily apply, and to which *A* did and did not two-way possibly apply).

Thus the principle proposed earlier about the incompatibility of relations $R_1$ and $R_2$ (i.e., if $R_1$ and $R_2$ are incompatible ways for a predicate to be related to a subject, then if $A\ R_1\ x$ and $A\ R_2\ y$, then $x \neq y$) is a very general principle of which the pairs of relations "applies/does not apply," and "necesarily applies/two-way possibly applies" are only two special cases. Others would be "one-way possibly applies/necessarily does not apply," "necessarily applies/(plain) does not apply," and so forth. And this is the principle that will determine, throughout the plain and modal syllogisms, where a negative assertoric conclusion is obtainable in the second figure. Aristotle may have overlooked this here in *An. Pr.* A.19 in part because of his previously formed conviction that two affirmative premises can establish nothing in the second figure.

## 6.14. THIRD-FIGURE SYLLOGISMS

Chapter 20 (third-figure moods with two two-way possibility premises) presents many small (and, by now, nagging) problems: Where has Aristotle used qualitative conversion, and where term conversion? Which uses are valid, and which not? Where a proof uses some illicit conversion, can an alternate, and valid, proof be found? What sorts of possibility conclusions do we get in the valid moods? Is ampliation necessary for the validity of any term conversion or syllogism found in this chapter? If so, will it be one- or two-way ampliation? Chapter 20 is a 40-line thicket of such questions. And while finding out the answers is not a trivial task, the field report would make for tedious reading and would include nothing differing in principle from what has gone before. Therefore, we pass on to the more interesting case of third-figure syllogisms with one problematic and one assertoric premise.[51]

Chapter 21 on third-figure mixed moods (one assertoric, one contingent premise) presents, besides the usual sorts of questions, one or two especially thorny problems. First, there is the question of which moods give a two-way possibility conclusion, and which a one-way only. Aristotle himself is vague about this. At the outset of chapter 20 he gave an overview of the third figure (ch. 20–22) and remarked that "when both premises signify ⟨two-way⟩ possibility (*endechesthai . . . sēmainōsin*), the conclusion also will be possible (*endechomenon*), and also (referring ahead to ch. 21) when one ⟨premise⟩ is possible, the other assertoric (*huparchein*)" (39a5–8). In fact, with two problematic premises, the conclusion will also be a two-way possibility proposition – but only with ampliation of the middle term. These pure two-way possibility premise pairs were the business of chapter 20. With regard now to chapter 21 (third figure, one problematic, one assertoric premise), the modality of the conclusion will in fact vary, *pace* Aristotle's summary statement of chapter 20, between assertoric (and, *a fortiori*, one-way possibility) and two-way possibility. One might shrug this off as a mere slip, except that he goes on to say (still in the "preview" section of ch. 20) that with one necessity premise, "if the necessary premise should be affirmative there will be no conclusion, either necessary or assertoric, but if it should be negative, there will be a syllogism of not applying (*tou mē huparchein*), *just as in the earlier cases*. The possibility (*to endechomenon*) in these conclusions must be taken similarly" (39a8–12; emphasis added). The import of the last sentence is almost perfectly obscure. It is quite possible that, as Ross thinks, it means to claim that *certain* conclusions will signify only one-way possibility.[52] It would thus parallel similar remarks at 33b29–31 of chapter 15: "and the negative conclusions are taken not according to the definition of possibility [i.e., that of two-way possibility], but that of not belonging of necessity to any or not belonging of necessity to all [i.e., one-way possibly not applying to all or some]." Certainly it cannot mean that *all* conclusions in this figure pertain to *two*-way possibility, because Aristotle is well aware that that is not so (e.g., in the *reductio* proof for Bocardo $O_{pp}$, $A/O_p$ in ch. 21, 39b31–39). Nor can it mean that *all* the conclusions are *one*-way possibility propositions, for he seems equally clearly convinced that several of these moods do give two-way conclusions.

Ross takes the remark to apply only to the immediately preceding lines: "when the major premise is problematic, all these syllogisms will be perfect and possibility will pertain to the stated definition. But when the minor is problematic, all will be imperfect.'' Thus, on Ross's view, Aristotle is saying merely that a combination of a negative apodeictic (necessity) with a problematic premise yields only a one-way possibility conclusion. But this cannot be right either, because the cases Aristotle singles out are those with problematic minor and necessary major, not those containing a negative necessity premise. Ross also suggests that ''similarly'' means ''as with the corresponding combinations in the second figure.''[53] This is a way of supporting his previous suggestion, but this does not hold up either, for some of these moods with a negative necessity premise actually give an *assertoric* conclusion (see $E_n, A_{pp}/O$, $O_n, A_{pp}/O$, $E_n, I_{pp}/O$ in ch. 22, 40a25–32, 40b3–8, 40b3–8, respectively). Let us say for the moment that (a) Aristotle appears to claim at 33b25–31 (the opening of ch. 15) that in the valid moods of the third figure a two-way possibility major combined with an assertoric minor premise will entail a two-way conclusion, but an assertoric major combined with a two-way minor will give only a one-way possibility conclusion, and (b) in chapter 21 itself he gives no clear indication of when one should expect a one-way rather than a two-way possibility conclusion.

The question, then, of which moods do in fact give only one-way, and which give two-way (hence also one-way), possibility conclusions arises with the very first mood considered:

| Darapti *A, pp*/____ | Darapti *pp, A*/____ |
|---|---|
| *A* all *C* | *A pp* all *C* |
| *B pp* all *C* | *B* all *C* |
| ______ | ______ |

Wieland believes that *both* versions give both one- *and* two-way possibility conclusions.[54] Ross believes that the former mood gives only one-way, and the latter both sorts of conclusions. It is obvious enough that the second version gives a two-way (hence also a one-way) possibility conclusion, for we need use only the conversion of plain '*B* all *C*' to '*C* some *B*' in order to obtain '*A pp* all *B*' *via* the perfect mood Darii *pp, A*/*pp*. In the former case, if we reverse the premise order and convert the assertoric premise, we can validly obtain '*B pp* some *A*':

*B pp* all *C*
*C* some *A*
______
*B pp* some *A*

Aristotle would then convert the conclusion to get the desired '*A pp* some *B*'. The one hitch in all of this is that that final conversion is invalid.

By converting '*C* some *A*', then using *ekthesis*, we can obtain '*A i B pp*'. But this is not equivalent to '*A pp i B*'. A *reductio* taking '$A\ N\ a\ B \vee A\ N\ e\ B$' as the negation of '*A pp i B*' cannot, strictly speaking, be carried out within Aristotle's system. But this would be of no avail anyway, since the argument is invalid. Let $A$ = Rational, $B$ = In the Agora, $C$ (middle) = Human, in a situation in which there are only inanimate things in the Agora.

We could two-way ampliate the middle term ($C$), then convert '*B pp* all *C pp*' to '*C pp* some *B pp*':

$$\begin{array}{l} A \text{ all } C\ pp \\ \underline{C\ pp \text{ some } B\ pp} \\ A \text{ some } B\ pp \end{array}$$

Just as before, we now have an assertoric, hence a *one*-way possibility conclusion. So although we cannot obtain what Aristotle (or Ross or Wieland) imagined, we can, by use of ampliation, obtain other, unimagined results.

By contrast, one can see that the premise pairs $A_{pp}$, $A/$, $E_{pp}$, $A/$, $A_{pp}$, $I/$, and $E_{pp}$, $I/$ do not need to use ampliation to give two-way, hence one-way, possibility *A–B* conclusions, because each will need only an assertoric conversion of the minor premise to give the appropriate first-figure mood with two-way conclusion. The pair $I_{pp}$, $A/$ and its twin $O_{pp}$, $A/$ (treated separately by Aristotle at 39b26–31 and 39b31–39, respectively) cannot be proved in this way, because conversion of the minor premise would leave two particular premises, which would prove nothing. Putting aside momentarily the question of how to prove validity, Aristotle's view is that the former mood entails an $I_p$, and the latter an $O_p$, conclusion. Something is wrong with that view, however, because $I_p$ is not equivalent to $O_p$, whereas the premises of the two moods are precisely equivalent, differing only in that one has $I_{pp}$ where the other has the logically equivalent $O_{pp}$. The fact is that both moods yield a two-way particular conclusion, which entails that both will give both $O_p$ and $I_p$. This is readily shown by *ekthesis*:

$$\begin{array}{rl} & A\ pp\ i/o\ B \\ & \underline{C\ a\ B} \\ \text{To prove:} & A\ pp\ i/o\ C \end{array}$$

The major premise tells us that $A$ two-way possibly applies to some $B$, say $b$, and the minor that $C$ applies to every $B$, including $b$. Therefore, $A$ two-way possibly applies to something (namely, $b$) to which $C$ actually applies; that is, '*A pp* some *C*'. So each valid third-figure combination of two-way major and assertoric minor premise entails a two-way conclusion, where that major premise may be universal or particular.

The situation is quite otherwise with assertoric major and problematic minor. Here Aristotle again converts the minor premise to get a first-figure mood, but in so doing he must convert two-way possibility propositions. (This we saw earlier with regard to the specimen case of Darapti.) These do not convert on the sort of reading he seems to have in mind for two-way possibility (whenever we get good evidence of what reading he does have in mind). Moreover, these moods, like Darapti: *A, pp* __, are simply *invalid* on a term-thing reading. To invalidate the eight moods (four pairs of twins, really) with affirmative assertoric major and problematic minor,

$$\frac{\begin{array}{l} A\ a/i\ B \\ C\ pp\ a/e,\ i/o\ B \end{array}}{A\ pp\ i\ C}$$

let $A$ = Animal, $B$ = Horse, and $C$ = Brown, in a possible situation in which all brown things are in fact cloaks. In that situation the premises would be true, and because Animal is necessarily inapplicable to all cloaks, the conclusion false.

For the eight moods with negative assertoric major and problematic minor,

$$\frac{\begin{array}{l} A\ e/o\ B \\ C\ pp\ a/e,\ i/o\ B \end{array}}{A\ pp\ o\ C}$$

let $A$ =Animal, $B$ = Cloak, and $C$ = Brown, in a situation in which all brown things are horses. Notice that the same counterexamples will work against these moods even if the conclusion is only a one-way possibility proposition.

Finally, notice that by two-way ampliating the middle term in both premises, one can, by converting the minor, derive assertoric conclusions with ampliated (logical) subject terms, just as we saw earlier with Darapti *A, pp/pp*.

One final note: Wieland finds Aristotle's *reductio* proof for Bocardo "alien to the system" on grounds that it depends on the implication '*A pp o B* → not: *A N a B*'.[55] (I have put the formula in my *cop* notation, but that does not affect the point at issue.) He claims that nowhere else does Aristotle use this principle, which one can integrate only into a system such as that of Theophrastus. Aristotle's proof runs as follows:

(1) *A pp o B*
(2) *C* all *B*

(3) *A p o C*

Suppose (4), '*A N* all *C*'. Then adding (2), '*C* all *B*', we would have, via Barbara *NAN*, (5), '*A N* all *B*'. But (5) is inconsistent with (1). Now it is certainly true (as Wieland agrees) that (5) and (1) are inconsistent, because '*A pp o B* → not: *A N a B*'. But (*pace* Wieland) this is not the first time Aristotle has used the principle, for it is nothing more than an application of the familiar Aristotelian position that two-way possibly not applying and necessarily applying are mutually exclusive. (Recall that $I_{pp}$ and $O_{pp}$ are incompatible equally with $A_n$ and $E_n$.) Of course, the principle does hold with one way possibility also: If '*A P o B*', then 'not: *A N a B*'; and this may be Theophrastean in the sense that Theophrastus' system apparently included only one-way (not two-way) possibility. But even that (one-way) principle would not be alien to Aristotle's system, either: In fact, it is recognized and used by Aristotle whenever he constructs a *reductio* proof for an argument with a one-way possibility conclusion, for there he rightly takes the appropriate necessity proposition as contradictory to the desired one-way conclusion.

## 6.15. A DAY IN THE SUN FOR *EKTHESIS*

The final chapter (ch. 22, on third-figure moods with one necessity, one problematic premise) harbors no surprises – aside from the example of the sleeping horses, the only compound term in all these chapters (40a37–38). Here again, free use is made of conversion principles laid down long before [e.g., of $A_{pp}$ to validate Darapti *N, pp/p* (40a11–16), of $A_n$ to validate Darapti *pp, N/A* and hence, *a fortiori*, Darapti *pp, N/p* (a16–18)]. On the whole, things are very much as they were in chapter 21, with one assertoric and one problematic premise – but with one interesting wrinkle.

With a contingent major and necessity minor, Aristotle converts the minor to reach a first-figure *pp, N/pp* mood. In chapter 21, conversion of the minor premise posed no problem, because there the minor was assertoric and its conversion valid. But in the present chapter these conversions are valid only with that premise read as strong *cop* necessity. Read that way, they do give a valid proof: Because '*B* $N_s$ all *C*' validly converts to '*C* $N_s$ some *B*', which entails '*C* some *B*', the minor will then serve to bring some *B*'s under the *C*'s in the usual way. Given also that the major asserts that *A* two-way possibly applies to every *C, A* will two-way possibly apply to some *B*:

$$\begin{array}{ll} A\ pp\ all\ C \rightarrow & A\ pp \text{ all } C \\ B\ N_s \text{ all } C \rightarrow C\ N_s \text{ some } B \rightarrow & \underline{C \text{ some } B} \\ & A\ pp \text{ some } B \end{array}$$

Using a strong *cop* minor does not, however, allow one to draw any *assertoric* or *necessity* conclusion. Let *A* = Sleeping, *B* = Animal, and *C* = Horse, where no animals are asleep, to show the premises consistent with '*A* no *B*' and hence show that no affirmative assertoric or necessity conclusion follows. To see that no negative assertoric or necessity conclusion follows, let the terms be the same, but suppose that all animals are asleep.

With a weak *cop* minor premise, the conversion proof breaks down. But one can still obtain a problematic conclusion by *ekthesis* proof. For example, with the pair

(1) $A\ pp$ all $C$
(2) $B\ N_w\ all\ C$

we know by (1) that '$A\ pp\ c$' (where *c* is some *C*), and by (2) that '$B\ N_w\ c$', hence that '*A pp* some *B*'. Or again, one could weaken the minor premise, using '$B\ N_w$ all $C \rightarrow B$ all $C$', and then convert the resulting assertoric minor.[56]

Two curiosities remain: As in chapter 21, Aristotle proves Bocardo by *reductio* and, as usual in such cases, claims only a *one*-way possibility conclusion. Conversion (of the minor premise) will not work, because that would give a particular premise; because the major is here also particular, nothing would follow. But *ekthesis* again works perfectly well to obtain a *two*-way conclusion:

$$\begin{array}{ll} (1) & A\ pp\ o\ C \\ (2) & \underline{B\ N\ \ a\ C} \\ (3) & A\ pp\ o\ B \end{array}$$

Premise (1) tells us that *A* two-way possibly fails to apply, say, to *c*; (2) tells us that *B* is (necessarily) applicable to all *C*, including *c*. So *A* is two-way possibly inapplicable to something to which *B* actually applies: *A pp o B*. Similar remarks apply to $I_{pp}$, $A_n/I_{pp}$, the twin by qualitative conversion to Bocardo, and the other third-figure mood (with contingent major, necessity minor) for which Aristotle claims only a one-way possibility conclusion. (That mood he "proved" by conversion of $I_{pp}$ to obtain

Darii *N, pp/p*, which at 36a40–b2 he had proved by *reductio*.) Here the system fails to provide the technical means for a *reductio* proof, and conversion cannot work in any case; but *ekthesis* saves the day. Thus, Aristotle (and his commentators) may have settled for a series of one-way conclusions simply for want of having tried *ekthesis*.

# *Chapter 7*

# Aristotle's perfect syllogisms

Aristotle's syllogistic encompasses several groups of syllogisms—the assertoric ones and six or so modal groups (the exact number would be a matter of contention and is not important here). Almost without exception[1] each type (e.g., those with necessary premises and conclusion) contains its own subset of "perfect" or "complete"[2] (*teleios*) syllogisms, perfect because they are not only valid, but obviously so just as they stand, and another set of imperfect ones whose validity must be demonstrated "by something additional," usually by reduction to one or another perfect syllogism. Aristotle says relatively little about this perfection, and for the most part the commentators follow him in this. Everyone agrees that he has in mind, at least in part, the psychological feature of *obviousness* of validity; but some would maintain – and here controversy begins to set in – that for the assertoric case, at least, he also has in mind a single logical principle that is itself obviously valid and that directly entails the perfect syllogisms Barbara, Celarent, Darii, and Ferio. The principle in question is the *dictum de omni et nullo*: Everything predicated (or not predicated) of all of some group is predicated (not predicated) of everything contained in that group. If Aristotle does have such a principle in mind, the question arises whether he meant it to apply also to perfect modal syllogisms and hence whether he had a single underlying principle that could ground the perfection of all his perfect syllogisms, plain or modal.

The whole issue has recently been reopened and addressed in an unusually detailed and stimulating way by Günther Patzig,[3] who has challenged the view that there is any such simple and intuitive reasoning behind all the perfect modal syllogisms, let alone all the perfect modal and non-modal syllogisms of *Prior Analytics* A. He has also worked out a complex positive account of how Aristotle did separate the modal sheep from the goats, but one on which Aristotle departed from his own admirable treatment of assertoric syllogisms in an arbitrary and intuitively unsatisfactory way. The following discussion takes the work of Patzig as a point of departure; after identifying what seem to be serious weaknesses in his interpretation, it will try to show that (a) Aristotle did have in mind a single criterion of perfection for all his perfect syllogisms, and (b) this principle is in fact a version of the *dictum de omni*, only generalized so as to take account of the various ways in which, according to Aristotle, a predicate can be related to its subject.

## 7.1. PLAIN SYLLOGISMS AND THE *DICTUM DE OMNI*

Within the system of plain syllogisms, Aristotle recognizes, in *Prior Analytics* A.4, four perfect moods in the first figure:

| | Barbara *I* | | Celarent *I* |
|---|---|---|---|
| $A\ a\ B$ | ($A$ applies to all $B$) | $A\ e\ B$ | ($A$ applies to no $B$) |
| $B\ a\ C$ | | $B\ a\ C$ | |
| $A\ a\ C$ | ($A$ applies to all $C$) | $A\ e\ C$ | ($A$ applies to no $C$) |
| | Darii *I* | | Ferio *I* |
| $A\ a\ B$ | | $A\ e\ B$ | |
| $B\ i\ C$ | ($B$ applies to some $C$) | $B\ i\ C$ | |
| $A\ i\ C$ | | $A\ o\ C$ | (There is some $C$ to which $A$ does not apply) |

Syllogistic perfection had been characterized in *Pr. An.* A.1:

> I call that syllogism perfect which needs nothing beyond what is stated to make plain the necessity. I call that syllogism imperfect, then, which stands in need of one or more ⟨things⟩ which are necessary because of the terms laid down, but are not taken by means of the premises. (24b22–26)

Later (A.5, 28a5–7) he will remark, in the midst of his discussion of the second-figure plain syllogisms, that these are all imperfect: "for all are perfected by certain things being additionally assumed, which are either contained in the terms or taken as an hypothesis, as when we demonstrate by *reductio ad impossibile*." As it turns out, the additional things needed to reduce (*anagein*, or *apagein*) imperfect to perfect syllogisms – or to perfect (*epiteleisthai*) them – are usually conversion principles, with occasional appeals to proof by *reductio ad impossibile*. It will not be necessary to go into those procedures here. What we shall go into is the precise meaning of a perfect syllogism's "needing nothing additional to make the necessity obvious."

Aristotle's description of plain Barbara offers a first clue: "if *A* is predicated of every *B* and *B* of every *C*, then *A* must be predicated of every *C*" (25b37–39); also, "if the last [term] is in the middle as a whole, and if the middle is in the first as a whole, then the last is in the first as a whole" (25b32–33). As Patzig observes, Aristotle apparently considers it important that the two occurrences of the middle term be adjacent. When using his own standard technical language of "applying to" or "being predicated of," Aristotle puts the major term first: "if *A* is ⟨predicated⟩ of all *B* and *B* of all *C*, then *A* is predicated of all *C*" (25b37–39). When using the language of "being in as in a whole," he instead places the minor term first: if "the last is in the middle and the middle in the first . . ." (25b32–34). Once again he reverses the standard order when using the copulative "is": "If all the planets do not twinkle, and what does not twinkle is near, then the planets must be near" (*Post. An.* A.13, 78a30–35).[4] The result is that whichever terminology he uses, he virtually always mentions the terms in the order necessary to bring the two occurrences of the middle term together. The reason for this, Patzig argues, is that "in the formulation of Barbara the logical fact on which its validity depends, namely the transitivity of the relation 'belongs to all' between the terms which satisfy the syllogistic assumptions, becomes supremely clear."[5] William Kneale cites Patzig's explanation with approval;[6] Lynn Rose cites Kneale with approval.[7]

One important preliminary correction to Patzig (which he might rather regard as a clarification): The arrangement of the terms in Barbara does not, strictly speaking, make the transitivity of the relation 'apply to' obvious. That fact we must know *already*, perhaps as part of our comprehension of the relation of 'applying to' itself. (We shall see that this "knowledge" needs certain modification.) In any case, what Aristotle's arrangement of the terms makes obvious is that the terms *A*, *B*, and *C* are related in the premises by the relation 'applying to' in precisely the way that satisfies the antecedent of the definition of a transitive relation:

| | |
|---|---|
| General definition: | For all $x, y, z$, if $x\,R\,y$ and $y\,R\,z$, then $x\,R\,z$. |
| Premises of Barbara: | *A* applies to all *B*, and *B* applies to all *C*. |

Let it be given that 'applies to all' is a transitive relation; then it is obvious that *A*, *B*, and *C* satisfy the antecedent of the definition of transitivity, hence that *A R C* (here, '*A* applies to all *C*'). So although the arrangement of the terms cannot of itself tell us that 'applies to all' is a transitive relation – nor reveal to us that it is *obviously* transitive – the fact does remain that no other arrangement of terms could make it quite so "supremely clear," given the transitivity of 'applies to all', that the premises of Barbara entail its conclusion. With this friendly amendment, we may agree that Patzig has identified at least part of what Aristotle meant by the perfection of plain Barbara.

Kneale and Rose incautiously take Patzig's remark about transitivity as a sufficient account of the perfection of *all* perfect moods in the first figure. But Patzig had already observed that Celarent, Darii, and Ferio use 'applying to none' (or 'to some' or 'not to some') and that these relations are *not* transitive.[8] Thus, in generalizing from Barbara, one arrives not at a single principle involving transitivity, but at the rather less exciting principle that in all four cases "the end of the . . . step from *A* to *B* and the beginning of the . . . step from *B* to *C* coincide. The two steps of the premises, we can say, follow one another without a break."[9] In each first-figure case, this still makes essential use of the ordering of the terms described earlier, but now that ordering is essentially all that we have by way of an explanation of how "the necessity is made obvious."

The case of Celarent will help us appreciate the limitations of this explanation and at the same time point us toward a somewhat different criterion. If we are given as valid a schema

If *A e B* and *B a C*, then *A e C*

and if we know that no snarks are sharks and that all boojums are snarks, then to make it maximally obvious that this entails that no boojums are sharks, we might put our initial information in the form

Shark *e* Snark, and Snark *a* Boojum

This arrangement makes it obvious that the information given in the premises satisfies the antecedent of the given schema; one can verify this by running through the terms "without a break." But where do we get the schema? In this case we have no *distinct* general schema to which we can match, in ways that are more or less obvious, the mood Celarent. That is, there is evidently nothing playing the role of the definition of transitivity to which we were able to "fit" Barbara in the most obvious way possible. The schema in the present case just *is* Celarent. But in the absence of a more general principle to which we can try to fit the three propositions of Celarent, we must be content with saying simply that it is easier to see the validity of Celarent when the presentations of the middle term are consecutive than when they are not.

But at this point one might suggest that the traditional *dictum de omni et nullo* can play something like the role we gave the definition of transitivity with respect to Barbara. More precisely, it is a general principle that is itself obvious and to which we can fit, in ways that are more or less perspicuous, the information given in various premise pairs. Moreover, it is the *first*-figure arrangement of terms that makes it "supremely clear" that Barbara, Celarent, Darii, and Ferio do fit the terms of that general principle. With Celarent, for example, we would have as the relevant part of the *dictum*, 'If *A* applies to none of the things to which *B* applies, and *B* applies to all the *C*'s, then *A* applies to none of the *C*'s'. Again, the first-figure arrangement of terms ('*A e B*', '*B a C*') makes it as clear as possible that the premises of Celarent satisfy the antecedent of the *dictum* and that the conclusion satisfies its consequent.

This is not the grandiose sort of claim one might make for the *dictum*. Kneale and others rightly dismiss as either too vague or too strong or too remote from Aristotle's intention the claim that it is "the essence of all syllogistic" or that from it "everything else (in Aristotle's system) is deduced."[10] What I wish to investigate here is its importance for the validity – and the obviousness – not of everything in the system, but of Aristotle's *perfect* syllogisms.

From a logical point of view, the principle clearly does entail, as David Ross and many others have observed, the validity of Barbara, Celarent, Darii, and Ferio. One may say that the major premise of each of those four perfect moods will be either '*A a B*' or '*A e B*', which simply asserts that *A* applies to (fails to apply to) the whole of *B* (i.e., to everything to which *B* applies). Meanwhile, the minor premise '*B a C*' or '*B i C*' simply asserts that all (or some) *C*'s are included in *B* (i.e., included among the things to which *B* applies). So let the general *dictum* be slightly reworded:

> If *A* applies (fails to apply) to everything to which *B* applies
> and
> *B* applies to all (or some) of the *C*'s,
> then *A* applies (fails to apply) to all (or some) of the *C*'s.

This gives us immediately the four perfect first-figure assertoric syllogisms.

Because, on the other hand, this schema does not apply directly to the syllogisms of the second and third figures (they do not, as they stand, fit the schema), but only to syllogisms to which or through which these others are reduced by way of "additional" operations, one may say that only the four valid first-figure moods are obviously valid or "complete."[11] Thus the *dictum* can be seen as expressing a necessary as well as sufficient condition of perfection. That is, the complete assertoric syllogisms are all of those and only those having precisely one of the four valid forms – Barbara, Celarent, Darii, Ferio – immediately entailed by the *dictum de omni et nullo.*

But just how Aristotelian is this use of the *dictum?* Does Aristotle actually formulate anything exactly or even approximately like this? More important, does he *use* any such principle to explain or show the validity – or the obviousness of the validity – of his perfect assertoric syllogisms?

In the secondary literature one often reads that plain Barbara, Celarent, Darii, and Ferio are "self-evident" and justified by nothing further whatsoever. Indeed the passages we have looked at from Aristotle can suggest this. This is true not only of discussions that do not mention the *dictum*[12] but also of some that do. Thus, Lynn Rose says that the *dictum* is not used to test for validity even in the first figure: "For Aristotle 'tests' the first figure moods in two ways: he rejects the invalid moods by using counterexamples, *not* pointing out that they fail to fit the *dictum;* and he accepts the valid moods because individually they are self-evident or perfect, *not* because they fit the *dictum.*"[13]

Rose is certainly right that Aristotle did not appeal to the *dictum* to show *invalidity*, but typically used instead the method of "contrasting instances."[14] On the other hand, it can be argued that Aristotle did use something like the *dictum* in establishing the validity of Darii and Ferio of the first figure:

> whenever the *universal* premise, affirmative or negative, is related to the *major* term, and the particular premise is affirmative and related to the minor, then there must be a perfect syllogism. But when the ⟨universal premise⟩ is the minor, *or the terms are disposed in any other way*, it is impossible ⟨that there be a syllogism⟩. (26a17–21; emphasis added)

Of course, one must add what Aristotle had said in the case of moods with two universal premises (including Barbara and Celarent):

> When the ⟨premises⟩ are universal, it is clear in this figure when there will be and when there will not be a syllogism, and that if there is a syllogism the terms must be disposed as we said, and if they are disposed in this way, that there will be a syllogism. (26a13–16)

What Aristotle had said was, in effect, that if things are disposed as in Barbara and Celarent, one gets a valid syllogism (sufficient condition), but not with any other arrangement involving two universal premises (necessary condition).

It is not so difficult to see why David Ross might want to elevate such comments, and similar general statements that occur at the beginning, end, or section breaks of most of the chapters 4–22 of *Pr. An.* A, to the status of "rules."[15] And if these are logical rules laying down necessary and sufficient conditions for validity, it looks as if the rule governing the first figure will be, in effect (if not in precisely so many words), the *dictum de omni et nullo.* Thus the *dictum*'s advocate.

The reply would be that even the devil may quote scripture. First, Aristotle's statement about moods with two universal premises comes at the end of the section on such syllogisms. Thus it occurs only after Barbara and Celarent have been otherwise validated, and the moods *AEA, AEE, EEA*, and *EEE* invalidated by the use of counterexamples. Aristotle does not, as he works through these moods, appeal to the *dictum* as a self-evident condition that certain moods meet and others fail to meet. It would be more accurate to say that this "rule" *summarizes* results achieved by quite other means. The same goes for the section on syllogisms with one particular and one universal premise: The invalid moods are invalidated as they arise, by the method of contrasting instances, and the valid moods are declared valid apparently without reference to any such general principle. So although Aristotle does in fact (loosely) formulate a necessary and sufficient condition for validity for each major group of first-figure moods, and might be thought to take for granted a unified principle for the whole figure, he does not use this condition as a tool of evaluation, but seems to derive the "rule" (better, the summary statement) from results achieved entirely independently.

It may yet be replied on behalf of the *dictum* that although the point about invalid moods is well taken, and although Ross's "rules" are not used to evaluate syllogisms, this still overlooks a critical feature of Aristotle's treatment of the valid moods, for Aristotle does have something else interesting to say about the perfection of his perfect syllogisms – something that arguably amounts to an invocation of the *dictum* and that *is* used to justify perfect syllogisms. With Barbara, first of all, he says

> For if *A* is ⟨predicated⟩ of every *B* and *B* of every *C*, then *A* must be predicated of every *C* [25b37–39] . . . for it was stated earlier what we mean by ⟨predicated⟩ of all (*proteron gar eirētai pōs to kata pantos legomen*). (25b39–40)

Aristotle has in mind chapter 1, 24b28–30:

> We say 'predicated of all' when none ⟨of the subject⟩ can be taken of which the other ⟨term⟩ is not said. And similarly for '⟨predicated⟩ of none' (*legomen de to kata pantos katēgoreisthai hotan mēden ei labein [tou hupokeimenou] kath' hou thateron ou lechthēsetai. Kai to kata mēdenos hōsautōs.*

This suggests that the definition of 'applies to all' is fundamental to the validity of Barbara and that the "obviousness" of the validity of Barbara derives from the two facts that (a) in order to see its validity one needs merely to comprehend the notion 'applies to all', and (b) the ordering of the terms makes it as easy as possible (in the way discussed earlier) to see that the *C*'s are among the things to all of which *A* applies. For Celarent, one would need to comprehend 'applies to none', and so forth.

Nor is this Aristotle's only such citation of the definition of 'applies to all/none': On the contrary, he repeats it numerous times, and in fact at almost every introduction of a new group of complete syllogisms right on through chapter 22 (30a2–3, 30a17–23, 32b38–33a5, *et al.*, are discussed later, in Section 7.3). Moreover, the very first paragraph of the *Prior Analytics* foretells the importance of this definition, for the defining of 'this being or not being in that as a whole' and 'being predicated of all or none' is listed as one of six basic items on the agenda of the *Analytics.* The others, in order, are (a) to say what the work is about and to what inquiry it belongs, to define (b) 'premise' and (c) 'term' and (d) 'syllogism', and (e) to say what sort of syllogism is perfect, and what imperfect (24a10–16). Those items, along with (f) the definition of 'applies to all', are taken up, in order, in chapter 1 of *Pr. An.* A.

At the same time, notice that the citation of the definition of 'applies to all', as important as it is, does not introduce anything new, anything not contained in the premises as stated. Rather, it only reminds us of what precisely *is* contained in the premises. Thus, although it is not correct that Aristotle says nothing more about Barbara than that its validity is obvious or self-evident, it remains true that Barbara needs nothing besides what is set forth explicitly in the premises "for its validity to be obvious."

As for Celarent, Aristotle says merely that it will be the same as with Barbara. Again, he could appeal now to the definition of 'applies to none', but he does not do so explicitly. This *is* just what happens, however, with Ferio a few lines further on:

> if *A* applies to no *B*, but *B* to some *C, A* must fail to apply to some *C*. For what we mean by '⟨predicated⟩ of none' has been defined; so that there will be a perfect syllogism. (26a25–28)

And similarly for Darii:

> For let *A* apply to every *B*, and *B* to some *C*. Then if 'predicated of all' is what was said in the beginning, then *A* must apply to some *C*. (26a23–25)

Insofar as these passages show Aristotle concerned to provide a logical foundation even for his complete or perfect moods, they show him appealing to the definition of the (universal) quantifiers. This revives the possibility of a primordial underlying principle, but one that happens here to be elaborated in the language of quantification. And indeed, just here defenders of the *dictum de omni et nullo* will point out that these definitions of universal quantification give, in effect, the first clause of the *dictum* ("if *A* applies to everything/nothing to which *B* applies . . ."). What one lacks is some reference to the second clause, which asserts that all/some *C* is included among the things (*B*'s) to all of which *A* applies. Only then, strictly speaking, would one have a version of the *dictum* obviously entailing the validity of these pefect syllogisms. But those who make the *dictum* the foundation of these moods presumably would contend that Aristotle finds the contribution of the minor premise so obvious that he simply takes it for granted. In fact, they might well argue that it *makes no sense* for Aristotle to appeal, as he does, *simply* to the definition of 'applies to all' or 'applies to none' – and maintain that this shows the syllogisms in question valid – unless he is taking for granted the contribution of the minor premise.[16]

I believe this is about as far as Aristotle's remarks on plain syllogistic will take us. The modal chapters have many interesting things to add, however – things that will help resolve our question about the basis of syllogistic perfection and the status of the *dictum de omni et nullo.*

## 7.2. PERFECTION OF PERFECT MODAL MOODS

The perfect modal cases might seem to follow the same lines as their assertoric brethren, as Aristotle, Alexander, and Ross affirm. But Patzig raises interesting questions both about what Aristotle was thinking and about what he ought to have been thinking in these often more complex modal cases. Patzig's basic procedure is to focus on the various first-figure modal syllogisms in order, first, to isolate any differences between these and the imperfect ones that would have led Aristotle to classify the former alone as perfect, and then to identify Aristotle's reasons for accepting precisely these differences as criteria of perfection. In discussing Patzig's position it will be useful to have these syllogisms before us, both in modal *copula* and in modal predicate form – the former because that is the way Aristotle presents them, the latter because that is the form in which these syllogisms are discussed by Patzig, Ross, and others.

## 7.2 Perfection of perfect modal moods

*Perfect modal syllogisms in Barbara*

(1) Barbara *NNN* (A.8, 29b36–30a5)

$$\begin{array}{l} A\,N\;a\;B \\ B\,N\;a\;C \\ \hline A\,N\;a\;C \end{array} \qquad \begin{array}{l} nA\;\;a\;\;B \\ nB\;\;a\;\;C \\ \hline nA\;\;a\;\;C \end{array}$$

(2) Barbara *NAN* (A.9, 30a17–23)

$$\begin{array}{l} A\,N\;a\;B \\ B\,a\;\;C \\ \hline A\,N\;a\;C \end{array} \qquad \begin{array}{l} nA\;\;a\;\;B \\ B\;\;\;a\;\;C \\ \hline nA\;\;a\;\;C \end{array}$$

(3) Barbara *ANA* (A.9, 30a23–32)

$$\begin{array}{l} A\,a\;\;B \\ B\,N\;a\;C \\ \hline A\,a\;\;C \end{array} \qquad \begin{array}{l} A\;\;\;a\;\;B \\ nB\;\;a\;\;C \\ \hline A\;\;\;a\;\;C \end{array}$$

(4) Barbara *pp, pp/pp* (A.14, 32b38–33a1)

$$\begin{array}{l} A\,pp\;a\;B \\ B\,pp\;a\;C \\ \hline A\,pp\;a\;C \end{array} \qquad \begin{array}{l} ppA\;a\;\;B \\ ppB\;a\;\;C \\ \hline ppA\;a\;\;C \end{array}$$

(5) Barbara *pp, A/pp* (A.15, 33b33–36)

$$\begin{array}{l} A\,pp\;a\;B \\ B\,a\;\;C \\ \hline A\,pp\;a\;C \end{array} \qquad \begin{array}{l} ppA\;a\;\;B \\ B\;\;\;a\;\;C \\ \hline ppA\;a\;\;C \end{array}$$

(8) Barbara *pp, N/pp* (A.16, 36a2–5)

$$\begin{array}{l} A\,pp\;a\;B \\ B\,N\;a\;C \\ \hline A\,pp\;a\;C \end{array} \qquad \begin{array}{l} ppA\;a\;\;B \\ nB\;\;a\;\;C \\ \hline ppA\;a\;\;C \end{array}$$

*Imperfect first-figure syllogisms*

(6) Barbara *A, pp/p* (A.15, 34a34–b2)

$$\begin{array}{l} A\,a\;\;B \\ B\,pp\;a\;C \\ \hline A\,p\;\;a\;C \end{array} \qquad \begin{array}{l} A\;\;\;a\;\;B \\ ppB\;a\;\;C \\ \hline pA\;\;a\;\;C \end{array}$$

(7) Barbara *N, pp/p* (A. 16, 35b38–36a2)

$$\begin{array}{l} A\,N\;a\;B \\ B\,pp\;a\;C \\ \hline A\,p\;\;a\;C \end{array} \qquad \begin{array}{l} nA\;\;a\;\;B \\ ppB\;a\;\;C \\ \hline pA\;\;a\;\;C \end{array}$$

Patzig then formulates three principles or operations by which imperfect syllogisms, such as (6) and (7), may be perfected (again I give this in both modal predicate and modal copula format):

| | | |
|---|---|---|
| (I) | $nA\ a\ B \rightarrow A\ a\ B$ | $(A\ N\ a\ B \rightarrow A\ a\ B)$ |
| (II) | $ppA\ a\ B \rightarrow ppA\ a\ ppB$ | $(A\ pp\ a\ B \rightarrow A\ pp\ a\ B\ pp)$ |
| (III) | $A\ a\ B \rightarrow pA\ a\ B$ | $(A\ a\ B \rightarrow A\ p\ a\ B)$ |

In the assertoric cases, as remarked earlier, the perfect syllogisms were the ones needing "nothing additional" to make their validity obvious; the imperfect stood in need of "one or more" operations. Of Aristotle's perfect modal syllogisms, Barbara (2) and (5) do fit this pattern: Their assertoric minor premises simply bring the *C*'s under the *B*'s, so that when their major premises attribute *nA* or *ppA* to every *B*, it is obvious, without need of any further steps, that *nA* or *ppA* will apply to every *C*. The curious fact is that (1), (3), (4), and (8), although classified by Aristotle as perfect, all need, on Patzig's account, one operation or another for their transformation into an argument in Barbara fitting that basic pattern of perfection. Syllogism (1) uses principle (I), which, when applied to the second premise, gives '*B a C*', thence the unproblematically perfect syllogism (2). Syllogism (3) applies (I) to its minor premise to give the assertoric version of Barbara. Syllogism (4) applies (II) (i.e., ampliation) to the major premise to give '*ppA a ppB*' (or, on a *cop* reading, '*A pp a B pp*'). Syllogism (8) applies (I) to its minor premise to get '*B a C*', thence syllogism (5). Thus, after a single application of (I), syllogisms (1), (3), and (8) have a minor premise '*B a C*' and (on a modal predicate reading) a major premise attributing *nA, A*, and *ppA*, respectively, to all the *B*'s. Syllogism (4) requires instead an application of (II), which brings the *C*'s under the two-way possibly *B*'s – to all of which, according to the major premise, two-way-possibly-*A* applies. Thus of all these syllogisms, only (2) and (5) require no operations at all to make their validity obvious.

By contrast, the imperfect (6) requires application of (III) *and* (II) to its major premise in order to give the following perfect syllogism:

$$\text{(6A)}\quad \begin{array}{l} pA\ a\ ppB \\ \underline{ppB\ a\ C} \\ pA\ a\ C \end{array} \qquad \begin{array}{l} A\ p\ a\ B\ pp \\ \underline{B\ pp\ a\ C} \\ A\ p\ a\ C \end{array}$$

Finally, (7) requires application of (I), (II), and (III) to its first premise, which then gives (6A) again.

Patzig then asks why, in these modal cases, Aristotle should have drawn the line differently than in the assertoric ones. That is, why not make (2) and (5), which require no operations, perfect, then declare the rest more or less imperfect, with degree of imperfection a simple matter of how many operations are required for perfection? This not only would follow the distinction drawn in the assertoric case between syllogisms needing *no* operations and those needing "one or more"; it also would preserve the intuitively appealing, if not expressly Aristotelian, point that there are *no* degrees of perfection, although there are degrees of imperfection.

In answer, Patzig reasons, first, that Aristotle must have deemed (1) as perfect as (2), and (4) as perfect as (5), because "he must have thought it absurd to suppose that the *identity* of the operators on *all* the propositions of a syllogism could *impair* its evidence, so that a syllogism with three necessary components like (1) could be less evident than the same syllogism with an assertoric minor (2), and a syllogism with three problematic propositions as premises and conclusion like (4) could be less evident than one with a problematic major and assertoric minor like (5)."[17] Patzig is right to regard such an assumption as "a mere prejudice, no doubt a seductive one";[18] whether Aristotle was in fact seduced by it remains to be seen.

Even accepting this explanation for (1) and (4), however, we would still lack an explanation for why Aristotle classed (3) and (8) as perfect. But elsewhere Patzig does indicate how his approach can be extended so as to explain why Aristotle regarded (3) and (8) as perfect, along with (1) and (2), and (4) and (5). Moreover, the extended principle still hinges, as one might hope, on the question of how many operations a given syllogism requires for its perfection. Patzig's considered position, then, is that because Aristotle allowed that (1) and (4) are perfect (this due to the prejudice mentioned earlier), and because (1) and (4) each require only one operation to make their validity clear, Aristotle decided to let modal syllogisms be perfect if they needed only one operation, and imperfect if more than one: "in his modal logic Aristotle allows perfection to syllogisms whose evidence is established by one of the operations we here described."[19] If this is Aristotle's reasoning, then there is (as Patzig says) a fair degree of arbitrariness in how he draws the line between perfect and imperfect, and he seems to have taken that first, fateful step away from the intuitively straightforward sort of division applicable in the assertoric cases (either no operations are needed or one or more are needed) because of a prejudice in favor of "modal conformity of premises and conclusion."

Patzig's discussion raises useful questions whose resolution will give us a better understanding of what Aristotle had in mind. For this reason it is an advance over Alexander and Ross, even though in the end I believe they are closer to the truth in asserting the continuity of Aristotle's treatment of assertoric and modal cases. To see why this is so, we need to look at each case in which Aristotle declares a syllogism valid but in which one or more of Patzig's principles must be applied to "perfect" the syllogism.

Let us start with syllogism (4), which required one application of principle (II) (ampliation). My objection to Patzig's analysis is simply that principle (II) is *not* "a further *implication* of [Aristotle's] modal logic."[20] Aristotle's idea is rather that the proposition '*A pp a B*' can be *read* or *interpreted* in two ways (*dichōs estin eklabein*, A.13, 32b26): as (a) (on the modal predicate reading favored by Patzig) '*ppA* applies to every *B*' ('two-way-possibly-*A* applies to everything to which *B* applies'), or as (b) (the ampliated modal predicate version) '*ppA a ppB*' ('two-way-possibly-*A* applies to everything to which two-way-possibly-*B* applies'). Patzig at first has this right when he speaks of "a new definition of *endechesthai huparchein* . . ." and of the (ampliated) "meaning," '*ppA a ppB*' of '*ppA a B*'.[21] But he then backslides by setting out principle (II) officially as an "implication" from '*ppA a B*' to '*ppA a ppB*'.[22] Logically speaking, the difference is far from trivial, for the implication represented by (II) is not even valid: The fact that possibly-and-possibly-not-A applies to every actual *B* does not entail that that same predicate applies to every possibly-and-possibly-not-*B*. [Let *A* = Walking and *B* = White Thing on the Mat, and let all actual *B*'s be cats, but some *ppB*'s be cloaks. Then two-way-possibly-walking does truly apply to all white things on the mat (all of which are cats); but two-way-possibly-walking does *not* apply to all two-way-possibly-white-things-on-the-mat, because some of these are cloaks, which cannot possibly walk. Likewise, on a *cop* reading, the fact that *A* two-way possibly applies to every *B* does not entail that *A* two-way possibly applies to everything to which *B* two-way possibly applies.]

The point is critical for understanding why Aristotle classified (4) as perfect, for if ampliation specifies one *interpretation* or reading of '*ppA a B*', then the ampliated reading of syllogism (4) will be, from the start, perfect:

| | | |
|---|---|---|
| *ppA a ppB* | or | *A pp a B pp* |
| *ppB a C* | | *B pp a C* |
| *ppA a C* | | *A pp a C* |

This just is the way the major premise is to be understood, so that *no* operations are needed to display this syllogism's validity. Therefore, it is not correct to say that the *unampliated* version of this mood is declared perfect by Aristotle on grounds that it needs only one operation (i.e., ampliation) to make its validity obvious. On the contrary, the *unampliated* version [syllogism (4)] is *invalid* (let $A$ = Walking, $B$ = Thing on the Mat, and $C$ = Cloak, in a situation in which all actual $B$'s are cats). One should say, rather, that the *ampliated* version is perfect because it needs *nothing at all* – no conversion of a premise, no *reductio*, no *ekthesis*, nor anything else beyond what is expressly stated in the premises – to make its validity apparent.

The other three allegedly odd cases, (1), (3), and (8), were all, on Patzig's interpretation, transformed by application of (I) to derive '$B\ a\ C$' from '$nB\ a\ C$'. Again, the acceptance of syllogism (1) as perfect was explained by an alleged Aristotelean prejudice in favor of modal conformity between premises and conclusion. But I think there is a much more plausible explanation, and one that becomes especially clear on the modal copula reading, in the fact that a proposition explicitly asserting that $B$ *necessarily applies* to every $C$ will thereby assert, and just as explicitly, that $B$ *applies* to every $C$ or that the $B$'s are among the $C$'s. Thus, although Aristotle did hold that 'necessarily applies' entails 'applies', and although one *could* use that implication [by way of principle (I)] to transform syllogism (1) into (2), (3) into assertoric Barbara, and (8) into (5), it would be perfectly natural for Aristotle to declare syllogisms (1), (3), and (8) obviously valid *just as they stand* – that is, on the basis of what is explicitly and directly stated in the premises. Put another way, the *plain* premise '$B$ applies to all $C$', derived in these cases *via* principle (I), does *not* tell us anything that is not already *explicitly* stated in '$A$ necessarily applies to all $B$'. Thus, transforming the minor premises of (1), (3), and (8) by use of (I) in order to reveal their validity would be, in Aristotle's view, completely superfluous. And hence all three of these syllogisms are valid on exactly the same criterion as applied in the assertoric case and in the case of modal syllogism (4): That is, none of them needs anything additional (i.e., nothing beyond what is explicitly stated in the premises, and in the way in which it is stated) to make their validity obvious.[23]

Finally, Aristotle's *imperfect* modal syllogisms will be imperfect *not* because they require more than one of Patzig's operations, but because they require the use of conversion or *reductio ad impossibile* or *ekthesis* – just as Aristotle says in the relevant texts – and, again, just as with the imperfect assertoric syllogisms.

## 7.3. 'APPLIES TO ALL/NONE' AGAIN

The parallel between the modal and assertoric cases holds in yet another important respect in that in both contexts Aristotle consistently explains or shows (gives an *apodeixis* of) the obvious validity of his perfect syllogisms by appeal to the definition of 'applies to all'. For example, Aristotle spends little time on the pure necessity syllogisms of chapter 8: Things are, he says, very much as they were with assertoric syllogisms. The main difference is that now one will have 'necessarily applies (fails to apply)' where before one had plain 'applies'. "For," he says, "the negative [i.e., '*A N e B*'] converts in the same way, *and we define 'being in the whole of' and '[predicated] of all' in the same way*" (30a2–3; emphasis added). With '⟨predicated⟩ of all' defined as in the assertoric case, it would read "nothing ⟨of the subject⟩ can be taken of which the other ⟨term⟩ is not necessarily said." Given this, the four pure necessity counterparts to plain Barbara, Celarent, Darii, and Ferio will be obviously valid, and for the same reasons as their assertoric brethren.

In the first-figure moods combining assertoric with necessity premises [chapter 9; our preceding syllogisms (2) and (3)] he is slightly more expansive:

> . . . for example if *A* is assumed to necessarily belong or not belong to *B*, but *B* only to belong to *C;* for with the premises taken in this way *A* will of necessity belong or not belong to *C. For since A necessarily belongs or does not belong to every B, and C is some of the B's (to de C ti tōn B esti), it is obvious that ⟨A⟩ will of necessity ⟨apply to⟩ C in one of these ways.* (30a17–23; emphasis added)

Here, again, the term *phaneros* signifies the *obvious* validity of the mood in question, and here Aristotle invokes, in the italicized sentence, something even more like the *dictum de omni et nullo* than before to explain their obvious validity, for here we have not only the reference to 'applies to all (none)' but also explicit mention of the premise bringing (every) *C* under *B*.

The case is virtually the same for mixed Darii and Ferio, with one small, but important, variation:

> First let the universal ⟨premise⟩ be necessary, and let *A* apply of necessity to *B*, and *B* only apply to some *C*. Necessarily, then, *A* will apply of necessity to some *C*. For *C* is under *B*, and ⟨*A*⟩ applied of necessity to all *B;* and similarly if the syllogism should be negative [as with Ferio]. *For the proof will be the same (hē gar auto estai apodeixis*). (30a37–b2)

Besides the earlier sort of explanation of validity in terms of 'applies to all', and the additional remark about *C*'s being under *B*, we find now a reference to this explanation as a "demonstration" (*apodeixis*). This is no doubt a broad use of the term: Aristotle does not seem to see himself as giving a prior deduction whose conclusion is that Darii or Ferio is valid. One difference between the *apodeixis* (the second sentence just quoted) and the statement of Darii itself (the first sentence) is the explicit addition in the latter of the word "all," as if to say that one need only recall how 'applies to all' was defined in order to see why this mood is valid. There is a second difference, in the use of *hupo* in the *apodeixis*, as if to underline explicitly that (some) *C*'s are among the *B*'s to all of which *A* applies. Taken together, these two details consolidate and emphasize the role of the definition of 'applies to all' as a conscious element in Aristotle's laying of the foundations of his modal as well as non-modal syllogistic, and they point to the use of the *dictum de omni* as a basis of the complete syllogisms of *Pr. An.* A.9.

Our next perfect syllogism is the pure two-way possibility mood in Barbara of chapter 14 [our syllogism (4) in the preceding list], to which Aristotle appends an unusually lengthy appeal to the quantificational definition:

> Whenever, then, *A* possibly applies to all *B* and *B* to all *C*, there will be a perfect (*teleios*) syllogism that *A* possibly applies to all *C*. This is obvious (*phaneron*) from the definition. *For this is what we mean by possibly applies to all* (*endechesthai panti huparchein*). Similarly if *A* possibly fails to apply to *B*, but *B* ⟨possibly applies⟩ to all *C*, ⟨there will be a syllogism⟩ that *A* possibly fails to apply to all *C*. *For A to possibly fail to apply to that to which B possibly applies, was for nothing to be left out of the things to which B possibly applies.* (32b38–33a5; emphasis added)

The final clause clearly recalls the definition of 'applies to all' from the very first chapter of the *Prior Analytics*. And Aristotle comments on Darii later in the chapter: "this is obvious (*phaneron*) from the definition of 'possibly applies ⟨to all⟩' " (33a24–25). Similarly, he will immediately add, regarding Ferio: "the proof is the same" (*apodeixis d' hē autē*) (33a25–27). Once more the perfection of these perfect moods is itself given an *apodeixis* amounting to an invocation of the definition of 'applies to all'; and again Aristotle explicitly mentions the minor premise (bringing all/some *C*'s under the *B*'s) as well.

Only two types of perfect syllogisms remain: the mixed possibility/assertoric ones of chapter 15 [including our preceding (5)] and the mixed possibility/necessity moods [including our preceding (6)] of chapter 16. In the former case we encounter again both key phrases of the *dictum de omni:*

> For let *A* two-way apply to all *B*, and *B* apply to all *C*. Since *C* is under (*hupo*) *B* and *A* possibly applies to *B*, it is obvious (*phaneron*) that ⟨*A*⟩ possibly applies to all *C* also. Indeed there comes about a perfect (*teleios*) syllogism. Similarly if the *AB* premise is negative. . . . (33b33–36)

Here, for a change, the explicit emphasis is on the minor premise, the one bringing the *C*'s under the *B*'s: For once, it is the definition of 'applies to all' that is taken for granted. When, later in the chapter, the minor premise is particular, Aristotle says that there "will be a perfect (*teleios*) syllogism, just as when the terms are universal. The proof (*apodeixis*) is the same as before" (35a35).

Chapter 16 adds two small but interesting details. First, with regard to the perfection of Barbara *pp*, *N/pp* [our preceding syllogism (8)], Aristotle says only that it is "perfected immediately through the initial premises (*euthus gar epiteleitai dia tōn ex archēs protaseōn*)" (36a6–7). Insofar as this bears on the matter under discussion, it confirms that no principle not contained directly in the initial premises is needed to complete the syllogism: Even if Aristotle considered the validity of the mood to rest upon the definition of 'applies to all', and even if he considered the definition itself something distinct from and prior to any affirmative or negative universal premise, and even if he considered the citation of that definition as some kind of *apodeixis*, the fact remains that nothing except what is expressed in the premises is needed to perfect any of his perfect moods. Rather, one only needs to understand what is being expressed, and, in particular, the concept 'applies to all', to see their validity. The chapter's second contribution is more significant and will be taken up in a moment.

This survey of texts concerning the perfect modal syllogisms has confirmed not only a conscious role for the definition of 'applies to all/none' at the base of the system but also the unity of Aristotle's principle of perfection for assertoric and modal syllogisms: They are all "self-perfecting" (*epitelountai di' autou*, said of plain Darii and Ferio, 29b6–8) in the sense that they need nothing more than what is expressly stated in the premises, in the way it is stated, to make their validity obvious. But the content of the general underlying principle at work will have to go far beyond what we found in the assertoric cases alone:

| | | |
|---|---|---|
| If *A* applies<br>or necessarily applies<br>(or possibly applies) | to everything | to which *B* applies<br>(or possibly applies) |
| and *B* applies<br>(or possibly applies) | to everything/something | to which *C* applies<br>(or possibly applies) |
| then *A* applies<br>or necessarily applies<br>(or possibly applies) | to everything/something | to which *C* applies<br>(or possibly applies) |

'Applies' covers all cases of applying, whether necessary or accidental.

The loose-limbed appearance of this formulation may suggest, as before in the assertoric case, that the basic principle is simply a *composite* or summary statement of all the individually obvious modal and non-modal cases of Barbara surveyed in chapters 4–22 of *Pr. An.* A. But the assertoric cases (ch. 4–7) plus the pure necessity cases (ch. 8) and the mixed necessity/assertoric cases (ch. 9) can at least be joined in more economical fashion:

> If *A* applies in a certain way to everything to which *B* applies, and *B* applies to everything/something to which *C* applies, then *A* applies in that same way to everything/something to which *C* applies.[24]

This will not be general enough, however, because 'two-way possibly applying' is mutually independent of '(actually) applying in some way'. One would thus need something more general than a relation of *actual* application (or actual failure to apply), whether accidental or necessary, between *A* and *B*. But further generality is easily supplied: if $R_1$, $R_2$, and $R_3$ stand for ways in which a predicate may relate to a subject, we have

> If *A* is (predicatively) related in way $R_1$ to everything to which *B* relates by $R_2$, and *B* relates by $R_2$ to everything to which *C* relates by $R_3$, then *A* relates in way $R_1$ to everything to which *C* relates by $R_3$.

The passage from *Pr. An.* A. 16 referred to a moment ago reflects just such a general principle. Speaking of Celarent *pp*, *N/pp*, Aristotle says

> Again, let the affirmative premise be necessary and let *A* possibly fail to apply to *B*, but *B* apply of necessity to all *C*. The syllogism will then be perfect (*teleios*), but not of applying but rather of possibly not applying. For this is how the premise was taken, the one related to the major term. . . . (36a17–22)

That is, because what the first premise asserts is not that *A fails to apply*, or *necessarily fails to apply*, to every *B*, but that *A possibly fails to apply* to every *B*, then (given that all the *C*'s are *B*'s) the conclusion will be that *A possibly fails to apply* to every *C*. The implication is that had the first premise asserted that *A* was related *in some other way* (*actually* fails to apply, *necessarily* fails to apply) to every *B*, then given that all the *C*'s are *B*'s, the conclusion would have been that *A* was related in *that* way (does not apply, necessarily does not apply) to every *C*. And this is just what one did get in texts concerned with those other syllogisms. Recall, for example, Aristotle's comment on the mixed assertoric/necessity moods: "since *A necessarily* belongs or does not belong to *B*, . . . it is obvious that *A* will of necessity ⟨relate to *C*⟩ *in one of these ways*" (30a21–23; emphasis added). The implicit contrast in this case is with that in which *A* simply *applies* to all *B*, hence simply *applies* to all *C*.

I would suggest that Aristotle's remarks at 36a7–22 and 30a1–23 point precisely to the general principle to which our consideration of perfect modal syllogisms has already led, for it expresses a single principle governing all predicative relations, a principle then to be elaborated in as many specific ways as there are ways for one thing to be predicatively related to another.[25] Once again, this can be seen as a version of the *dictum de omni*, now in a relatively compact formulation, and generalized so as to cover various sorts of predicative relations. At the level of abstraction represented by *Pr. An.* A.1–22, these will be the kinds of relations expressed by Aristotle's various modal copulae [i.e., plain, necessarily, and (one-way and) two-way possibly applying]. And these are, of course, designed to express the ways in which species, genera, propria, and accidents may be related predicatively to their subjects (i.e., all of the possible ways a predicate can relate predicatively to its subject).

# *Chapter 8*

## Principles of construction

Aristotle was interested in "what followed, . . . certain other things being the case." He was not, however, interested in identifying all cases of one thing's following from another. Even putting aside the difficult question of whether or not he believed that every non-syllogistic inference could be captured syllogistically, the fact remains that he did not investigate many types of inferences that were clearly expressible categorically. The largest single example would be that of syllogisms with at least one one-way possibility premise.[1] Barbara *pAp*, for example (on the "term-thing" reading of one-way possibility), would be just as obviously valid as Barbara *NAN*:

$$\begin{array}{c} A\ p\ a\ B \\ \underline{B\ a\ C} \\ A\ p\ a\ C \end{array}$$

So, also, for Barbara *pNp*. But as we have often observed, the only text (aside from the possibly inauthentic 34b2–6) that considers any syllogisms with one-way possibility *premises* is *Pr. An.* A.15, where Aristotle speaks, in effect, of pure one-way possibility syllogisms as parallel to pure necessity ones. And even there the former are mentioned only for the role they will play in validating syllogisms containing no one-way possibility premises. But at least one can say that Aristotle believed he had identified those syllogisms whose premises expressed the basic kinds of (predicative) relations that can obtain between predicate and subject and that are reflected in the doctrine of the "four predicables."

This stands in striking contrast to modern systems of modal logic, which typically begin with an undefined notion of "logical necessity," this being interdefinable with a notion of *one-way* (logical) possibility. Aristotle's starting points and subsequent focus were so different because they reflected the metaphysical and epistemological motivations of the system: Again, necessity and *two*-way possibility were the modalities corresponding to basic kinds of relations between predicables and their subjects. And so Aristotle diligently investigated what follows from the various combinations of premises asserting plain, necessary, or two-way possible relations between predicate and subject. One-way possibility propositions did turn up in the course of that investigation, but only insofar as the sorts of premises in which Aristotle was mainly interested sometimes entail only a one-way possibility *conclusion* (e.g., Barbara $A$, $A_{pp}/A_p$, 34a34–b2, Barbara $A_n$, $A_{pp}/A_p$, 35b38–36a2), or a one-way proposition enters into a *reductio* proof (as for the conversion of $E_n$ at 25a27–32), or he needs certain pure one-way syllogisms as part of his "at worst false and not impossible" proof technique (as in chapter 15).

Aristotle's own conclusions have been compiled and usefully charted more than once.[2] I would like to concentrate in this final chapter on a basic feature of Aristotle's system that is at best only partially revealed (and is sometimes obscured) by such charts, namely, the manner in which it is built up step-by-step. From the point of view of this study, there is a prior issue deriving from the fact that Aristotle did not realize that each of his modal propositions admitted of two distinct readings, each based on Aristotelian essentialist views about reality. This does not imply that his modal logic must remain forever inconsistent. Because we have seen that the basic ideas underlying his modal logic are consistent, it remains for us to assign the two readings their Aristotelian definitions and then track their logical effects on the issues of conversion (both term and qualitative), contradictoriness, and syllogistic validity. This will reveal that some conversions and syllogisms are valid on one reading, and others on another. Consistency, then, requires not that one or the other reading – along with all logical principles based on that reading – be eliminated from the system, but that only the valid principles, framed explicitly in terms of whatever reading makes them valid, be retained. Nor does it entail that we shall end up with two separate modal systems, for we have on many occasions seen that our various readings (each being always unambiguously identified) can be safely and even profitably mixed. One just has to be careful about trying this at home. To return to the larger point at hand: Accommodating both readings will multiply the *kinds* of premise pairs included in the system. Corresponding to *Pr. An.* A.8, for example, one will have pure $N_w$ moods, pure $N_s$ moods, and various combinations of the two, rather than just one set of necessity moods. And when necessity premises are mixed with plain ones, as in *Pr. An.* A.9–11, there will be mixtures with $N_w$ and $N_s$ premises to consider. And so on.

Needless to say, all these readings will be categorical ones employing modal copulae rather than modal predicates or *dicta*. Still, it is worth noting that aside from the difficulties we have surveyed with those two traditional approaches to Aristotle's modal logic, there is the additional problem that any modern representation in terms of quantified S4 or S5, say, would be much stronger than Aristotle's system, in that it would contain not only the syllogisms he regarded as valid but also some he deliberately did not include. To begin with, these systems are built on modern first-order predicate calculus, which already extends far beyond Aristotle's assertoric syllogisms.[3] An analogous point would hold for the modal portions of the system. This might not seem particularly disturbing. After all, one could regard Aristotle's system as a fragment, interesting from certain historical or philosophical points of view, of a more powerful modern logic. Indeed, this would make possible an interesting comparison of the strength of Aristotle's system with contemporary modal logics,[4] as Aristotle's plain syllogistic is often translated as a fragment of first-order logic.

As worthwhile as that may be – again, overlooking for a moment problems with carrying out the required translations – the drawback is that it obscures the manner in which Aristotle actually built up his system (and thought of himself as building up the system), and with it the way in which he built in portions that interested him and left out portions that did not. So how did Aristotle proceed? Probably what leaps first to the eye is his introduction of various kinds of syllogisms, one by one, into the system, the kinds being identified by the modality of their premise pairs, followed by identification, wherever possible, of some members of each kind as complete or perfect, with provision of the means necessary to reduce the incomplete syllogisms of each type to perfect ones. That allowed him to introduce all and only the kinds of premise pairs in which, for logical or extra-logical reasons, he might have an interest.

This description can, however, create the impression that the system consists of a series of free-standing columns, as it were, each with a base of (four) complete first-figure syllogisms and a number of imperfect syllogisms resting upon that base. This certainly is the situation in the assertoric case (ch. 4–7) (Figure 8.1).

The second- and third-figure syllogisms reduce to first-figure ones by term conversion or *reductio.* (Some can be reduced by either method, and by *ekthesis* [with a reducing syllogism] too; note the three proofs for Datisi at 28b11–15.) Aristotle proceeds as much as possible in exactly parallel fashion with every new group of syllogisms. Thus, in *Pr. An.* A.8 he declares that pure necessity syllogisms are just like their plain counterparts

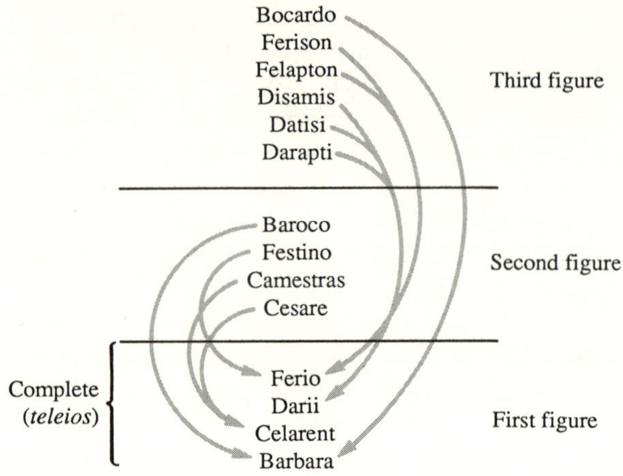


Figure 8.1

except that one has 'necessarily applies' where before one had plain 'applies'.

Aristotle also believes that his necessity propositions convert exactly as their plain counterparts, so that conversion proofs exactly parallel to the plain cases will now validate his necessity syllogisms (Figure 8.2). The only difference he notes is that Baroco and Bocardo cannot (by the means available to him) be proved by *reductio.* Because they cannot be proved by conversion, either, they must be validated by *ekthesis.* As we remarked earlier, this does not impugn the rigor of the proofs; it does, however, cut the direct epistemological (as well as the logical) tie to those first-figure syllogisms whose validity is "obvious" just as they stand. Still, as Aristotle remarks, Baroco and Bocardo are proved by *ekthesis* using syllogisms "in their own figure" (Camestres and Felapton, respectively, 30a13–14), and the reducing syllogisms are in turn reduced by conversion to perfect ones (Celarent and Ferio). This means that column II in Figure 8.3 is still free-standing in the sense that it need not lean on any other column for support.

The surprising fact is that of the remaining seven columns recognized by Aristotle, only one is entirely free-standing: column V, where both premises involve two-way possibility. There he declares all the second-figure syllogisms invalid (we saw that, and why, this was mistaken). He then reduces all the third-figure moods except Bocardo to first-figure syllogisms via term conversion; Bocardo is equivalent, *via* qualitative conversion, to Disamis.

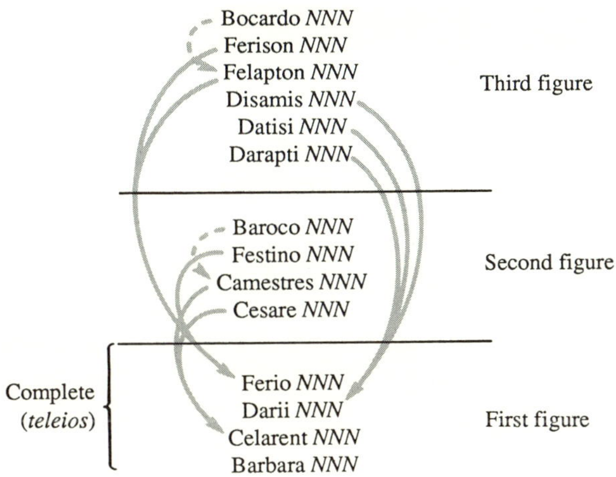


Figure 8.2

Column VII (assertoric major, two-way possibility minor) looks at first glance like a free-standing column in that, with the exception of Disamis, its second- and third-figure syllogisms reduce to its own first-figure ones. But the situation is more interesting than that; in fact, it happens to be one of only two columns (the other being IX, necessity major, two-way possibility minor) containing *no* complete syllogisms. Thus its own versions of Barbara, Celarent, Darii, and Ferio reduce to imperfect syllogisms of column III (necessity major, plain minor), which in turn reduce to perfect syllogisms of groups I, III, and IV.

The introduction of groups VII (ch. 15) and IX (ch. 16) thus underlines the important fact that Aristotle did not proceed exclusively by introducing sets of perfect syllogisms and their dependents. Rather, his most basic principle involved a specific sort of comprehensiveness: All the kinds of premise pairs constructed from assertoric, necessity, or two-way possibility propositions were to be investigated. Of course, groups VII and IX ultimately depended on perfect syllogisms from other groups. The point is that they are not introduced for that reason, but in their own right, as having the sorts of premise pairs Aristotle intended to investigate. Again, his decision to investigate these sorts of premise pairs, and his relative neglect of one-way possibility premises, had to do with extra-logical considerations about the purposes for which he was devising a modal logic in the first place. While all nine columns are essential to the larger plan of including certain kinds of premise pairs, there would occasionally be a

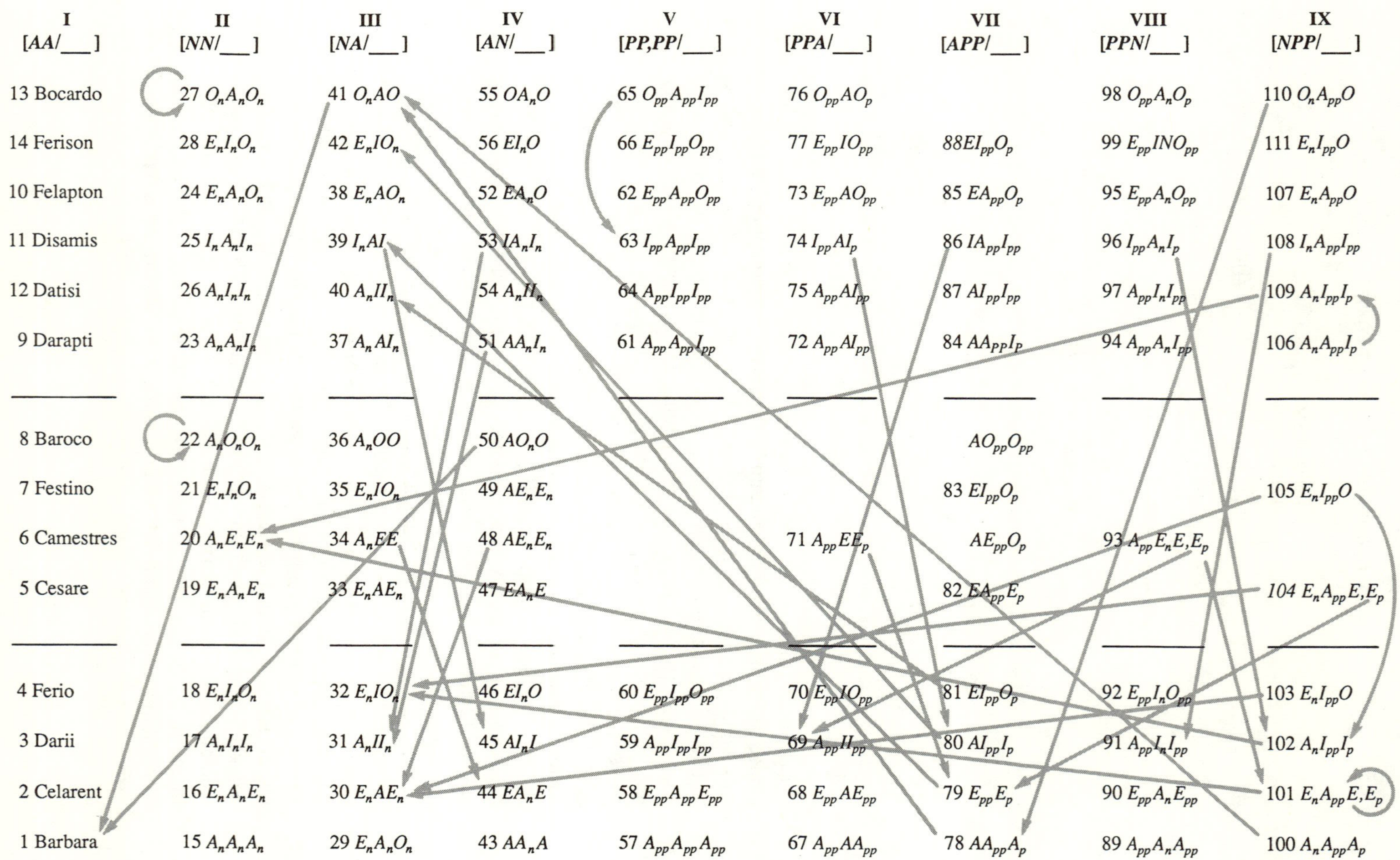
I [$AA$/___]
II [$NN$/___]
III [$NA$/___]
IV [$AN$/___]
V [$PP,PP$/___]
VI [$PPA$/___]
VII [$APP$/___]
VIII [$PPN$/___]
IX [$NPP$/___]
13 Bocardo | 27 $O_nA_nO_n$ | 41 $O_nAO$ | 55 $OA_nO$ | 65 $O_{pp}A_{pp}I_{pp}$ | 76 $O_{pp}AO_p$ | | 98 $O_{pp}A_nO_p$ | 110 $O_nA_{pp}O$
14 Ferison | 28 $E_nI_nO_n$ | 42 $E_nIO_n$ | 56 $EI_nO$ | 66 $E_{pp}I_{pp}O_{pp}$ | 77 $E_{pp}IO_{pp}$ | 88$EI_{pp}O_p$ | 99 $E_{pp}INO_{pp}$ | 111 $E_nI_{pp}O$
10 Felapton | 24 $E_nA_nO_n$ | 38 $E_nAO_n$ | 52 $EA_nO$ | 62 $E_{pp}A_{pp}O_{pp}$ | 73 $E_{pp}AO_{pp}$ | 85 $EA_{pp}O_p$ | 95 $E_{pp}A_nO_{pp}$ | 107 $E_nA_{pp}O$
11 Disamis | 25 $I_nA_nI_n$ | 39 $I_nAI$ | 53 $IA_nI_n$ | 63 $I_{pp}A_{pp}I_{pp}$ | 74 $I_{pp}AI_p$ | 86 $IA_{pp}I_{pp}$ | 96 $I_{pp}A_nI_p$ | 108 $I_nA_{pp}I_{pp}$
12 Datisi | 26 $A_nI_nI_n$ | 40 $A_nII_n$ | 54 $A_nII_n$ | 64 $A_{pp}I_{pp}I_{pp}$ | 75 $A_{pp}AI_{pp}$ | 87 $AI_{pp}I_{pp}$ | 97 $A_{pp}I_nI_{pp}$ | 109 $A_nI_{pp}I_p$
9 Darapti | 23 $A_nA_nI_n$ | 37 $A_nAI_n$ | 51 $AA_nI_n$ | 61 $A_{pp}A_{pp}I_{pp}$ | 72 $A_{pp}AI_{pp}$ | 84 $AA_{pp}I_p$ | 94 $A_{pp}A_nI_{pp}$ | 106 $A_nA_{pp}I_p$
8 Baroco | 22 $A_nO_nO_n$ | 36 $A_nOO$ | 50 $AO_nO$ | | | $AO_{pp}O_{pp}$ | |
7 Festino | 21 $E_nI_nO_n$ | 35 $E_nIO_n$ | 49 $AE_nE_n$ | | | 83 $EI_{pp}O_p$ | | 105 $E_nI_{pp}O$
6 Camestres | 20 $A_nE_nE_n$ | 34 $A_nEE$ | 48 $AE_nE_n$ | | 71 $A_{pp}EE_p$ | $AE_{pp}O_p$ | 93 $A_{pp}E_nE,E_p$ |
5 Cesare | 19 $E_nA_nE_n$ | 33 $E_nAE_n$ | 47 $EA_nE$ | | | 82 $EA_{pp}E_p$ | | 104 $E_nA_{pp}E,E_p$
4 Ferio | 18 $E_nI_nO_n$ | 32 $E_nIO_n$ | 46 $EI_nO$ | 60 $E_{pp}I_{pp}O_{pp}$ | 70 $E_{pp}IO_{pp}$ | 81 $EI_{pp}O_p$ | 92 $E_{pp}I_nO_{pp}$ | 103 $E_nI_{pp}O$
3 Darii | 17 $A_nI_nI_n$ | 31 $A_nII_n$ | 45 $AI_nI$ | 59 $A_{pp}I_{pp}I_{pp}$ | 69 $A_{pp}II_{pp}$ | 80 $AI_{pp}I_p$ | 91 $A_{pp}I_nI_{pp}$ | 102 $A_nI_{pp}I_p$
2 Celarent | 16 $E_nA_nE_n$ | 30 $E_nAE_n$ | 44 $EA_nE$ | 58 $E_{pp}A_{pp}E_{pp}$ | 68 $E_{pp}AE_{pp}$ | 79 $E_{pp}E_p$ | 90 $E_{pp}A_nE_{pp}$ | 101 $E_nA_{pp}E,E_p$
1 Barbara | 15 $A_nA_nA_n$ | 29 $E_nA_nO_n$ | 43 $AA_nA$ | 57 $A_{pp}A_{pp}A_{pp}$ | 67 $A_{pp}AA_{pp}$ | 78 $AA_{pp}A_p$ | 89 $A_{pp}A_nA_{pp}$ | 100 $A_nA_{pp}A_p$

narrower purpose pertaining to one or another column in particular (e.g., the pure necessity group, for its connection to scientific demonstration).

So the first-figure members of groups VII and IX will reduce to other syllogisms – either directly to a complete syllogism of some other group, or first to some other incomplete syllogism that in turn reduces to a complete syllogism. When one adds up all the cases from groups VII, IX, and elsewhere in which (in Aristotle's presentation) an incomplete syllogism is reduced *via* a syllogism belonging to some group other than its own, the resulting "reduction chart" includes quite a few intercolumnar reductions. With this, the image of a series of nine free-standing columns collapses. (Unless otherwise indicated in Figure 8.3, second- and third-figure syllogisms are reduced by Aristotle to first-figure arguments of their own groups).

Notice that syllogisms containing 1, 2, or 3 two-way possibility propositions will be equivalent to 1, 3, or 7 other syllogisms, respectively, *via* qualitative conversion. (Aristotle occasionally discusses such equivalents separately.) Also, some individual syllogisms must hop about for some time before finding a perfect base on which to rest. [Trace the reduction history of Darapti $A_n A_{pp} I_p$ (column IX), for example, or Disamis $I_{pp} A I_p$ (column VI).]

Our own results could be arrayed in this same style, only with quite a few more columns. Where Aristotle had one column of pure necessity syllogisms, we would already have four: one for pure strong *cop* premise pairs, one for pure weak *cop*, one for strong *cop* major premise and weak *cop* minor, and one for weak *cop* major and strong *cop* minor. Besides the question of which syllogisms are in fact valid, we have seen that one must also determine which could be validated by the sorts of proofs given by Aristotle. It turns out that the strong *cop* term conversions are valid, so that his conversion proofs work, and his *reductio* proofs for Baroco and Bocardo also go through on this reading. Thus the pure strong *cop* column would look exactly like Aristotle's pure necessity column (II).

By contrast, some weak *cop* pure necessity syllogisms were in fact

---

Figure 8.3. I have retained Ross's nine groups and his numbering of syllogisms, so that this chart might readily be used as a supplement to his. Nos. 22 and 24 are validated within Aristotle's system only by *ekthesis*, hence not by reduction using any complete syllogism. Nos. 20, 65, 105, and 106 reduce to other incomplete syllogisms within their own groups, thence to complete syllogisms. Again, unless otherwise noted, each incomplete syllogism reduces to a complete one within its own group.

invalid, and *none* could be validated by Aristotelian conversion proofs, because the requisite weak *cop* conversions are invalid. As we saw, the valid weak *cop* syllogisms could still be validated by *reductio* or *ekthesis* proofs of a sort Aristotle gives elsewhere. It turned out also that with two-way ampliation, all the pure weak *cop* syllogisms were valid – although once again it was necessary to use a different sort of proof from Aristotle's in the third figure.

Obviously one could map out an extensive system covering all combinations of premise types, with and without ampliation in each case, and with one- and two-way ampliation for each ampliated premise. Thus one would have a multiple of four columns even within the realm of pure necessity premise pairs, depending on how many of those logical possibilities one wanted to pursue. This would quickly take us far beyond anything Aristotle was interested in either for logical or non-logical reasons. Nevertheless, we shall see in a moment how the greater part of such a system could be erected on a fairly restricted basis of Aristotelian principles.

Meanwhile, it is worth remarking that other ancient or modern categorical logicians could use Aristotle's approach to construct a wide variety of different systems. For example, one might want to include plain syllogisms and, for scientific purposes, the strong *cop* column. Because *reductio* proofs for strong *cop* conversions and for syllogisms with strong *cop* conclusions will involve arguments with one-way premises – and, more specifically, one-way "term-term" propositions – one might well want to construct a new column, with its own complete syllogisms, for pure one-way possibility deductions, and two more for the mixed necessity/one-way possibility versions. Some scholars have, in effect, suggested that Theophrastus' modal logic was of this sort, including only the "*de dicto*" modalities of necessity and one-way possibility. But it would be possible also to limit a system to assertoric propositions, weak *cop* necessity, and one-way (term-thing) possibility, thus giving a simple *de re* system. In short, it is clear that Aristotle's basic approach is flexible and can be adapted to serve different logical and non-logical purposes.

We also looked in some detail at the effects of combining weak and strong *cop* premises. The most interesting results were that (i) in the second figure one could derive a strong *cop* conclusion from one weak and one strong *cop* premise, given that the strong *cop* proposition was the minor premise, and (ii) in the third figure two weak *cop* premises could entail a strong *cop* conclusion. These mixed weak/strong cases were of interest also because their validation involved semantic considerations bearing on the intended application to relations among genera, species, propria, and accidents. This allowed us to verify that a given type of syllogistic inference could have no counterexamples and was therefore to be included in the system, even if it was neither "self-evident" nor formally derivable within the system from self-evident principles.

In addition to the possibility of adding quite a few more columns to Aristotle's system, it would also be possible to go beyond his text in the opposite direction, as it were, by reducing some of his basic principles to others. Specifically, one could presuppose the validity of a rather small number of his complete syllogisms, then validate the rest by use of various intermodal principles (along with the usual term and qualitative conversion principles, *reductio ad impossibile*, and *ekthesis*).[5] Given '$N_w \to A$' (i.e., '*A N a/e/i/o B*' → '*A a/e/i/o B*'), for example, presuppose the validity of weak *cop* Barbara *NAN*, and then derive Barbara *NNN*. Now, Aristotle explicitly classifies the latter as perfect, and the former is obviously perfect as well, for exactly the same sort of reason he gives for the cases of Barbara *A*, *A*/*A*, *N*, *N*/*N*, and *pp*, *A*/*pp*. But Barbara *NNN* could be reduced to *NAN* by applying the principle '$N \to A$' to its minor premise. (This would show, by the way, a distinction between a non-derivable inference rule or axiom of a system – i.e., one that is not derivable within that system – and the Aristotelian concept of a complete mood that *needs no* supporting derivation, whether or not we may be able to construct one within the very system for which it serves as an "obviously" valid inference scheme.) Assertoric Darii and Ferio are further examples, as discussed later.

To go one step further, let us presuppose only the first-figure moods *AAA*, $N_wAN_w$, $N_sN_sN_s$, *pp*, *pp*/*pp* (with ampliation), and *pp*, *A*/*pp*. Then, from *AAA*, along with the intermodal principle

$$N_s \to N_w \to A \to p$$

we could derive all the first-figure moods in $N_wN_wA$, $N_wAA$, $AN_wA$, $N_sN_wA$, $N_wN_sA$, $N_sAA$, and $N_sN_sA$ – that is, those each with an assertoric conclusion and at least one necessity premise. From $N_wAN_w$ plus that same principle, we obtain $N_wN_wN_w$, $N_sN_wN_w$, and $N_wN_sN_w$; from *pp*, *A*/*pp* plus that intermodal principle come *pp*, $N_s$/*pp* and *pp*, $N_w$/*pp*. So we need only five (sets of four) complete syllogisms plus the intermodal law to generate a broad array of complete syllogisms, including counterparts to Aristotle's groups I–VI and VIII, plus a number of others he did not investigate.

On the other hand, one could remain somewhat closer to Aristotle by invoking less inclusive principles ('$N_w \to A$', '$pp \to p$', and the like) so as to generate some portion of his system without overshooting the mark. (For example, '$N_w \to A \to p$' would capture a large so-called *de re* portion of the system.[6]) But the fact is that despite his foundationalism, Aristotle does not seem consistently interested in finding the most economical system possible, at least in the sense of finding the smallest set of starting points still adequate for his purposes. (The *only* case in which he explicitly tends toward this is the demonstration in chapter A.7 that the plain, first-figure, perfect syllogisms Darii and Ferio reduce via *reductio ad impossibile* to second-figure Camestres and Cesare, respectively, where the latter pair had been shown in A.5 to reduce to first-figure Celarent. Thus, given his reduction in A.5 and A.6 of all second- and third-figure syllogisms to first-figure perfect ones, he has shown that all his valid assertoric syllogisms can be reduced to Barbara and Celarent. He did not, however, take the further step – which he might well have taken, given his belief in '$N \rightarrow A$', '$A \rightarrow p$', '$pp \rightarrow p$' – of reducing some perfect syllogisms to others *via intermodal* principles.[7])

Still, Aristotle maintained throughout the *Analytics* an interest in certain uses of logic along with the goal of basing all syllogisms needed to serve those ends on a smaller number of deductions whose validity was "obvious." So although he was investigating the logical issue of what follows from what and manifestly had considerable interest in that question in its own right, his system of plain and modal syllogistic and, above all, the elements at the base of the system, were shaped in critical ways from the start by extra-logical considerations. The resulting system thus left to one side many logically possible developments. It also fell short in some respects as a scientific or philosophical organon adequate for his own purposes. But for all of that, it is still well enough developed that one may appreciate the power of its underlying principles and their impressive development in *Pr. An.* A.1–22, along with a wealth of interesting and often highly ingenious discussions of more local issues.

# *Appendix*

# Categorical propositions and syllogisms

I. Assertoric (plain) statements

1. *A a B* (or, *A* all *B*): *A* applies to every *B*
2. *A e B* (or, *A* no *B*): *A* does not apply to any *B*
3. *A i B* (or, *A* some *B*): There is some *B* to which *A* applies
4. *A o B* (or, *A* not-some *B*): There is some *B* to which *A* does not apply

II. Propositions of necessity

A. Neutral *cop* formulations

5. *A N a B: A* necessarily applies to every *B*, or '*necessarily applies to all of* relates *A* to *B*'.
6. *A N e B: A* necessarily fails to apply to any *B*, or '*necessarily applies to none of* relates *A* to *B*' (equivalent to: There is no *B* to which *A* possibly applies)
7. *A N i B:* There is some *B* to which *A* necessarily applies
8. *A N o B:* There is some *B* to which *A* necessarily fails to apply (equivalent to: There is some *B* to which *A* does not possibly apply)

B. Weak *cop* (9–12 are given here in terms of the underlying semantics; they can also be read in the same way as 1–4, Chapter 2, Section 7)

9. *A* $N_w$ *a B: A* is entailed by the essence of each *B*
10. *A* $N_w$ *e B: A* is incompatible with the essence of each *B*
11. *A* $N_w$ *i B: A* is entailed by the essence of some *B*
12. *A* $N_w$ *o B: A* is incompatible with the essence of some *B*

C. Strong *cop*

13. *A* $N_s$ *a B: A* is entailed by the essence of each *B*, and *A* is entailed by *B* itself (Recall that 'being entailed by an essence *E*' is a broadening of the concept of 'being in the definition of an essence *E*'; cp. 5–8, Chaper 2, Section 7)

14. *A $N_s$ e B: A* is incompatible with the essence of each *B*, and *A* is incompatible with *B* itself
15. *A $N_s$ i B:* For some *C, B $N_s$ a C & A $N_s$ a C*
16. *A $N_s$ o B:* For some *C, B $N_s$ a C & A $N_s$ e C*

D. *De re* necessity using modal predicates

17. Necessary-*A a B* (or, *nA a B*): Necessary-*A* applies to every *B*
18. Possibly-*A e B* (or, *pA e B*): Possibly-*A* does not apply to any *B*
19. Necessary-*A i B* (or, *nA i B*): Necessary-*A* applies to some *B*
20. Possibly-*A o B* (or, *pA o B*): There is some *B* to which possibly-*A* does not apply

N.B.: 'Necessary-*A e B*' asserts that necessary-*A* does not apply to any *B*, or that possibly-not-*A* applies to every *B*, and so is not the counterpart to '*A N e B*'.

'Necessary-*A o B*' asserts that there is some *B* to which necessary-*A* fails to apply, or that possibly-not-*A* applies to some *B*

E. *De dicto* necessity

21. Nec: *A a B:* It is necessarily true that *A a B*
22. Nec: *A e B:* It is necessarily true that *A e B*
23. Nec: *A i B:* It is necessarily true that *A i B*
24. Nec: *A o B:* It is necessarily true that *A o B*

III. One-way possibility

A. Type I, or "term-term" *cop*

25. *A P a B: A* is compatible with *B* itself
26. *A P e B: A* is not entailed by *B* itself
27. *A P i B:* For some *C, B $N_s$ a C* and *A P a C*
28. *A P o B:* For some *C, B $N_s$ a C* and *A P e C*

B. Type II, or "term-thing" *cop* (lowercase *p*)

29. *A p a B: A* possibly applies to each *B;* or, *A* is compatible with the essence of each *B* (contradictory to 12)
30. *A p e B: A* possibly fails to apply to every *B;* or, *A* is not entailed by the essence of any *B* (contradictory to 11)
31. *A p i B: A* possibly applies to some *B;* or, for some *B, A* is compatible with the essence of that *B* (contradictory to 10)
32. *A p o B: A* possibly fails to apply to some *B:* For some *B, A* is not entailed by the essence of that *B* (contradictory to 9)

IV. Two-way possibility *cop* (contingent or problematic propositions)

A. Type I, or "term-term" *cop*

33. *A PP a B: A* is neither entailed by nor incompatible with *B* itself
34. *A PP e B: A* is neither incompatible with nor entailed by *B* itself
35. *A PP i B:* For some *C*, $B\,N_s\,a\,C$ and *A PP a C*
36. *A PP o B:* For some *C*, $B\,N_s\,a\,C$ and *A PP e C*

B. "Term-thing" *cop*

37. *A pp a B:* For every *B, A* is neither entailed by nor incompatible with the essence of that *B* (or '*possibly applies to and possibly does not apply to* relates *A* to each *B*'; or, '*possibly applies to all of and possibly applies to none of* relates *A* to *B*')
38. *A pp e B:* For every *B, A* is neither incompatible with nor entailed by the essence of that *B*
39. *A pp i B:* For some *B, A* is neither entailed by nor incompatible with the essence of that *B*
40. *A pp o B:* For some *B, A* is neither incompatible with nor entailed by the essence of that *B*

V. Ampliated propositions

In principle, all of Aristotle's propositions, plain or modal, could be ampliated, although he mentions ampliation only in connection with two-way possibility. Notice, in addition, that any ampliated proposition may be ampliated using either one- or two-way possibility. Two examples will suffice to show how any statement could be ampliated:

41. *A pp a B pp: A* two-way possibly applies to everything to which *B* two-way possibly applies
42. *A pp a B p: A* two-way possibly applies to everything to which *B* one-way possibly applies

Only by using one-way ampliation (as in 42) do we literally obtain "ampliation," i.e., an extension that is at least as large as, and typically larger than, that of the given subject term. But Aristotle has no word for ampliation, nor does he make it clear whether he intends one- or two-way ampliation.

VI. Syllogistic figures and moods

Aristotle's syllogisms fall into one or another "figure" (*schēma*), where a figure is defined by the disposition of its terms. Let $P$ = major term (the predicate of the conclusion), $S$ = minor term (the subject of the conclusion), and $M$ = middle term. Then the first three figures are as follows:

$$\begin{array}{ccc} \text{I} & \text{II} & \text{III} \\ P\,M & M\,P & P\,M \\ \underline{M\,S} & \underline{M\,S} & \underline{S\,M} \\ P\,S & P\,S & P\,S \end{array}$$

(The hoary question of the "fourth figure" has nothing to do with modal logic in particular and will not concern us here.) The medievals supplied names for the valid syllogisms, with each name containing three vowels signifying the quantity and quality of a given syllogism's constituent statements. For example, because Barbara belongs to the first figure, and contains three occurrences of *a*, it will go like this:

$$\begin{array}{c} P\ a\,M \\ \underline{M\,a\,S} \\ P\ a\,S \end{array}$$

$$\begin{array}{cc} \text{Second-figure Cesare} & \text{Third-figure Ferison} \\ M\,e\,P & P\,e\,M \\ \underline{M\,a\,S} & \underline{S\,i\,M} \\ P\ e\,S & P\,o\,S \end{array}$$

The "mood" of a syllogism is determined by the type (*A*, *E*, *I*, or *O*) of its three constituent statements. Given its figure and mood, one can construct any syllogism in the system. To remember which names go with which figures, one used to memorize the following bit of doggerel:

Barbara, Celarent, Darii, Ferio *que prioris*
Cesare, Camestres, Festino, Baroco *secundae*
*Tertia,* Darapti, Disamis, Datisi, Felapton, Ferison, Bocardo . . .

Extrapolating to modal syllogisms, we have, for example, Barbara *A, pp/p:*

$$\begin{array}{c} P\ pp\ a\,M \\ \underline{M\,pp\ a\,S} \\ P\ p\ \ a\,S \end{array}$$

Barbara *pp, pp/pp* (with two-way ampliation in the major premise):

$$\frac{\begin{array}{l} P\ pp\ a\ M\ pp \\ M\ pp\ a\ S \end{array}}{P\ pp\ a\ S}$$

Finally, the initial letter of the name of each incomplete or imperfect syllogism indicated the perfect syllogism (if any) by which it was to be reduced, either through *reductio* or conversion. Cesare and Camestres reduced to Celarent, Baroco and Bocardo were validated via *reductio* using Barbara, and so on.

I abbreviate the modal syllogisms as 'Barbara *pp, A/pp*' (contingent major premise [with lower case '*pp*' indicating the "term-thing" reading, IVB above], assertoric minor, contingent conclusion) and the like, but sometimes omit the comma and slash (as in 'Barbara *NAN*') where the result is still clear.

# Notes

## CHAPTER 1

1. See, for example, Saul Kripke on the origins of living things (e.g., a human being's having *these* particular parents) as essential to them, or Hilary Putnam on basic physical makeup, as described by contemporary physical sciences, as essential to physical objects: Saul Kripke, "Naming and Necessity," in *Semantics of Natural Languages*, ed. G. Harman and D. Davidson (Dordrecht: Reidel, 1972), pp. 253–355, with appendix, pp. 763–9; Hilary Putnam, "The Meaning of 'Meaning'," in *Minnesota Studies in the Philosophy of Science VII*, ed. K. Gunderson (Minneapolis: University of Minnesota Press, 1975). A large literature has now grown up around these issues.
2. Kripke again: "Semantical Considerations on Modal Logic." *Acta Philosophica Fennica* 16(1963):83–94; reprinted in *Reference and Modality*, ed. L. Linsky (Oxford University Press, 1971).
3. Jan Łukasiewicz's sophisticated and systematic treatment in *Aristotle's Syllogistic*, 2nd ed. (Oxford University Press, 1957), is a landmark in this regard.
4. Günther Patzig, *Aristotle's Theory of the Syllogism*, trans. J. Barnes (Dordrecht: Reidel, 1968), pp. xvi–xvii, 194f.
5. I shall use "syllogism" as a translation of *syllogismos*, even though Aristotle may intend to give that term (in his definition at 24b18–22) a broader sense than that requiring exactly two premises and three terms arranged in one of a small number of "figures" (*schēmata*). I shall not be worried here about exactly how much he has built into the definition of *syllogismos*, since the issue does not concern Aristotle's modal logic in particular. The reader should consult especially John Corcoran, "Aristotle's Natural Deduction System," in *Ancient Logic and Its Modern Interpretations*, ed. John Corcoran, pp. 85–131, for a defense of "deduction" as a translation of *syllogismos* and a view of "syllogisms" as deductive structures. [See also Timothy Smiley, "What Is a Syllogism?" *Journal of Philosophical Logic* 2(1973):136–54; Robin Smith, *Aristotle, Prior Analytics*, translated, with introduction, notes, and commentary (Indianapolis: Hackett, 1989) (hereafter

cited as *Notes*), p. 106 (on 24a12) and esp. pp. 109f. (on 24b18–22); and Jonathan Barnes, "Proof and the Syllogism," in *Aristotle on Science: The Posterior Analytics*, ed. E. Berti (Padova: Antenore, 1981), pp. 17–59.] For defense of a narrower construal, see Michael Frede, "Stoic vs. Aristotelian Syllogistic," *Archiv für Geschichte der Philosophie* 56 (1974): 1–32. This issue is related to the view of "incomplete" syllogisms – as opposed to "complete" (*teleios*) ones – as deductive structures requiring certain steps (conversions of premises or conclusion, use of *reductio ad impossibile*) to make them into valid deductions. (For discussion, see the works by Corcoran, Smiley, and Smith just cited.)

6. Thus Martha and William Kneale, in *The Development of Logic* (Oxford University Press, 1962), p. 85, are right to suggest that "Aristotle was probably determined by metaphysical considerations to make contingency, rather than possibility, the leading notion in his theory of problematic syllogisms; for in his metaphysics the distinction between the necessary and the impossible on the one hand and the merely factual on the other is of fundamental importance." They are unduly alarmed, however, by the fact (as they see it) that a problematic statement will be a "disguised conjunctive statement" rather than a simple one. We shall see that this is not quite accurate and that there is no insuperable obstacle to the inclusion of such propositions in syllogisms, or to Aristotle's ability to test such syllogisms for validity.

There are at least three substantial problems, however, with fitting scientific demonstrations into the general system Aristotle produces in the *Prior Analytics:* (1) Scientific demonstrations, and the need to require a *per se* link – as in *Posterior Analytics* (*Post. An.*) A.4 – between terms, rather than just propositions of necessity, are discussed in Chapter 3. (2) Syllogisms involving "by nature" or "for the most part" (*hōs epi to polu*) propositions, classed as two-way possibility statements in *Pr. An.* A.13, are taken up in Chapter 6, along with the views of Gisela Striker, Jonathan Barnes, and others on how Aristotle might have thought such propositions could be strengthened, as they must be, for scientific service. The *Posterior Analytics* may address this problem, too, by associating such propositions with *per se* connections of the fourth type listed in A.4 (73b11–16). (3) The *Posterior Analytics* seems to recognize that some scientific propositions will involve terms standing for items that are not of a sort to be predicated of one another (point and line, for example), whereas all the syllogistic propositions of the *Prior Analytics* involve predicative relations between terms. This recognition seems to be reflected not only in the discussion of *per se* connections, but also, as I suggest in Chapter 2, note 11, in the definition of universal quantification at A.4, 73b25ff.

All three points suggest that the *Posterior Analytics* was written later than the modal chapters of the *Prior Analytics*, for Aristotle would hardly have offered those chapters as the basis of scientific demonstration (and the first sentence of the work seems to say that he will be offering such a basis in the *Prior Analytics*) if he had already come to a realization of points (1) and (3) [and perhaps (2)] listed above.

7. To cite only one example that will not be discussed in what follows, but that shows a clear connection, the proof in *Post. An.* A.19 that demonstrations must be finite [on which, see Jonathan Lear's excellent discussion, *Aristotle and Logical Theory* (Cambridge University Press, 1980), ch. 2] rests on the assumption that chains of predication are finite – an assumption that *Post. An.* 22 defends by appeal, in Lear's words, "to a structure implicit in nature." Lear is in basic agreement with D. W. Hamlyn ["Aristotle on Predication." *Phronesis* 6(1961):110–26], J. Barnes [*Aristotle's Posterior Analytics* (Oxford University Press, 1975), pp. 166–73], R. Demos ["The Structure of Substance According to Aristotle." *Philosophy and Phenomenological Research* 5(1944–5):255–68], and D. Ross [*Aristotle, Prior and Posterior Analytics: A Revised Text with Introduction and Commentary* (Oxford University Press, 1949), pp. 573–83] in saying that "chains of predication are not abstract mathematical entities; they reflect an order possessed by a subject and its predicate. This order is reflected by the structure of a proof and restricts the proof to finite length. A study of nature can therefore reveal an important property of proofs" (*Logical Theory*, p. 30). (For the key role of epistemology in Aristotle's argument, see 82b37–83a1.)
8. "Necessity proposition" is not an attractive phrase, and I shall occasionally substitute the still not altogether happy "proposition of necessity." For the most part I avoid "necessary proposition" as a general term for Aristotle's proposition of necessity because that phrase is now standardly used of necessary truths, to express what is called *de dicto* necessity (which will be discussed later).
9. Note that '*A N e B*' does *not* mean 'Necessarily-applies is a relation that relates *A* to none of the *B*'s': That would be equivalent to 'possibly-does-not-apply relates *A* to each *B*'. Three additional ways (besides the one given above) of saying what it *does* mean would be '*Possibly-applies* is a relation that relates *A* to no *B*'s', '*Necessarily-does-not-apply* relates *A* to all of the *B*'s' and '*Necessarily-applies-to-none-of* relates *A* to *B*'. For fuller discussion of Aristotle's copulae, and of why the last alternative should be preferred, see Chapter 2, Section 2.1.
10. Fate, too, had a hand in this, as the history of the loss and recent recovery of Stoic logic will testify.
11. See the quotation from al-Farabi in Nicholas Rescher, "Aristotle's Theory of Modal Syllogisms and Its Interpretation," in *The Critical Approach to Science and Philosophy: Essays in Honor of Karl Popper*, ed. M. Bunge (New York: Collier-Macmillan, 1964), pp. 153f.
12. Jan Łukasiewicz, *Aristotle's Syllogistic*, p. 133. For other choice polemical tidbits, see the passages quoted (but not endorsed) by Storrs McCall, *Aris-*

*totle's Modal Syllogisms* (Amsterdam: North Holland, 1963), pp. 2–4. Nicholas White, though not expressing himself in quite so extreme a fashion, alleges a widespread confusion on Aristotle's part concerning the placement of modal operators – confusion that "seems to plague Aristotle's modal syllogistic in a thoroughgoing way" ["Origins of Aristotle's Essentialism." *Review of Metaphysics* 26(1972–3):57–85, quotation on 60f.] Patzig laments that Aristotle's modal logic "is still a realm of darkness" (*op. cit.* p. 86, n. 21).

13. Albrecht Becker, *Die aristotelische Theorie der Möglichkeitsschlüsse: Eine logisch-philologische Untersuchung der Kapitel 13–22 von Aristotle's Analytica Priora I* (Berlin: Junker & Dünnhaupt, 1933); see, e.g., pp. 39–42, 64 (the distinction comes up in many places). Becker himself (p. 2) seems to have seen his main contribution as the application of a Jaegerean developmental approach to the text of *Pr. An.* A.3, 8–22. Thus he attempts to resolve many issues through postulation of textual additions by a "later hand," or later additions by Aristotle himself. Becker's remarks along these lines are consistently interesting, even if he is sometimes too quick to revise the text. All his textual decisions still result, however, in an Aristotle who unwittingly alternates between two conceptions of modality.
14. See William Kneale, "Modality *de re* and *de dicto*" in *Logic, Methodology, and Philosophy of Science*, ed. E. Nagel, P. Suppes, and A. Tarski (Stanford University Press, 1962), p. 624 (hereafter cited as "Modality").
15. I shall use "statement" and "proposition" interchangeably, except where context requires that a distinction be observed.
16. One could, of course, understand a modal operator as applying to a whole sentence in a *de re* manner, as asserting that some *fact* obtains necessarily. But because Aristotle does not appear to recognize any such *res* as a *fact* to which modality might apply (for our purposes, his ground-level entities are things falling into one of the *Categories'* ten kinds of things that there are), I leave that possibility aside here.
17. For useful discussion of several characterizations of the *de re/de dicto* distinction, see R. Sorabji, *Necessity, Cause, and Blame: Perspectives on Aristotle's Theory* (Ithaca, NY: Cornell University Press, 1980), ch. 12. G. Forbes [*The Metaphysics of Modality* (Oxford University Press, 1985)] suggests a further criterion that will become relevant in Chapter 6.
18. "Conversion" here refers to the interchange, or turning about (*antistrophē*), of the two terms of a categorical proposition. Thus the converse of 'A applies to some *B*' is '*B* applies to some *A*'. The universal affirmative '*A* all *B*' entails not '*B* all *A*' but only '*B* some *A*', and so it is said to convert to a particular (*Pr. An.* A.2, 25a7–9). '*A e B*' entails its converse, '*B e A*'; the particular negative '*A o B*' does not entail '*B o A*' (or any other '*B* – *A*' proposition) and so is said "not to convert."
19. Such syllogisms are those needing nothing beyond the premises laid down to make manifest the necessity of the conclusion's following from the prem-

ises. I shall translate *teleios* indifferently as "complete" or "perfect," because one standard meaning of "perfect" (as a translation of *teleios*) is simply "not missing any parts." What will matter for present purposes is Aristotle's definition of completeness and, above all, the way in which he demonstrates the completeness or perfection of his perfect syllogisms. Chapter 7 investigates Aristotle's criterion of perfection, especially as applied to modal syllogisms.

20. As Michael Ferejohn points out, Aristotle validates the "mixed" moods Cesare, Camestres, Festino, and Felapton *NAN* by converting necessity premises – which requires a *de dicto* reading – to effect a reduction to certain perfect mixed first-figure moods – whose validity depends on a *de re* reading. In these cases the problem is especially acute, since within a single argument he must switch from one sort of necessity to another ("Essentialism in Aristotle's Organon." Ph.D. dissertation, University of California at Irvine, 1976, esp. pp. 219ff.).

21. Again, the best-known modern treatment in this vein is that of Albrecht Becker (see note 13); for an especially pointed formulation of the alternatives, see pp. 39f. I. M. Bochenski's treatment in *Ancient Formal Logic* (Amsterdam: North Holland, 1963) agrees with Becker on most essentials. Peter Geach works out this sort of interpretation (alternation between *de re* and *de dicto* modalities) for the chapters on necessity syllogisms and mixed assertoric-necessity syllogisms, which Becker had not treated thoroughly, in his unpublished (and unfinished) commentary on *Pr. An.* A. See also R. Sorabji, *Necessity, Cause, and Blame*, p. 202, and J. Hintikka, "On Aristotle's Modal Syllogistic," ch. 7 of *Time and Necessity: Studies in Aristotle's Theory of Modality* (Oxford: Clarendon Press, 1973). William Kneale points out that this approach can be traced back through several centuries of medieval logic ("Modality," pp. 625f.).

22. For the hyphenated modal predicate, see W. Kneale, "Modality," p. 623, and R. Sorabji, *Necessity, Cause, and Blame*, p. 187. Sorabji, like Kneale, believes (p. 202) this conception necessary for making sense of the "mixed" assertoric-necessity syllogism just formulated. On the modal predicate approach, see also David Wiggins, "The *de re* 'Must': A Note on the Logical Form of Essentialist Claims," in *Truth and Meaning*, ed. G. Evans and J. McDowell (Oxford University Press, 1976); Peter Geach, unpublished commentary on *Pr. An.* A; D. Ross, *Commentary;* Günther Patzig, *Aristotle's Theory*, pp. 61ff.; Robin Smith, *Notes*, p. xxvii.

Non-historical discussions of the *de re/de dicto* distinction normally formalize these notions using modern modal predicate logic. A great many formulae can get into the act, but let two suffice to represent *de re* and *de dicto* versions of Aristotle's universal affirmative proposition of necessity:

*de re:* $(x)(Bx \rightarrow \Box Ax)$
*de dicto:* $\Box(x)(Bx \rightarrow Ax)$

In discussions of Aristotle, and where there is concern for investigating how, in detail, Aristotle conceived of his modal principles and proofs, most commentators cast the *de re* version, at least, in categorical terms, using (hyphenated) modal predicates. (Virtually none attempt to give a categorical version of the *de dicto* reading; those of which I am aware use two modal predicates. This sort of approach is discussed in Chapter 6.) For this reason, most of the discussion here of the *de re/de dicto* distinction will be based on the now widespread modal predicate representation of *de re* modality – and, of course, on more informal characterizations of the intuitive content of the distinction. Chapter 6 will, however, examine one quite sophisticated and textually sensitive approach that abandons the categorical framework altogether, yet tries to capture the details of Aristotle's thought in terms of modal predicate logic.

23. It should be added that still other major figures (e.g., Peter of Spain, Ockham, Scotus) took the *de dicto* reading as fundamental. See Ernest Moody's summary remarks in "Medieval Logic," in *Encyclopedia of Philosophy*, vol. 4, ed. Paul Edwards (New York: Macmillan, 1967), p. 533. If memory serves, I was first made aware of the venerable *cop* approach by Charles Kahn during the course of a seminar taught by Peter Geach on *Prior Analytics* A at the University of Pennsylvania in 1973.
24. One finds this sort of shift to *de dicto* modality in I. Bochenski, "Notes historiques sur les propositiones modales." *Revue des Sciences Philosophiques et Theologiques* 26(1937):675. For an intriguing suggestion as to how Aristotle, following common Greek usage, might have thought of the copula either as standing before the two terms of a proposition or as coming between them, see Robin Smith, *Notes*, pp. 108f., on 24b16–18. The passage is discussed here in Chapter 2, Section 2.1.
25. See, e.g., W. Kneale's (apparently intentional) shift from describing Abelard's view in terms of "modification or qualification of the propositional link between things" ("Modality," p. 624) to a characterization of that same sort of statement as one in which "the modal word belongs in effect to the predicate and may therefore be said to express a mode or manner in which the subject is characterized."
26. Most representations in terms of contemporary modal predicate logic are not well suited for that task either, despite their considerable interest in other respects. More will be said about that in Chapter 6, but certainly we shall not proceed here by recasting Aristotle's propositions as statements of modal predicate logic and then carrying out demonstrations of validity within, say, S4 or S5. Readers who are comfortable only within an established formal framework may be uncomfortable reading this book, which will, in effect, fashion the constituents of such a system step-by-step.
27. It should be stressed that the modal predicate approach represents only one variety of *de re* modality and that a proposition with a modal copula could be read *de re* – i.e., as saying *of some thing* that some attribute applies to it

in a specified way – without absorbing modality into the predicate proper. Although I would have no objection to classifying weak *cop* necessity propositions as *de re* ones in a broad sense, there are still reasons (having to do with the further analysis of weak *cop*, as developed in Chapter 2) to distinguish even weak *cop* necessity from most common understandings of *de re* modality. As remarked earlier, the modal predicate version of *de re* modality will be important in this study not because it is the only categorical version of *de re* modality (for it clearly is not) but because among categorical formulations it is by now dominant in the secondary literature. (See note 22.)

28. On term conversion, see note 18. In qualitative conversion (valid only for two-way possibility propositions), the quality of a proposition is switched from affirmative to negative, or vice versa. For example, the qualitative converse of '*A PP i B*' is '*A PP o B*'. (Ross calls this "complementary" conversion, for reasons that are not entirely clear to me. I shall call it "qualitative" conversion, because it consists in switching the apparent quality of a proposition.) Both sorts of conversion – but especially the first – are important, in ways to be discussed at length later, for "reduction" of imperfect to perfect syllogisms.

29. This ageless pair are still teasing logical intuitions. Barbara *NAN* asserts that if *A* necessarily applies to all *B*, and *B* applies to all *C*, then *A* necessarily applies to all *C;* Barbara *ANN* asserts that if *A* applies to all *B*, and *B* necessarily applies to all *C*, then *A* necessarily applies to all *C*. Are both valid (as Lukasiewicz argues)? Both invalid (Theophrastus)? One valid, one invalid (Aristotle)? What difference does it make how their constituent propositions are read? Chapter 4 addresses these questions, using them as an entry to discussion of several competing approaches to Aristotle's modal syllogistic in general.

30. With ampliation, a proposition no longer reads, say, '*A* two-way possibly applies to all *things to which B applies*', but '*A* possibly applies to all *things to which B possibly applies*'. The function and technical implementation of this device raise a large number of interesting problems. Aristotle introduces the concept only in connection with two-way possibility propositions, but we shall consider its effects in certain other cases.

## CHAPTER 2

1. Aristotle's usual term for both sorts of possibility is *endechetai*, with *dunatai* occurring occasionally. Usually the context makes it clear which sort (one- or two-way) is intended, but we shall encounter a few cases in which this is not so. In translating, I have, where Aristotle's meaning is clear, sometimes added "one-way" or "two-way" in diamond brackets; where there is any question about which is meant, I have simply translated "possible."

   Incidentally, the nearly ubiquitous term *endechetai* has to be translated in general as "it is possible" rather than as "it is contingent," for the latter translation would directly and unjustifiably introduce many falsehoods and contradictions into Aristotle's text. (For example, wherever he asserts or presupposes that necessity entails possibility, or that the possible includes the necessary, etc.: Substituting "contingency" for "possibility" makes all such statements false and places them in contradiction to other passages.)

With regard to *huparchein*, it should be noted that Aristotle sometimes uses this and its modal variants metalogically, simply to indicate the modality of a sentence: A sentence *tou huparchein* or *tou huparchontōn* or *en tōi huparchein* is an assertoric proposition (at least); one *tou anangkaiou* or *tōn anangkaiōn* is a necessity proposition (see, e.g., A.3, 25a27; A.12, 32a6ff.).

2. Ackrill translation. See his notes for a variety of difficulties about the interpretation of these passages from *De Interpretatione*, in Ackrill, *Aristotle's Categories and De Interpretatione* (Oxford: Clarendon Press, 1963). I have restricted myself to relatively secure claims that do not, in any case, rest essentially on the evidence of *De Interpretatione*.
3. Peter Geach has emphasized this to me in correspondence.
4. We can postpone to "some other time," as Socrates might say, a full discussion of this issue, because none of the main theses of this study depend on the notion that Aristotle's syllogistic propositions all consist of two unadorned terms plus a complex copula indicating both quality and quantity of application. One could, for example, regard quantification as an independent component of the propositional frame, indicating to how much of the subject (some, all, none, not all) the predicate applies or necessarily applies, etc. Still, the question is worth pursuing here, because, as we shall see in a moment, it does have implications for the reading of modal propositions in particular.
5. Aristotle does not here appeal to the later "traditional" concept of subalternation, on which each universal proposition entails the corresponding particular one (*A* entails *I; E* entails *O*). Still, this conversion *kata meros* of *A* propositions shows their existential import: Neither term of a true *A* proposition is empty.
6. Aristotle's system does not contain the traditional operations of obversion (wherein one changes the quality of the proposition and of the predicate term proper; *A, E, I*, and *O* statements are all equivalent to their obverses) or contraposition (wherein one switches the positions of the terms, as in conversion, and also changes the quality of both terms; *A* and *O* are necessarily equivalent to their contrapositives, but *E* and *I* are not). This is probably due to the fact that Aristotle does not here take up the possible use of negative terms. Using conversion plus those further operations, one could reduce *all* the valid plain non-first-figure moods. But without negative terms – hence without obversion or contraposition – Aristotle must use either *reductio ad impossibile* or *ekthesis* to validate Baroco and Bocardo. [Aristotle does else-

where recognize negative terms, both in *De Int.* (esp. ch. 10) and in *Pr. An.* A.46.]

7. Becker sees that this rules out a *de dicto* reading of the conversion – because on that reading it is plainly invalid (*Möglichkeitsschlüsse*, p. 16) – just as he sees that the other two-way possibility conversions, along with Aristotle's necessity and one-way possibility conversions, work if read *de dicto*, but not if read *de re* (pp. 41, 63).
8. Aristotle had in fact warned us earlier about this anomaly (25b16–18).
9. The circularity was noted by Ross (*Commentary*, p. 295), Becker (*Möglichkeitsschlüsse*, p. 90), Wieland [''Die aristotelischen Theorie der Konversion von Modalaussagen.'' *Phronesis* 25(1980):109–16] (hereafter, ''Conversion''), and others. Becker wished to excise lines 25a29–34, in part because of the circularity. Ross found it plausible that Aristotle might have written the entire passage. A second serious problem is that, as both Ross and Becker noticed, Aristotle's *reductio* proof at 25a40–b3 appears to use '*A N e B*' as the (unique) contradictory of *both* '*A P i B*' and '*A PP i B*'. (The *en hapasin* of a40 indicates that he has in mind both sorts of possibility.) But he shows clearly in ch. A.17 and elsewhere a realization that the contradictory of '*A PP i B*' is in effect '*A N a B* or *A N e B*'. Becker's excision is intended to remove this blemish as well (p. 90).
10. Jacques Brunschwig, *Aristotle Topiques I. Livres* I–IV (Paris: Budé, 1967).
11. While this emphasizes one basic strand of continuity between the *Topics* and the *Prior Analytics* (and common also to the *Posterior Analytics* and the *Categories*), I do not wish to suggest that the former's treatment of certain formal points is on the same level as that of the latter. Nor, as Robin Smith points out, do *protasis* and *problēma* have quite the same use in the *Topics* as in the *Prior Analytics*. In the *Topics*, both ''carry with them more of a suggestion of argumentative role'' (*Notes*, p. 148, on 42b27). One party to a dialectical encounter presents a *protasis* as a *question* (''proposed in the manner of a contradiction,'' and so admitting a yes-or-no answer) in the course of an argument; the demonstrator then takes a ''premise'' as the basis for a proof. (This last step does, however, provide a strong link to the logical concept of a premise.)

    Meanwhile, a *problēma* in the *Topics* is ''the proposition under discussion (which one party to the debate undertakes to defend and the other to attack)'' (*Notes*, p. 148, on 42b27). In the *Prior Analytics*, a *problēma* is one of the four basic types of categorical sentences and may well, because the term usually refers to a type of statement to be proved, have the primary connotation of a type of categorical proposition to be proved. So although the terminological fit is not perfect, it is close, and the differences may well be due to the *Prior Analytics'* broader view in which *protaseis* and *problēmeta* play not only dialectical roles but also scientific and perhaps other roles. [For a more fully blown ''genetic'' approach to these and related points, see

Jonathan Barnes, "Proof and the Syllogism," in *Aristotle on Science: The Posterior Analytics*, ed. E. Berti (Padova: 1981), pp. 17–59.]

Nor do I mean to say that Aristotle is, in the *Topics*, intending directly to give a semantics for the system of the *Prior Analytics*, for in that case (as Robin Smith points out in correspondence) he presumably would have followed the latter's treatment of, say, quantification and propositional negation. My suggestion is just that the subject and predicate terms of the "premises" and "problems" constituting the arguments of both works are related as genus to species, species to accident of that species, etc., and that these "predicables" provide the semantic background for the syllogistic system of the *Prior Analytics*.

12. On the distinction between the *Topics'* list of ten "categories" of predicates and the *Categories'* list of ten ultimate kinds of being, see M. Frede, "Categories in Aristotle," in *Essays in Ancient Philosophy* (Minneapolis: University of Minnesota Press, 1987), pp. 29–48. Cf. Aryeh Kosman, "Aristotle's First Predicament." *Review of Metaphysics* 20 (1967):483–506.

13. Aristotle assumes in the *Analytics* that essentially applying entails necessarily applying. Elsewhere he seems to recognize that not all things sharing some definable essence need actually possess all the essential properties of that natural kind (e.g., sheep with three legs). This need not affect our treatment of the logical system of the *Analytics*, but raises some interesting questions about the relation of scientific demonstration to his modal syllogistic (see Chapter 6).

14. Barnes's translation [*Aristotle's Posterior Analytics* (Oxford University, Press, 1975)]. The *Posterior Analytics'* examples of point and line, line and triangle, suggest that Aristotelian science, at least in the case of mathematics, will include propositions in which neither term is predicated of some/all of the other, hence in which the terms are necessarily related, but not as genus/species or differentia/species, etc. For example, certain geometric assertions relating points and lines will not even assert, whether truly or falsely, either that some/all points are lines or vice versa. [The relation between point and line may still be one of "belonging to" or "pertaining to" (*huparchein*) in some broad sense: see note 16 on *Pr. An.* A. 36.] There are no such examples in the *Prior Analytics'* development of the logical system, and insofar as Aristotle indicates in that work the intended application of his formal logic, he always has in mind the four predicables, and propositions in which some predicate is predicated of some/all of some subjects. So it may be that the *Post. An.* shows awareness of one way in which scientific demonstration will go beyond the strict confines of the logic of *Pr. An.* A.1–22.

This may also explain an unobtrusive but significant difference between his definitions of *kata pantos* in the two works. At *Pr. An.* A.1, 24b28–30, he defines "being predicated of all" (*kata pantos katēgoreisthai*) as meaning that there is nothing [of the subject] of which the predicate is not said (*kath hou . . . ou lechthēsetai*). By contrast, at *Post. An.* A.4, 73a28–34, *katēgoreisthai* has dropped out of the *definiendum*, so that he defines simply *kata pantos* (rather than *kata pantos katēgoreisthai*), as "what is not *epi* some but not others (*ho an ei mē epi tinos men, tinos de mē*), or at some times but not others. For example, if animal ⟨is predicated⟩ of every human, then if it is true to say that *this* is a man, it is true also ⟨to say that⟩ this is an animal, and if now the one, now the other; and similarly if point is in every line (*en pasēi grammēi*)." He seems in the *Posterior Analytics* to be conscious of defining universal quantification so as to include not only predicative relations, where one thing is "said of" (predicated of) another (*lechthēsetai; Pr. An.* A.1, 24b30), but also cases in which the one term is "in" (*en*) every instance of the other, but is not predicated of the other – as with the examples he there gives of point and line, line and triangle; *epi* seems deliberately chosen as a relatively colorless and generic term suitable for covering both sorts of cases. If so, this would call for a friendly amendment to Barnes's statement (*Aristotle's Posterior Analytics*, p. 113, note on 73a28) that "Aristotle means 'of every' to have the same sense he gave it in *APr A.1*.") It would also call for an investigation, not undertaken by Aristotle, of syllogisms asserting two non-predicative *per se* connections in the premises, and of the effects of mixing predicative and non-predicative connections in the premises.

15. If Aristotle wishes to leave open the possibility that the same property might be related accidentally to one subject but necessarily to another (e.g., as White is treated in some examples in the *Prior Analytics* with respect to humans and swans or snow, respectively), then he should be taken as defining here '*P* is an accidental property of subject *S*' rather than just '*P* is an accidental property'. As Aristotle also says, an accidental property of a subject will be something that applies to that subject, but not as a species, genus, differentia, or proprium of the subject (102b4). This is less satisfactory not only for the reason Aristotle gives (i.e., that it presupposes the notions of genus, species, etc.) but also because what he wants is something that will define, say, White as accidentally related to Socrates, whether it actually belongs to Socrates or not. (That this is what he wants is clear from his handling of examples in which it is quite irrelevant whether or not a given accident actually does apply to the subject in view.) On the other, preferred, definition (204b4–7), White would be an accident relative to Socrates just in case it might and might not actually apply to him.

There remain some stray oddities about Aristotle's usage, probably due to the fact that much of his technical terminology was just then being forged. Thus he says at *Topics* 101b23–24 that he will "follow common usage" (among Academic philosophers?) in calling the remaining sort of predicable an *idion.* Later he introduces a qualified sense of *idion* that covers a predicate that does not apply necessarily to a subject, but does temporarily belong only to that subject (as sleep may apply to a man, 102a22ff.). In the strict sense, an *idion* is *necessarily* counterpredicable with all and only that of which it is an *idion;* in the qualified sense it is an *idion* only *pros ti kai pote.*

Occasionally in the *Posterior Analytics* he appears to refer to *idia* (in the strict sense) as *sumbebēkota kath' hauta.* [I agree on this reading with M. Ferejohn, *The Origins of Aristotelian Science* (New Haven: Yale University Press, 1991), pp. 124f.] Presumably the phrase *kath' hauto* captures the necessity of the connection – due to the connection between the *idion* and the essence of that of which it is an *idion* – while *sumbebēkota* captures the fact that the *idion* is not part of the essence proper of the subject in question (so it is neither genus nor species nor differentia of it). I mention this terminological fluidity mainly to observe that it presents no obstacle, so far as I can see, to the very close association of Aristotle's concept of an accident in the *Topics* with two-way possibility as defined and illustrated in the *Prior Analytics.*

If the *Topics* sometimes seems an endless source of difficulties, I think that only one more need be noted here. The relation between genus and differentia is somewhat obscure. Although Aristotle conveniently lumps the former with the latter (101b18–24, quoted earlier) on grounds that both are "generic" or "genus-like" (*genikon*), he does not elaborate, and he does in fact (as one reader of this manuscript pointed out) note several differences between genera and differentiae (see *Topics* IV.2, 122b12–14, IV.5, 126b13, and VI.6). This is significant, but makes no difference to my present series of points, which could all be made in terms of the "five predicables" if need be.

16. One must agree with Ross (*Commentary*, p. 407) that when in *Pr. An.* A. 36 Aristotle notes that *huparchein* may stand for various relations, some of which are not predicative, and that the relation asserted in the conclusion must match that asserted in the premises, he points to a large area for potential logical theorizing that he himself did not develop. Again, all the connections defined and illustrated in *Pr. An.* A, ch. 1–22, are predicative ones. Recall, however, note 14.

17. In terms of the Aristotelian relation of "signification" (*semainein*) between a general term and the corresponding property [discussed by Michael Ferejohn in "Aristotle on Necessary Truth and Logical Priority," *American Philosophical Quarterly* 18(1981):285–93], weak $A_n$ could be formulated as "The (non-linguistic) predicable signified by 'A' applies necessarily to everything to which the predicable signified by 'B' applies (whether necessarily or otherwise)." ['Applies to' could then be further analyzed in terms supplied by the *Categories* ('inheres in' and 'is said of') or the *Posterior Analytics* ('is predicated *per se*' and 'is predicated accidentally'), etc.]

This brings out the general point that both terms in the syllogistic propositions of *Pr. An.* A.1–22 are, strictly speaking, predicates, even though one is standardly referred to in the secondary literature as the "subject" term. The subject term is just the predicate that picks out the ontological subject(s). This is made clear in Aristotle's text from time to time – for example, in his discussion of ampliation in A.13, where he says that a possibility proposition will assert, for example, either that "*A* possibly applies to everything to which *B* applies" or that "*A* possibly applies to everything to which *B* possibly applies." Similarly, one could add, the assertoric counterpart would assert "*A* applies to everything to which *B* applies," again both terms functioning as predicates. This bears on an objection (which we must leave aside on the present occasion) raised by Peter Geach to any system that purports to use exactly the same term both as predicate in one sentence and as subject in another, without change of sense: "History of the Corruptions of Logic," in *Logic Matters* (Berkeley: University of California Press, 1972), pp. 44–61.

18. The use of "cannot" in a definition of "contrariety," which is in turn essential to the definition of "necessity," will naturally suggest a question about which modality is primitive in Aristotle's system. Logically speaking, one could make either possibility or necessity primitive, defining the one in terms of the other. But Aristotle neither addresses the question directly nor adopts any uniform practice one way or the other. We have seen that he defines contingency, or two-way possibility, in terms of necessity and impossibility: It is that which "is not necessary and which, being assumed to obtain, results in nothing impossible." (*Pr. An.* A.13, 32a18–20). This suggests that one might take necessity as primitive and define the one-way possible as what is "not necessarily not." But it would be just as well to adopt a suggestion from Sarah Waterlow Broadie that we simply regard the whole family of modal concepts as being interdefinable and having no further nonmodal definitions [*Passage and Possibility: A Study of Aristotle's Modal Concepts* (Oxford University Press, 1986), p. 16]. I would add that one must, in any case, give specific *Aristotelian* meanings to these terms (whether or not one wishes to designate a single basic primitive) by reference to his essentialist metaphysics, and my suggestions will differ significantly from hers, although in a way that complements rather than competes with her definitions of temporally relativized modalities.

19. It is worth asking whether strong *cop* propositions need to be defined conjunctively or whether their definitional components will *entail* their weak *cop* counterparts. The matter is complicated by Aristotle's use of *leukos* ("white") and by some of his examples involving White as a predicate. First, he does not syntactically distinguish between the adjectival and nominative uses of such terms. As a noun, it would be truly "said of" (in the strict sense of the *Categories*) any subspecies of White Color (Colonial White, Soft White) and of any particular occurrence of White Color (if there are such in the ontology of the *Categories*) and would be part of the definition of any particular subspecies of White Color. As an adjective, it would be truly predicated rather of things colored white (Socrates or Coriscus, i.e., things in which White Color inheres). In the *Prior Analytics*, Aristotle himself *always* uses such terms as *leukos* (i.e., ones having both a nominative use and an adjectival use) adjectivally. So let us simply observe that to

include such statements as 'White $N_s$ *a* Antique White', Aristotle should distinguish 'White' as the name of a kind of color from 'white' as an adjective. This would allow him to represent the results of a science of color, and also to block such fallacies as 'Color *a* White [color] and White some Cloak, therefore Color some Cloak' (i.e., some cloak is a color).

On the nominative use, the definitional components (5)–(8) of strong *cop* necessity propositions would entail their weak *cop* counterparts (1)–(4). So, too, with any adjectives (e.g., 'rational') that always apply essentially to their subjects. But with 'Colored $N_s$ *a* White Colored' and 'Horse $N_s$ *a* Winged Horse', the weak *cop* counterparts will be dubious or worse. One could try to formulate restrictions on terms to be allowed into strong *cop* propositions; or, alternatively, one could review the evidence on what Aristotle means by a true "definition" and argue that one has *that* only where the weak *cop* analogue is true. But it seems more sensible for present purposes to simply add conditions 1–4 to 5–8 in defining strong *cop*. This having been said, there are still serious complications, to be discussed in just a moment, arising from the adjectival occurrences themselves of *leukos*.

20. Ignacio Angelelli raises this question, and then responds by saying that in general we can take Aristotle's examples seriously. He does not give any reason, however ["The Aristotelian Modal Syllogistic in Modern Modal Logic," in *Konstruktionen versus Positionen*, ed. K. Lorenz (Berlin: Walter de Gruyter, 1979), pp. 176–215, esp. p. 181]. Perhaps he would regard some examples as slips (e.g., 'White *pp a* Animal'). Certainly there are some slips in Aristotle's examples, and he himself says with regard to one of them that he will have to choose his terms better [which remark shows, as Gisela Striker has observed, "Notwendigkeit mit Lücken." *Neue Hefte für Philosophie* 24/25(1985):146–64, a concern about his examples].

21. See Jeroen van Rijen, *Aspects of Aristotle's Logic of Modality* (Dordrecht: Kluwer, 1989), pp. 133ff., for this and other quotations and for further discussion.

22. D. Balme, "Aristotle's Use of Division and Differentiae," in *Philosophical Issues in Aristotle's Biology*, ed. A. Gotthelf and J. Lennox (Cambridge University Press, 1987), pp. 69–89, esp. pp. 74f. In keeping with this response, one could alter the definition of '$A\ N_s\ i\ B$' so as to require directly a *per se* connection between $A$ and $B$:

$$A\ N_s\ i\ B \quad \text{iff} \quad A\ N_s\ a\ B \quad \text{or} \quad B\ N_s\ a\ A$$

This has some independent support in Aristotle's discussion of the first two sorts of *per se* connection in *Post. An.* A.4, where he gives as examples of the second sort, in which $B$ belongs in the account of what $A$ is, such cases as Odd and Even belonging to Number, Straight and Curved belonging to Line. Obviously '$A\ N_s\ i\ B$' holds in these cases, as well as in such cases as Number belonging to Odd, or Animal belonging to Human. Against a background, then, of relations among genus, species, etc., it is an easy move from these remarks of *Post. An.* A.4 to the definition of '$A\ N_s\ i\ B$' given just above. Let us note, however, that Aristotle seems to have a broader range of cases in mind there, including some non-predicative ones (e.g., Line belonging to Triangle, and Point to Line). Still, that broader discussion of *per se* connections can be applied to the more limited context of predicative *per se* relations among genus, species, and so on.

Besides explicitly requiring a *per se* link between *A* and *B*, this definition has the added effect of making the conversion of '$A\ N_s\ i\ B$' quite trivial. Notice that such a definition would be unjustified in the case of '$A\ N_w\ i\ B$', or '$nA\ i\ B$', or any *de re* version of the proposition. There the conversion is, as we have seen, invalid, whereas in the case of a *per se* connection we can show, on the basis of the underlying relations of genus, species, and so on – rather than by simply appealing to the commutativity of '$p \vee q$' – that the conversion holds. Indeed, on Balme's suggestion '$A\ N_s\ i\ B$' entails '$A\ N_s\ a\ B \vee B\ N_s\ a\ A$'; but '$A\ N_w\ i\ B$' clearly does not entail '$A\ N_w\ a\ B \vee B\ N_w\ a\ A$'.

23. Among these I would include Nicholas Rescher, Storrs McCall, Wolfgang Wieland, Fred Johnson, and Paul Thom; for detailed discussion of their interpretations, see Chapter 4.

24. Again, "being entailed by" being *B* is used broadly to cover not only things included in the definition of *B* (its essence in a narrow sense) but also the thing's *propria* (*idia*), which are present because of the thing's essence but are not mentioned in the thing's definition.

25. Hintikka's remark that "just because both the *de re* reading and the *de dicto* one are very natural interpretations of the verbal formula '*A* necessarily applies to all *B*' they were easily run together by Aristotle" ("On Aristotle's Modal Syllogistic," p. 145) now appears a bit too simple. Nonetheless, our discussion here of the common syntax and underlying semantic connections between strong and weak *cop* gives considerable plausibility to that general type of explanation: From Aristotle's point of view, our two modal *copula* readings are entirely natural and very closely connected.

A. Becker suggested (*Möglichkeitsschlüsse*, pp. 2, 83, 90) rather that our text of the modal chapters (3, 8–22) is sketchy or fragmentary and that with further work on the subject Aristotle would have discovered the distinction between *de dicto* and *de re* readings of necessity propositions (p. 42). Again, something like the latter clause may well be true and is entirely compatible with Hintikka's suggestion, but I cannot agree that this section of the *Prior Analytics* is "sketchy," even if it is in some respects unfinished. Generally speaking, and despite a number of rough and obscure passages, it is in one important sense "finished." In particular, it works quite methodically through a long series of permutations of various types of premises, along with accompanying arguments and issues, in accordance with a clear and comprehensive plan. If the plain syllogistic of chapters 4–7 reaches us in a more polished state than some of the later chapters, that is most likely because the plain syllogisms do not raise certain formal and philosophical difficulties that bedevil the modal syllogistic from the start. (Incidentally, this means that I can see no basis *here*, at least, for dating the modal logic late.)

26. Notice again that weak *cop* necessity, as defined here, *allows* this but does not require it. Thus 'Animal $N_w a$ Human' is true. One could define weak *cop* so as to *require* a non-essential relation between subject and predicate terms so that 'Animal $N_w a$ Human' would be false. The definitions adopted here have the important advantage of emphasizing a strong similarity or underlying kinship between the two readings of necessity propositions, which in large part explains how Aristotle failed to distinguish them clearly.

    One might try to reduce strong *cop* to weak *cop* by using two explicit modal operators in each proposition. '$A$ $N_s$ all $B$' might be replaced by '$A$ $N_w$ all $B$ $N_w$' (or 'nec: $A$ all $B$' by '$nA$ all $nB$'). Although this might give results parallel to those reached on a strong *cop* (or on a *de dicto*) reading, it does not reflect the way in which Aristotle thought of these modal propositions, nor, therefore, how he thought of them behaving within proofs. Also, it seems to miss Aristotle's intention to express, in some contexts, at least (e.g., that of scientific demonstration), a direct connection between two natures. Certainly one can show, as we saw earlier, that if $A$ and $B$ both apply necessarily to the same subject(s), then they will normally stand in some necessary relation to one another (but see p. 46 above). But again, Aristotle would prefer to say, in such situations, that $A$ necessarily applies to everything to which $B$ necessarily applies *because* of some relation between the natures $A$ and $B$: $A$ might be related to $B$ as genus to species, or species to *idion*, etc.; or $A$ is part of what it is to be a $B$, or $A$ is included in the definition of $B$, and the like. For discussion of some recent proponents of this "doubly modalized" necessity proposition, see Chapter 4, note 53.

27. Those who don't mind this can consult my "Conversion Principles and the Basis of Aristotle's Modal Logic." *History and Philosophy of Logic* 11(1990):151–72. There are some warts on that treatment, however.

28. See Aristotle's *Parts of Animals* I.2–3 and Balme, "Aristotle's Use of Division and Differentiae," esp. p. 73. From that point of view, the present diagram is simplistic. It may represent Aristotle's thinking in the *Prior Analytics*, but in any case it can illustrate the point I wish to make. (Notice also that one could represent White as (a differentia?) connecting with this tree from "outside," or use 'White-Bird'. See Section 2.8.)

29. One final comment on a curious and stubborn textual question: The passage in which Aristotle first begins to discuss possibility propositions (*Pr. An.* A.3, 25a37–b14) is, as Ross remarks, a "very difficult" one. Aristotle says, first, that possibility is said in several ways (*pollachōs legetai to endechesthai*, a37–38), for we call the necessary, the not necessary, and the potential possible (a38–39). He then maintains that, in *all* cases, the positive possibility propositions convert in the same way (as do their plain and necessity counterparts): Both universal and particular affirmatives convert to a particular

affirmative. Although he does not yet *clearly* separate one- from two-way possibility, he seems to have all positive propositions of *both* kinds in mind; later on he will appeal to all these conversions.

He then says that the negative propositions do not all convert in the usual way: That depends on how "possible" is taken. And here he seems, in lines 25b4–14, to intend, and to illustrate by example, one-way possibility (see Becker, *Möglichkeitsschlüsse*, p. 84, and Ross, *Commentary*, pp. 295f.): "But negative ⟨possibility propositions⟩ do not ⟨convert⟩ in the same way. Those which are said to be possible by virtue of necessarily (not?) applying or by virtue of not necessarily (not?) applying do ⟨convert⟩ similarly – for example, if someone should say that a man is possibly not a horse or that white possibly does not apply to any cloak: for of these, the one necessarily does not apply, the other does not necessarily apply, and the protasis converts in the same way ⟨as in the plain and necessity cases⟩."

Ross believes (p. 296), with Becker (pp. 86f.), that only by omitting the first questionable "not," and printing the second, can one give coherent sense to the passage. The manuscript evidence is roughly evenly divided on both points (as Ross remarks). But I would adopt the opposite solution. Aristotle is here talking about negative one-way propositions and, in effect, says (printing the first *mē*, and deleting the second) that there are two sorts of cases covered: (1) that in which a predicate is necessarily inapplicable to a subject (*ex anangkēs mē huparchein*); (2) that in which a predicate is not necessarily applicable to a subject (*mē ex anangkēs huparchein*). In both sorts of cases one may (rightly) say that the predicate one-way possibly fails to apply to the subject. He then illustrates both cases: Humans are necessarily not horses, hence Horse possibly fails to apply to Human; White is not necessarily applicable to any cloak, hence White possibly fails to apply to all Cloak. As he himself comments on his examples, "for of these the one necessarily does not apply, the other does not necessarily apply" (25b7–8). This is perfectly coherent. I believe that the reason Becker and Ross do not perceive this possibility is that they fail to appreciate that Aristotle is, throughout 25b3–9, describing and then illustrating cases in which *negative* one-way possibility propositions obtain. They seem to think, probably in view of the *endechesthai* at b4, that Aristotle is trying to clarify the nature of one kind of case of (positive) possibility (namely, that in which *A* necessarily applies to all *B*, or *A* is not necessarily inapplicable to all *B*). This could well lead them to suppose also that the text sometimes gets garbled because commentators are misled by the ensuing (negative) examples of possible inapplicability. But the sort of case Aristotle is clarifying in b4–5 is precisely that of possible *non-application*, so that the negative examples are directly pertinent to this point as well as to the point about non-convertibility of negative one-way possibility propositions.

I see no problem, moreover, in taking the *endechesthai* of b4 to refer to cases in which a predicate possibly does *not* apply; indeed, the context strongly supports this. Aristotle has just finished commenting on the positive one-way possibility *and* two-way possibility propositions (a39–40). Now he turns to negative propositions (b3); among these, he says, are those that are said to be possible "because of [some predicate's] necessarily not applying. . . ." Here the "possible" propositions are clearly *negative* one-way propositions.

## CHAPTER 3

1. I agree with Robin Smith that there is no need to think of that which is set out as a sensible particular (*Notes*, p. 121, on 30a12–13). But I also do not see why (*pace* Smith) *hoper* need have here its admittedly frequent sense of indicating the essence of a thing. The thing set out need not "be essentially a so-and-so" (which would limit the applicability of *ekthesis* in an unnecessary and lamentable way); rather, the *hoper* underlines the fact that the thing set out was "by hypothesis" *taken precisely to be* some part of *C*. Hence *hoper* has here its usual intensifying function, but not its technical Aristotelian one of indicating an essence.
2. Becker, among others, has pointed this out (*Möglichkeitsschlüsse*, pp. 64f.). Indeed, Aristotle's determination or predisposition to see things work out everywhere, as far as possible, just as in the assertoric chapters 4–7, was, according to Becker, one factor in Aristotle's failure to distinguish between two basic readings of modal propositions. If so, I believe it was a minor factor. (More important considerations having to do with the underlying unity of the two readings emerged in Chapter 2.)
3. *Reductio ad impossibile* for *any* mood having a necessity conclusion would require use of a syllogism with a one-way possibility premise, and Aristotle does not treat such moods within his system. (For a single, locally employed exception, see Chapter 6, Section 6.4.) Robin Smith objects to this sort of explanation of why Aristotle does not here use a *reductio* proof on the ground that "when Aristotle does get around to this case [A.16, 36a2–7; that of a syllogism with a possible major and a necessary minor yielding a possible conclusion] he tells us that it is a complete deduction: why then does he not appeal to that fact here?" (*Notes*, on 30a3–14). The answer is that *there* he is speaking of a syllogism with a *two*-way possibility major premise, whereas here a *reductio* argument would require a syllogism with a *one*-way possibility premise. Ross also overlooks this when he says (*Commentary*, p. 317) that the *reductio* would use one "problematic" (i.e., two-way possibility) premise and one necessity premise, and then explains Aristotle's failure to use a *reductio* here by the fact that he has "not yet examined the conditions of validity in mixed syllogisms."
4. Patzig once charged (*Aristotle's Theory*, p. 23) that Aristotle's language blurs the distinction by implying that *any* conclusion will be itself necessary given

the premises (recall the phrase, *tinōn ontōn anangkaion*). He prudently withdraws from this position in the Preface to the second edition of *Aristotle's Theory* (p. xvi).

5. The point is made by Jonathan Lear, *Aristotle and Logical Theory*, pp. 6f.
6. Celarent $N_w N_w N_w$ is the first-figure mood

| | |
|---|---|
| $A\ N_w\ e\ B$ | (*A* necessarily fails to apply to every *B*) |
| $B\ N_w\ a\ C$ | (*B* necessarily applies to every *C*) |
| $A\ N_w\ e\ C$ | (*A* necessarily fails to apply to every *C*) |

7. *Post. An.* A.15 says that the primary premises of demonstration include negative "immediate" predications. For a discussion of why negative predications should be included, see M. Ferejohn, *The Origins of Aristotelian Science*, p. 131. See also the biological demonstration using "no reptiles give milk" as a premise at *Parts of Animals* 692a10–14. (Thanks, for this example, to James Lennox by way of Ferejohn.)
8. There is one more interesting twist to the story of our second-figure pure weak *cop* syllogisms. If we here introduce *one*-way possibility ampliation, we obtain the following version of Cesare $N_w N_w N_w$:

$$\frac{\begin{array}{c} B\,N\,e\,A\,p \\ B\,N\,a\,C\,p \end{array}}{A\,N\,e\,C\,p}$$

The major premise now reads '*B* is necessarily inapplicable to everything to which *A* is one-way possibly applicable', which entails '*B* no *A p*' (rather than just '*B* no *A*'). And this in turn entails '*B p e A p*' ('*B* is possibly inapplicable to everything to which *A* is possibly applicable'). We can now combine this with '*A p* some *C p*', the negation of the conclusion to be proved, to get '*B p o C p*'. But this conclusion contradicts our original minor premise, '*B N* all *C p*'. So the (one-way) ampliated version of this mood is valid and can be validated by a *reductio* proof.

As remarked earlier, Aristotle does not introduce ampliation until chapter 13, and then only in connection with *two*-way possibility premises. There he *seems* to have in mind two-way possibility ampliation of such premises. However, the *reductio* argument just given breaks down with two-way ampliation. Moreover, as we shall see in a moment, the syllogism is in fact invalid with that sort of ampliation. Chapter 6 will say more about possible logical or philosophical motives behind the concept of ampliation. For now, notice that one-way possibility ampliation can save a number of otherwise invalid second-figure syllogisms, for Camestres, Cesare, Festino, and Baroco are also valid with one-way ampliation of the major and/or the minor term – as can be easily shown by *reductio* proofs.

With two-way ampliation, however, none is valid. Consider Camestres as representative:

$$\frac{\begin{array}{c} B\,N\,a\,A\ pp \\ B\,N\,e\,C\ pp \end{array}}{A\,N\,e\,C\ pp}$$

The problem here is similar to, but more complicated than, the one we found with completely *unampliated* premises. There our counterexample exploited a situation in which *A* and *C* themselves applied accidentally to their denotata, and were accidentally related to each other, even as their relations to *B* entailed that the essences of the *A*'s and the *C*'s were mutually incompatible. At first glance one might suppose the same counterexample would work here, because with two-way ampliation we are working from the start with *A* and *C* which are incidentally related to their denotata. But this appearance is misleading: Let *B* (middle) = Human, *A* = Awake, and *C* = White, in a situation in which everything that is two-way possibly awake is (necessarily) a raven and everything that is two-way possibly white is a rock. Then the premises are both true. [Notice that if all animals are two-way possibly awake, the major premise ($B\ N\ a\ A_{pp}$) rules out the existence of any animals except ravens. This would not be an implication of the unampliated assumption that Raven necessarily applied to all (actual) Awake.] But if Awake *can* apply only to animals, then the conclusion ($A\ N\ e\ C_{pp}$, or Awake *N e* $\text{White}_{pp}$) will be true, for the things to which *C* two-way possibly applies will then have to be non-ravens, hence non-animals, so that Awake $N\ e\ C_{pp}$ will come out true. So this attempted counterexample fails.

A certain amount of futile casting about (which exercise I leave as an option for the reader) leads eventually to the realization that we should look for *A* and *C* that apply accidentally to some of their subjects, and necessarily to others – more specifically, an *A* that is two-way possibly applicable to the *B*'s but necessarily applicable to some *C*'s, and a *C* that is necessarily applicable to the *B*'s but two-way possibly applicable to some non-*B*'s – even as *A* and *C* are related only accidentally to one another. So let *B* = Swan, *A* = Swimming, and *C* = White, where all things two-way possibly swimming are swans and all the two-way possibly white things are cloaks and sharks. (It's a small world.) Suppose, finally, that sharks are necessarily swimming (swimming is part of their essence; they die if they stop swimming; this is true of the great white shark, I believe, although not asserted by Aristotle). Given, then, that Swan is necessarily inapplicable to cloaks and sharks, the premises will be true and the conclusion false. Similar terms will show Cesare, Festino, and Baroco *NNN*, with *two*-way ampliation, invalid.

Similar but less convoluted considerations will show that the first-figure moods in Barbara, Celarent, Darii, and Ferio are valid with one-way ampliation, but invalid with two-way ampliation.

I see no tremendous advantage in pursuing the matter further. Still, this brief digression into ampliated necessity moods not only points to a very large number of unexplored modal syllogisms but also underlines the seriousness of Aristotle's neglect of the subject. After introducing ampliation in chapter 13, he fails to say not only where it should be used but also which sort of ampliation (one- or two-way) should be used in a given circumstance. Because some syllogisms work only *with* ampliation, and some without it, and, of the former, some work only with one-way ampliation, and others only with two-way ampliation, no uniform policy is possible – except, perhaps, "apply as needed."

9. On a strong reading we may also validate the mood by an *ekthesis* proof utilizing Cesare rather than Camestres, for if we pick an appropriate 'some *C*', say *D*, we have

$$B\,N_s\,a\,A$$
$$B\,N_s\,e\,D$$

Converting the minor premise and inverting the premise order gives

$$D\,N_s\,e\,B$$
$$B\,N_s\,a\,A$$

This is just Celarent $N_sN_sN_s$, which gives $D\,N_s\,e\,A$, which converts to $A\,N_s\,e\,D$, from which, because *D* is (by definition) some *C*, we get finally $A\,N_s\,o\,C$.

10. For a more general discussion of *ekthesis*, see the treatments by Patzig (*Aristotle's Theory*, pp. 156–68) and Robin Smith, "What Is Aristotelian *Ecthesis?*" *History and Philosophy of Logic* 3(1982):113–27, along with the relevant sections of the works of van Rijen, Johnson, and Thom to be discussed in Chapter 4.
11. Patzig, *Aristotle's Theory*, p. 160.
12. Alexander of Aphrodisias, *In Aristotelis Analyticorum priorum librum I commentarium*, ed. NT. Wallies, CAG II, 1 (Berlin, 1883), 99, 28–32. Patzig, *Aristotle's Theory*, p. 160.
13. For discussion of this last passage, see Allan Baeck, "Philoponus on the Fallacy of Accident." *Ancient Philosophy* 7(1987):131–46, and Smith's *Notes* on the passage.
14. As Robin Smith points out (*Notes*, p. xxv).
15. As Robin Smith has observed in another connection (*Notes*, p. 136).

## CHAPTER 4

1. This *peiorem* rule applied to quantity and quality, as well as modality (124.11–17). In the present case it is (correctly) assumed that a necessity proposition is stronger than (entails but is not entailed by) its assertoric counterpart.
2. Ross, *Commentary*, p. 43.

3. Lukasiewicz, *Aristotle's Syllogistic*, p. 184.
4. *Ibid.*, p. 186.
5. Becker, *Möglichkeitsschlüsse*, p. 40.
6. Geach, Commentary on *Prior Analytics*, 1.9.1.
7. This is the view of Becker (*Möglichkeitsschlüsse*, p. 42), Geach (Commentary on *Prior Analytics*, 1.9.1), Sorabji, W. Kneale (see Chapter 1, note 22), and others.
8. As an anonymous reader of this manuscript has pointed out, actually working through Łukasiewicz's wire model in detail raises a series of extremely messy problems. For the present occasion, we may, in the words of Mark Twain, draw over these a veil of charity.
9. For a more elaborate formulation incorporating "attaches by a string" to model "applies but does not necessarily apply," see Patterson, "The Case of the Two Barbaras: Basic Approaches to Aristotle's Modal Logic." *Oxford Studies in Ancient Philosophy* 7(1989):1–40. The central criticism there is the same as here.
10. Here the question of ampliation is again of some interest, for those second-figure weak *cop* moods that we just found invalid without ampliation turn out valid with one-way ampliation. Moreover, they can even be validated through conversion proofs of the usual Aristotelian style. So let us briefly consider the effect of ampliating the necessity premise of Cesare, first with one-way ampliation:

    Cesare *NAN*

    $$\frac{\begin{array}{c} B\ N\ e\ A\ p \\ B\ a\ C\ p \end{array}}{A\ N\ e\ C\ p}$$

    It is not immediately obvious what effect, if any, this will have on this mood's validity. If we ask first about the conversion of ampliated $E_n$ – i.e., whether or not '*A N e B p*' entails '*B N e A p*', things start to look very hopeful:

    $$A\ N\ e\ B\ p$$

    Suppose

    $$B\ p\ i\ A\ p \qquad \text{(the contradictory of '}B\ N\ e\ A\ p\text{')}$$

    It obviously follows from these two propositions that

    $$A\ N\ o\ A\ p$$

    But it is impossible that *A* necessarily fails to apply to something to which *A* possibly applies. So if '*A N e B p*' is true, '*B p i Ap*' is false, so that '*B N e A p*' is true.

    Equipped with this result, we can then convert the major premise of Cesare, which will reduce to the perfect one-way ampliated first-figure mood Celarent:

$$\frac{\begin{array}{c} A\,N\,e\,B\,p \\ B\,p\;a\,C\,p \end{array}}{A\,N\,e\,C\,p}$$

Notice that we made use of the intermodal rule that $A \rightarrow P$ (each assertoric proposition entails its one-way possibility counterpart) in order to get '*B p a Cp*' as a minor premise, rather than '*B a Cp*'. Also, we *needed* one-way rather than two-way ampliation, because the latter would have yielded (at best)

$$\frac{\begin{array}{c} A\;N\;e\;B\,pp \\ B\;a\;\;C\,pp \end{array}}{}$$

But now the minor premise will not yield '*B pp a C pp*', which would be needed to make the syllogism work: For all the premise says, *B* might apply *necessarily* to all the things to which *C* two-way possibly applies.

Thus, ampliation saves an unampliated mood that Aristotle had, at least on a weak *cop* reading (the one needed to make his first-figure moods valid), erroneously declared valid. Had he seen the invalidity of these (unampliated) moods, he then might also have realized that they could be salvaged by one-way ampliation. But even if that speculation were correct, the question would remain whether or not Aristotle would have any independent reason to one-way ampliate a necessity proposition, or any other proposition. (A possible scientific motive for one- or two-way ampliating contingent propositions is discussed later, in Chapter 6.)

Ampliation of the necessity premise of Camestres *ANN* has interestingly different effects, and ampliation of the *assertoric* premises of these moods produces further (mild) surprises. But let this suffice for an additional glimpse of logical possibilities that Aristotle has left unexplored.

11. Geach, Commentary on *Prior Analytics*, 10.3.
12. Despite the fact that Aristotle later accepts Bocardo *NAN* (using it as a reducing syllogism at 34a38–40), he actually gives a counterexample to it at 32a4–5. There he had just given counterexamples to Bocardo *ANN* (31b40–32a1) and Ferison *ANN* (32a1–4), both of which would have $O_n$ conclusions. In the counterexamples, these conclusions are both 'Awake *N o* Human', which, as Aristotle observes, is false. His terms for Bocardo *NAN*, then, are Two-footed, Moving, and Animal. Here he merely lists the terms without further comment, and he may have mistakenly thought that the conclusion would be 'Moving *N o* Animal', which would be false in the same way as the conclusions of the two previous counterexamples. But in fact no arrangement of these terms will give such a conclusion (along with true premises). So here I regard the declaration of invalidity as a minor slip.

13. Rescher, "Aristotle's Theory of Modal Syllogisms."
14. McCall, *Aristotle's Modal Syllogisms.*
15. *Ibid.*, pp. 5, 22ff.
16. *Ibid.*, p. 10. McCall, like Rescher, and like Lukasiewicz before him, uses *X* for a plain proposition, and *L* for a necessary one.
17. McCall, *Aristotle's Modal Syllogisms*, p. 10.
18. *Ibid.*, p. 11.
19. *Ibid.*
20. The basic analysis agrees with Ross, *Commentary*, p. 319.
21. Jaakko Hintikka, "An Aristotelian Dilemma." *Ajatus* 22(1959):87–92.
22. Geach, Commentary on *Prior Analytics*, 9.4.
23. Hintikka, "An Aristotelian Dilemma," p. 91.
24. Rescher, "Aristotle's Theory of Modal Syllogisms," p. 161.
25. *Ibid.*, p. 153.
26. *Ibid.*, p. 165.
27. Rescher claims that for contingent (two-way possibility) propositions there is no contradictory "in the framework." This is right insofar as Aristotle does not provide for disjunctive propositions. (or disjunctive *copulae*). Still, Aristotle, in effect, recognizes that the contradictory of '*A pp* all *B*' is '*A N* some *B* or *A N*-not some *B*'.
28. Rescher, "Aristotle's Theory of Modal Syllogisms," p. 167.
29. *Ibid.:* "The modality of the conclusion follows from that of the major premise."
30. *Ibid.*, p. 168.
31. *Ibid.*, p. 172.
32. *Ibid.*, p. 171. In Aristotle's example, the middle term should be "distant," not "twinkling thing": The stars' distance from the earth would explain their twinkling, not vice versa. (Aristotle actually states the matter in terms of the planets' nearness to the earth as an explanation of their not twinkling, at *Post. An.* A.13, 78a31–39.)
33. McCall, *Aristotle's Modal Syllogisms*, p. 25.
34. *Ibid.*
35. *Ibid.*, p. 26.
36. *Ibid.*, p. 25.
37. Rescher maintains that "by providing an axiomatization of Aristotle's modal syllogistic in the manner of Łukasiewicz's well-known axiomatization of Aristotle's assertoric syllogistic, McCall demonstrates the internal coherence and consistency of Aristotle's theory." But whether or not this is so depends on what one means by "Aristotle's theory." Certainly if one means it narrowly, as "Aristotle's formal theory," putting aside all questions about the motivation and semantic interpretation of the theory, Rescher and McCall are right. That is, if we do not ask what Aristotle *meant* by necessity and (one- and two-way) possibility, but just accept his base of complete syllogisms and conversion principles as uninterpreted formulae having this or that

structure, we can then derive other formulae in the way Aristotle indicates. [The point is noted by Paul Thom, "The Two Barbaras," *History and Philosophy of Logic* 12(1991):144; this was not made clear in my own "The Case of the Two Barbaras" article.]

38. McCall, *Aristotle's Modal Syllogisms*, p. 38.
39. *Ibid.*
40. *Ibid.*
41. van Rijen, *Aspects of Aristotle's Logic of Modalities.*
42. *Ibid.*, p. 214f.
43. *Ibid.*, p. 211.
44. *Ibid.*, p. 177.
45. *Ibid.;* fuller and more technical definitions are found on p. 179.
46. *Ibid.*, pp. 201, 215.
47. *Ibid.*, p. 201.
48. *Ibid.*, p. 208.
49. This follows van Rijen's treatment on pp. 208f.
50. van Rijen, *op. cit.*, p. 215.
51. *Ibid.*, p. 215. What he has in mind is that Aristotle's validations of these moods *via* various reductions appeal to conversion of assertoric premises. But such premises may contain heterogeneous terms. Thus, even if prior to conversion their subject terms did name the objects of discourse by reference to a non-accidental characteristic, we could not be sure that the new subject terms would do so after conversion. So we are no longer guaranteed homogeneity of the terms appearing in the premises such that they entail a conclusion whose terms must be homogeneous. On van Rijen's view, Aristotle overlooks this fact and mistakenly declares these moods valid.
52. Jonathan Barnes's translation. Barnes floats the possibility that this is a reference to Barbara *NAN*, but hesitates to affirm it definitely, preferring a reading on which both premises are non-necessary (i.e., Barbara AAN) (see Barnes, p. 125, on 74b29). Either way, the objection to van Rijen will be the same.

    [Incidentally, I do not agree with Barnes that at 75a4 ("one can deduce a necessity from a non-necessity") Aristotle has in mind Barbara *AAN*. That mood would be plainly invalid, and also in direct contradiction to Aristotle's correct claim in *Pr. An.* A.12 that derivation of a necessity conclusion requires at least one necessity premise. What Aristotle must mean here is that two plain propositions can entail, say, '*A* all *C*' (*via* pure assertoric Barbara *AAA*), where it happens that *A* does apply necessarily to all *C*. For example, 'Animal all Moving' and 'Moving all Human' entail 'Animal all Human'. The premises do not, however, entail 'Animal *N* all Human', even though that necessity proposition is in fact true. Thus, Aristotle draws the parallel to deriving a true conclusion from false premises. (We could supply the example that 'Two-legged all Cat' and 'Cat all Human' entail 'Two-legged all Human'.) The conclusion may in fact be *true*, but the argument does not *show* that it is true unless we know that the premises are *true*. Similarly, two assertoric premises can entail a proposition that is in fact necessary, but they do not *show* that it is necessary (i.e., they do not entail the *assertion that* such-and-such a connection is necessary) unless they themselves – or at least the right one of them – are stated as necessity premises.]

53. van Rijen, *op. cit.*, p. 201. Aryeh Kosman has independently hit upon a strategy similar to, but less fully worked out than, van Rijen's. [See especially Kosman, "Necessity and Explanation in Aristotle's *Analytics*," in *Biologie, Logique, et Metaphysique chez Aristote* (Paris: Editions du CNRS, 1990), pp. 349–64. Cf. Kosman, "Aristotle on Incontrovertible Modal Proposition." *Mind* NS 79(1970): 254–8.] Like van Rijen, he attempts to understand certain key aspects of the *Prior Analytics* – including the mixed Barbaras – in light of Aristotle's theory of science in the *Posterior Analytics*. But Kosman rightly refrains from meddling with assertoric propositions – including the minor premise of Barbara *NAN*. Still, he cannot keep his hands off Barbara's major premise. Kosman formulates clearly the standard problem of explaining how Aristotle could have accepted both Barbara *NAN* – which requires a *de re* reading, such as '$(x)\ (Bx \rightarrow NAx)$', "with the modal operator within the quantified proposition" ("Necessity and Explanation," p. 352) – *and* the conversion of universal affirmative propositions of necessity – which requires a "modal operator outside the entire quantified proposition" (p. 353). As a solution, he proposes that because the system of the *Prior Analytics* is constructed with scientific demonstration in mind, and because that aims at explaining a necessary connection between the terms of one's (scientific) conclusions, which in turn requires necessary connections between the middle term and each of the extremes, the necessary propositions involved should be understood not along the lines of '$(x)\ (Bx \rightarrow NAx)$' but rather in terms suggested by A. Becker, as '$(x)\ (NBx \rightarrow NAx)$'. Kosman is right, of course, that the former formulation (a counterpart to weak *cop*) does not guarantee any necessary connection between the terms (natures) *A* and *B* themselves. The latter does guarantee such a connection. (Kosman does not argue for this, but it is clear, in terms used earlier, in Chapter 3, that because *B* and *A* will both belong *kath' hauto* to the *B*'s, they will be part of a common essential tree.)

Why is this a solution? Well, if we can read the necessity propositions of the *Prior Analytics* in that way, we then have versions of universal affirmatives that do convert and that can serve as major premises and conclusions of Barbara *NAN*, without having to shift the position of the modal operator from outside the whole proposition (to accommodate conversion) to inside (to accommodate Barbara *NAN*) (see p. 353).

There are some problems, however, with this ingenious proposal. First, how does one justify reading '*A* necessarily applies to all *B*' everywhere as '$(x)\ (NBx \rightarrow NAx)$'? Kosman simply asserts that this is "the standard type of proposition that Aristotle has in mind and [it exhibits] the structure of modal necessity invoked throughout the *Analytics*'' (p. 358). But we have seen that there are concrete examples of *de re* proposition in the *Prior Analytics* that do not have this structure – ones that have concerned numerous commentators and that van Rijen, for example, has recognized and tried to counter.

Second, there is no firm basis for doubly modalized necessity propositions. Kosman cites the fact that Aristotle recognizes in *Pr. An.* A.13 doubly modalized possibility propositions. [This is the ''ampliation'' passage that yields Becker's '$(x)$ (Pos $Bx \rightarrow$ Pos $Ax$)' (Möglichkeitsschlüsse, p. 353).] But (as Hintikka notes, ''On Aristotle's Modal Syllogistic,'' p. 145) Aristotle never mentions ampliation or any other sort of double modalization in connection with necessity propositions. (In fact, he doesn't *need* it to make Barbara *NAN* valid, whereas he *does* need it to preserve Barbara *PP, PP/PP.*)

Third, reading Barbara in Kosman's way does not clearly preserve its validity: Let $A$ = Animal, $B$ = Human, and $C$ = In the Agora ($B$ middle):

> Anything that is necessarily human is necessarily an animal
> Anything that is in the Agora is human
> ———
> Anything that is necessarily in the Agora is necessarily human

There may be some deformities here (How would we represent this categorically with three terms?), but let it suffice to note that in a situation in which everything in the Agora is human, the premises will be true and the conclusion false. (I take it that no one will want to argue that Aristotle would regard the conclusion as *vacuously* true.) But Kosman has apparently not noticed these problems with his version of Barbara *NAN*.

54. Wolfgang Wieland, ''Die aristotelischen Theorie der Syllogismen mit modal gemischten Prämissen.'' *Phronesis* 20 (1975):77–92.
55. Wieland, ''Die aristotelischen Theorie der Notwendigkeitsschlüsse.'' *Phronesis* 11(1969):35–60, p. 52.
56. *Ibid.*, p. 57.
57. *Ibid.*, pp. 54ff.
58. *Ibid.*, p. 44.
59. *Ibid.*, p. 43.
60. See the last portion of ''Notwendigkeitsschlüsse.''
61. Ross, *Commentary*, p. 325.
62. *Ibid.*
63. I agree with Robin Smith, *Notes*, p. 124, about the emphasis of A.12.
64. ''Notwendigkeitsschlüsse,'' p. 49, note 20.
65. Becker, *Möglichkeitschlüsse*, p. 125.
66. *All* the concrete examples of pure assertoric syllogisms contain at least one proposition based on a necessity relation, and several (26a8–9, 11–12; 27a19–20, 22–23; 27b5–8; 28a32–35) are composed *entirely* of such propositions.
67. Wieland, ''Notwendigkeitsschlüsse,'' p. 54 (emphasis added).

68. Fred Johnson, "Models for Modal Syllogisms." *Notre Dame Journal of Formal Logic* 30(1989):271–83.
69. *Ibid.*, p. 272.
70. *Ibid.*, p. 271.
71. *Ibid.*
72. Johnson remarks that he has "so far been unable to give a satisfactory interpretation of [Aristotle's] contingency operator," p. 283; Thom does not comment in his paper on any syllogism involving contingency.
73. Johnson, "Models for Modal Syllogisms," p. 274.
74. Thom, "The Two Barbaras," p. 136.
75. Thom questions Johnson's assumption (axiom A2) that for every term there are things of which that term is essentially predicated. This is complicated, however. Terms in "categories" other than "Substance" will be part of the "what-it-is" of items in their own category; so it would seem that, for example, 'Color' is part of the essence of 'White Color', and 'White Color' of all concrete instances of white color.
76. Thom, "The Two Barbaras," p. 138.
77. *Ibid.*, p. 142.
78. *Ibid.*, p. 146, note 60.
79. *Ibid.*, p. 146, note 61.
80. *Ibid.*
81. All these objections apply to Thom's simplified semantics (p. 149), on which $I_n$ and $E_n$ are given essentially the same definitions as before.

## CHAPTER 5

1. As Wieland maintains at, e.g., "Möglichkeitsschlüsse," p. 136.
2. *Ibid.*, p. 126, note 6.
3. See Wieland, *op. cit.*, pp. 126, 136.
4. R. Smith appreciates this point: *Notes*, p. 126 (on 32b4–22).
5. Thus it is clearly correct, as Hintikka ["Aristotle's Different Possibilities," in *Time and Necessity: Studies in Aristotle's Theory of Modality* (Oxford: Clarendon Press, 1973), pp. 34f.] and Striker ("Notwendigkeit mit Lücken," p. 150) have maintained, to regard Aristotle as defining a modality that covers two sorts of cases (and claiming that certain properties apply to both cases, even if for different reasons), rather than as defining two meanings of "possible" (then claiming that each meaning is such as to imply the application of certain properties).
6. Cf. Hintikka, "Aristotle's Different Possibilities," p. 35, and J. Barnes, "Sheep Have Four Legs," in *Proceedings of the World Congress on Aristotle* (Athens: Ministry of Culture and Sciences, 1981–3), p. 117. As G. Striker points out ("Notwendigkeit mit Lücken," p. 150), Theophrastus and

Eudemus apparently rejected qualitative conversion for just this reason (Ps.-Amm. in *Pr. An.* 45.42–46.2).

7. For this reason, A. Becker doubted the authenticity of the passage linking two-way possibility to science (*Möglichkeitsschlüsse*, p. 77). The usual examples are White/Human, White/Animal, Moving/Animal, and the like (33b7, 34b11–18, 35b10, 36a30, 36b14f, *et al.*).
8. The things "entailed by" the definition of *A* will cover any propria *A* may have.
9. Sarah Waterlow Broadie, *Passage and Possibility*, esp. ch. 2.
10. Wieland, "Möglichkeitsschlüsse," p. 127.
11. This seems to me more likely than Martha and William Kneale's suggestion that he had in mind an argument involving an inference from " 'It is contingent that-*p*' entails 'It is contingent that-*q*' " to " '*p*' entails 'q'." (*The Development of Logic*, p. 87).
12. Becker, *Möglichkeitsschlüsse*, pp. 36f.
13. *Ibid.*
14. This is not a happy solution, because as it stands, *kath' hou to B huparchei to A endechetai* does look like one of the alternative readings contrasted in $c_1$–$c_5$, namely, the unampliated one. To fit into the argument of the passage precisely where it occurs, however, it should be taken as parallel to 32b32–34: "first, let us say what syllogism, and of what sort, there will be if *B* possibly applies to that to which *C* applies, and *A* ⟨possibly⟩ applies to that to which *B* ⟨possibly⟩ applies." The next sentence says that this will mean we start with pairs of premises of the same sort – i.e., both pertaining to possibility. Then comes our $e_1$, which ought to have said "But ⟨when *B* applies to that to which *C* applies, and⟩ *A* ⟨possibly⟩ applies to that to which *B* applies, one premise pertains to ⟨plain⟩ applying, the other to possibly applying." This would make perfect sense, but requires that one supply the phrase in diamond brackets.

    As for Becker's excision of 34–37 ($d_2$, $e_1$, $e_2$), one might just as well excise only $e_2$. This would still clearly make the point that here, as before, we should consider syllogisms with unmixed premise pairs before those with mixed (necessity/possibility, assertoric/possibility) pairs.
15. Hintikka argues that when Aristotle says that possibility propositions can be taken in two ways (*dichōs*), he means that they cover two sorts of cases (rather than that they have two different meanings). He concludes that 'it is possible for *A* to apply to all *B*' should in general be read as 'it is possible for *A* to belong to everything to which *B* either in fact applies or two-way possibly applies', which is equivalent to 'it is possible for *A* to belong to everything to which *B* one-way possibly applies' ("Aristotle's Different Possibilities," pp. 38f.). This is a possible reading, although I think the argument from *dichōs* is by itself quite inconclusive. Hintikka also maintains that Aristotle "never seems to use [the unampliated proposition] but only the [one-

way ampliated version] in his subsequent discussion of syllogisms from possible premises'' (p. 39). Again, this could be right, but I simply do not know how it is supposed to be demonstrated textually or in any other way. [His citation of a later passage, from *Pr. An.* A.29 (45b31–34), does not seem to me to establish the point.] Our investigation in Chapter 6 of two-way possibility syllogisms will strongly suggest that Aristotle at least sometimes had in mind ampliation using two-way possibility.

16. There is a long-standing controversy about the authenticity of 32a21–29. I did not comment on this earlier because it does not appear to have significant implications for Aristotle's main concerns in this chapter. I do believe, however, that Becker (*Möglichkeitsschlüsse*, pp. 11–13) and Ross (*Commentary*, pp. 327f.) are right to suspect the passage. Granted, it does make perfectly good sense (as Becker points out) as a comment on one-way possibility: 'not possible', 'impossible', and 'necessarily not' are ''all the same or follow from one another'' and are also the contradictories of 'possible', 'not impossible', and 'not necessarily not', all of which are the same or follow from one another (*akolouthei allēlois*, a24). But the remark about ''being the same or following from one another'' does not appear in any way to make ''clear'' (*phaneros*) either the nature of two-way possibility, whose definition had just been given (32a18–20), or the fact that ''the necessary is called possible homonymously'' (32a20–21). On the contrary, that remark is not even true of two-way possibility; nor is the two-way possible called necessary, even by homonymy.

By contrast, the last statement of the suspect passage (''the possible, then, will not be necessary, and the not necessary will be possible,'' a28–29) does make sense as a statement about two-way possibility, provided 'necessary' is taken broadly (as rightly suggested by Hintikka) to cover both necessarily applying and necessarily not applying. So this portion could be retained even if lines 21–28 were deleted. Still, what one would like to see at this point in the text is for Aristotle to point out that (1) on one use of 'possible' (one-way), the necessary will be called possible just because things that are necessary are also possible on that use of the term (the one on which 'possible', 'not impossible', and 'not necessarily not' are all the same or follow from one another, etc.), but (2) on the definition just given at 32a18–20, the necessary will not be possible, nor the possible necessary. This *would* help clarify, by contrast with one-way possibility, the use of 'possible' being defined here.

## CHAPTER 6

1. I say ''main copula'' because an ampliated proposition will, in effect, involve two modal copulae: *A possibly applies* to everything to which *B possibly applies.*
2. G. Patzig, *Aristotle's Theory*, p. 63.

3. Darii (33a23–25) and Ferio *pp, pp/pp* (a25–27) are explicitly recognized as complete, again on the basis of the "definition of possibility" (33a25). Actually, two manuscripts have "definition of possibly applying to all" (*horismon tōn kata pantos endechesthai*), which is, strictly speaking, the more correct formulation – and should be adopted – because it is the definition of 'possibly applies to all', not the narrower definition of the modality in question ('possibly applies'), that he here, as elsewhere, wishes to invoke. Aristotle sometimes puts the point in an even more abbreviated way (e.g., "obvious from the definition," 32b40), presumably with the quantificational definition in mind.

   The syllogisms considered so far all use a type II reading of contingency. This will continue to be the main focus here, because it seems to be what Aristotle has in mind in *Pr. An.* A.14–22. This is indicated by the passage on ampliation, which has implications for non-ampliated propositions as well. Specifically, the most natural reading of '*A* possibly applies to everything to which *B* possibly applies' would be that the predicable *A* is two-way possibly related not to *B* itself, but to items that might or might not be *B*'s. Similarly, the non-ampliated '*A* possibly applies to all *B*' would be read as '*A* two-way possibly applies to everything to which *B* actually applies', which would state that the predicable *A* itself is two-way possibly related not to *B* itself, but to all the actual *B*'s.

   In addition, we have seen that on the one occasion when Aristotle actually discusses a two-way conversion, he has in mind a reading of type (II) (see Chapter 5, Section 5.4). The type (I) reading will not be simply ignored, however: Its term conversion was discussed in Chapter 5, and its possible role in syllogisms will be taken up in this chapter wherever it appears to be of special interest.
4. G. Striker, "Notwendigkeit mit Lücken," esp. p. 160. See also her useful discussion of how the notion of a "necessity with gaps" can apply to both "hypothetical" and "pure" or "natural" necessity (pp. 160–2) drawing on John Cooper's discussion of the distinction between those two kinds of necessity in "Hypothetical Necessity," *Aristotle on Nature and Living Things. Philosophical and Historical Studies Presented to David Balme on His Seventieth Birthday*, ed. A. Gotthelf (Pittsburgh: Mathesis, 1985), pp. 151–67. I have not taken up the distinction here because I find no trace of it in the *Prior Analytics*, nor would it seem to have anything directly to do with the logic of necessity and possibility.
5. Striker, "Notwendigkeit mit Lücken," p. 154.
6. *Ibid.*, pp. 158f.
7. *Ibid.*, p. 160.
8. *Ibid.*, p. 159.
9. *Ibid.*, p. 163.
10. *Ibid.*, p. 153.
11. A. Becker, *Möglichkeitsschlüsse*, p. 33; G. Patzig, *Aristotle's Theory*, p. 63.

12. This approach, developed by Ulrich Nortmann, is discussed in Sections 6.6 and 6.7. Striker also mentions omnitemporality and its possible connection to necessity and to science (p. 159, note 8).
13. *The Origins of Aristotelian Science*, pp. 109ff. Thus, for example, we will have 'Dies applies by nature to all that has its throat cut'.
14. In a brief but quite useful discussion of some of the difficulties surrounding the concept of *to hōs epi to polu* (see "Sheep Have Four Legs") Jonathan Barnes notes that among the scattered passages bearing on its semantic interpretation there are some that seem to treat it as a kind of modal operator on a par with the phrase *ex anangkēs* [Barnes proposes (p. 116) *Physica* B5, 196b11; *Mem.* 2, 451b13; *Met.* E 2, 1026b27–35; *Poet.* A 2, 1357a31; *Rhet.* 7, 1450b30]. Barnes also cites passages that suggest a frequency interpretation, or a temporal one, or a combination of those two, in addition to our passages from *Pr. An.* that connect the concept with two-way possibility.

    This new "scientific" operator clearly should not be given a purely temporal or frequential sense. Among other things, this would immediately encounter a problem of diminishing frequency – a problem that would become especially acute with longer chains of syllogisms. For example, suppose that something about beards normally caused them to turn grey with old age, but that in a certain number of cases something about the beard (rather than some prior factor that caused both the whiskers and their greying) happened to thwart this normal outcome. If each of the normal processes (growing a beard and a beard turning grey) met interference only 30% of the time, it would already be true that only 49% of all male humans would eventually grow grey beards. Even without looking at longer chains, we are already out of the realm of "for the most part" connections (cf. J. Barnes, *Aristotle's Posterior Analytics*, note on 96a8, p. 229). In such cases, "applying by nature and for the most part" is not transitive. (This holds, of course, even if "applying for the most part," in a pure frequency sense, is not entirely definitive of our connective, but only – along with "by nature" – one necessary condition.)

    One might actually find that some conclusions that were less-than-for-the-most-part propositions in a purely frequential sense were still "by nature" true and hence scientific. The Aristotelian ground for this would be that in such cases (e.g., where 49% of adult male humans grow grey beards) the conclusion proper ("adult male humans by nature grow grey beards") is nonetheless true because the natural disposition toward growing a grey beard is still there; it fails to occur only because something has interfered at many points in the natural course of events. This would depend on our ability to determine the "natural" course of events – at least in *some* cases – independently of pure frequency of occurrence.

    Jonathan Barnes also observes that "according to pseudo-Ammonius, Aristotle's pupils took note of the non-[qualitative] convertibility of '$\Omega p$' ['for the most part: *p*'] with 'Ω(not-*p*)': recognising [this], they determined to abandon thesis (5) [*endechetai* that *p* if *endechetai* that not-*p* – a state of affairs is possible if its contradictory is possible], and to construct an alternative logic of *endechetai*. . . . That story implies that, in Peripatetic eyes, the primary function of problematic syllogistic was to provide an $\Omega$-logic; and that Aristotle's pupils took [the nonconvertibility of $\Omega$-propositions] as an objection . . . to their master's problematic logic. Far from abandoning the . . . interpretation [of *to hōs epi to polu* in terms of *endechetai*], they attempted to develop a logic of *endechetai* which would accommodate it.'' As Barnes also observes, although their attempt may have failed, it is an intelligible reaction to the problem raised by the nonconvertibility of $\Omega$-propositions (''Sheep Have Four Legs,'' pp. 117f.).

This seems to me a highly plausible reading of one fascinating episode in the history of Peripatetic modal logic. But from Aristotle's point of view it should be said that there are good reasons for developing a logic of two-way possibility, as that modality is defined in the *Prior Analytics*, even if it turns out that such a logic cannot accommodate the scientific concept of *to hōs epi to polu*. Consequently, a proper response would not be to abandon the use of *endechetai* that is convertible, but to enlarge the logical system by including an appropriate use (i.e., one suitable for use in science) that is not convertible – perhaps by introducing a new, suitably modalized copula, along the lines sketched just above.

15. It is essential to see that Aristotle's phrase ''that by which *B* exceeds *A*'' must be taken as ''that by which *B* exceeds *the two-way possibly A*.'' Ross (*Commentary*, p. 332) takes it that way, but without explicitly noting that Aristotle's literal statement (''that by which *B* exceeds *A*'') would not lead to a valid proof.

16. The attendant proof by counterexample is unusually complex. It does work, although there is one curious feature of the terms themselves. Making the indicated substitutions, one obtains as premises

Animal *pp* some/not-some White
White *pp* all/no Human/Cloak

The odd premise is, of course, the first: 'Animal *pp* some/not-some White.' What sort of white thing might Aristotle have in mind that is two-way possibly an animal? This *is* the question that immediately arises, because the chapter up to this point reads naturally in a straightforward ''term-thing'' manner. (The ''scare quotes'' are necessary for reasons given earlier, in Section 5.1.) *Perhaps* he is thinking of some white menses, or early fetus that might or might not become a human – or perhaps of some human body that might be a living body or only a corpse. The latter would not need to appeal to a doctrine of form as predicable of matter (whose interpretation is in itself somewhat controversial, and which *may* be a later development, if it *is* a development, in Aristotle's thought), as opposed, say, to a view on which souls may survive the bodies they inhabit. Thus, we might have here simply a familiar academic example, and not one whose truth Aristotle need himself accept, let alone stop to explain or justify on Aristotelian grounds. This is certainly true of some of his examples – e.g., number as a substance, at 27a18.

17. This analysis agrees with Robin Smith, *Notes*, p. 132.
18. M. Mignucci, *On a Controversial Demonstration of Aristotle's Modal Syllogistic: An Enquiry on Prior Analytics A.15* (Padua: Editrice Antenore, 1972).
19. *Ibid.*, p. 18.
20. A. Becker, *Möglichkeitsschlüsse*, pp. 56f.; M. Mignucci, *Demonstration*, pp. 31–6.
21. Mignucci says that Ross concedes the possibility of such a reading, but Ross does not concede this. What Ross says is that if the text had read *to A ou panti tōi B endechetai huparchein* (note the added *huparchein*), then it "might perhaps mean" what Becker (and Mignucci) want it to mean, but that as it stands, "I [Ross] do not think [it] can mean this" (*Commentary*, p. 338). In fact, even Ross's highly guarded concession seems to me overly generous.
22. Ross, *Commentary*, pp. 338ff.
23. A. Becker, *Möglichkeitsschlüsse*, p. 16.
24. M. Mignucci, *Demonstration*, p. 16.
25. Ross (who also rejects the authenticity of the passage), *Commentary*, p. 339. Ross notes, however, that both Alexander and Pacius have the passage.
26. *Ibid.*
27. E.g., G. Striker, "Notwendigkeit mit Lücken," p. 153.
28. P. Geach, Commentary on *Prior Analytics*, p. 15.7: "Aristotle is not saying here that all plain universal affirmatives must be construed omnitemporally, rather than as properly present-tensed; but only that such a construction is necessary for the validity of the syllogisms presently under consideration." Nortmann takes it to apply to all such premises in syllogisms involving two-way possibility (cf. Ross, *Commentary*, p. 340).
29. A. Becker, *Möglichkeitsschlüsse*, pp. 57ff.
30. The text has 'knowledge' (*epistemē*), but this must be read as 'knower' if the other premise, 'Knower *pp a* Man', is to be true.
31. Compare Sarah Waterlow Broadie, *Passage and Possibility.*
32. See Alexander of Aphrodisias, *On Aristotle's Prior Analytics I. 1–7*, trans. J. Barnes *et al.*, p. 79, note 157, and the reference there to Ammonius, in *Int.* 153, 13–15; 215, 11–14.
33. Geach, Commentary on *Prior Analytics*, p. 15.8. But see n. 51 below.
34. Peter Geach points this out (p. 15.7); cf. Ross (*Commentary*, p. 340) and Becker (*Möglichkeitsschlüsse*, p. 58).
35. Nortmann, "Ueber die Stärke der aristotelischen Modallogik," *Erkenntnis* 32(1990):61–82.
36. Hintikka had previously introduced tense-logical reconstructions of Aristotle's two-way possibility syllogisms ("On Aristotle's Modal Syllogistic").

He, too, relies heavily on the omnitemporality requirement of *Pr. An.* A.15. But he rightly worries that because (as he reads Aristotelian necessity) always being true implies being necessarily true, Aristotle's omnitemporally true assertoric propositions will not be distinguishable from apodeictic propositions (pp. 136, 141, 144). Nortmann will preserve a distinction, however, through the double modalization of necessity propositions. Still, I do not find grounds for making universal assertoric affirmatives into *any* variety of necessity proposition.

37. The formal proof and the semantic account of Barbara's validity just given are taken from correspondence.
38. Smith, *Notes*, p. xvif.
39. This argument, too, is taken from correspondence.
40. Nortmann must read *to A ou panti tōi B endechetai* (34a39) as 'poss: *A o B*', or 'it is possible that *A* fails to apply to some *B*'. But the Greek says that *A* does not possibly apply to all *B* – i.e., that *A* necessarily fails to apply to some *B*. Nortmann believes the Greek phrase expresses the idea that '*A o B*' is at worst false and not impossible, i.e., that it is possible. But Aristotle would express *that* using the phrase *pseudos kai ouk adunaton* (34a25–26, a27, a29, a37–38, 34b1).
41. Hintikka, "Aristotle on the Realization of Possibilities in Time," in *Time and Necessity*, p. 100.
42. See, e.g., van Rijen's discussion of the issue and of Hintikka in chapter 4 of *Aspects of Aristotle's Logic of Modalities*, pp. 59–72.
43. I. Angelelli also takes Aristotle's discussion of Barbara *A, pp/p* as a critical point of departure. He is puzzled that Aristotle does not acknowledge as valid the mood (note the assertoric conclusion)

$$\frac{\begin{array}{c} A\ a\ \ B\,p \\ B\,pp\ a\ \ C \end{array}}{A\ a\ \ \ C}$$

but instead argues "in a very roundabout way indeed for an '*A pp a C*' conclusion" ("The Aristotelian Modal Syllogistic," pp. 190f.).

He suggests that Aristotle must have had an "understanding of logical implication that was in one precise way stronger than ours" (p. 190). This Angelelli expresses in a "middle existential import" (MEI) condition: "Let *S* be a first figure form (*A* major, *B* middle, *C* minor term). We say that *A* ★ *C* (with ★ for any of: *a, e, i, o*, $N_a$, . . . , $P_a$, . . . ) satisfies the ⟩middle existential import⟨ condition (MEI) relative to *S*, iff for any *x* for which *A* ★ *C* (in conjunction with '*x* is *C*' or '*x* can be *C*', if *A* ★ *C* is universal) implies that *x* is *A* or is not *A*, *S* guarantees that *x* is *B*" (p. 190). Angelelli notes, however, that the mood in question actually violates MEI, and he responds by attributing to Aristotle two ways of establishing validity even where MEI is not met (Methods I and II, p. 192). He also tries to give MEI intuitive Aristotelian content in the form of a "middle term as cause" principle (MC, p. 198). I cannot see that the passage he quotes from Alexander really expresses such a condition; moreover (again, as Angelelli notes), it conflicts with Aristotle's acceptance of Barbara *NAN*. This calls forth further complications about the possibly non-uniform application of MC (p. 199).

The details of Angelelli's novel and sometimes exploratory paper are often of interest and will be enjoyed by all devotees of Aristotle's modal syllogistic. But I am afraid there is no support to speak of in the *Prior Analytics* for either MEI or MC (although the latter would apply in scientific contexts). Moreover, Angelelli has gotten the ball rolling by inadvertently misreading Aristotle's mood (Barbara *A, pp/*____). (The conclusion Aristotle has in mind is '*A p a C*', not '*A pp a C*'; but this is not the critical oversight.) If Aristotle had had in mind

$$\frac{\begin{array}{l} A\ a\ \ B\,p \\ B\,pp\ a\ C \end{array}}{}$$

as Angelelli supposes, it would indeed need explaining that he did not allow an assertoric '*A a C*' conclusion. But Aristotle is discussing the premise pair

$$\frac{\begin{array}{l} A\ a\ \ B \\ B\,pp\ a\ C \end{array}}{}$$

and this does not give an assertoric conclusion. With this premise pair, the *C*'s do not necessarily fall under the actual *B*'s, but only the two-way possibly *B*'s; hence the major premise will not guarantee that they fall under the actual *A*'s. Hence there is no mystery that needs to be cleared up by MEI.

44. A. Becker, *Möglichkeitsschlüsse*, p. 59.
45. W. Wieland, "Möglichkeitsschlüsse," p. 146, note 41.
46. Ross, *Commentary*, p. 341.
47. Finally, it remains to mop up the invalid moods. First, those in which the minor premise is a particular assertoric negative: $A_{pp}$, *O/*__ and $E_{pp}$, *O/*__. That these entail no conclusion is shown, of course, by the same terms that worked for $A_{pp}$, E/__ and $E_{pp}$, *E*. When the major premise is particular – regardless whether it is problematic or assertoric – no conclusion follows, whether the (universal) minor is affirmative or negative, problematic or assertoric, nor if both premises are particular, both indesignate, or one particular and one indesignate. By judicious selection of terms, one may show all these moods, and those with a particular major premise, invalid at one stroke. Aristotle gives us such terms at 35b18–19: for necessarily belonging, Animal, White, Human; for necessarily not belonging, Animal, White, Cloak (middle = White).
48. Ross, *Commentary*, p. 343.
49. The final aspect of the chapter calling for special notice is the closing proof "by terms" of the invalidity of all moods with two particular premises, or

two indesignate premises, or one of each. To show that such pairs entail no conclusion at all, Aristotle gives triples of terms purporting to show their premises consistent with both '*A N* all *C*' and '*A N*-not all *C*'. Now there are 24 first-figure moods to be considered, and Aristotle suggests a single pair of triples that, he believes, will work for all 24 moods, namely, Animal, White, and Human (to show the consistency with '*A N* all *C*', 36b14), and Animal, White, and Inanimate (to show the consistency with '*A N*-not all *C*', 36b15). "For," he explains, "animal is necessarily applicable to some White and necessarily inapplicable to some White, and so also White to some Inanimate. And similarly with respect to two-way possibility [i.e., Animal is possibly applicable to some White and possibly inapplicable to some White, etc.]. Thus the terms work in all cases" (36b15–18). (Notice, by the way, the clearly term-thing or type II use of the Animal–White relationship.) The one factor he does not mention, but which is also necessary if the same terms are to dispense with all these moods at one blow, is that White must *necessarily apply* to some human, and *necessarily not apply* to some human, and must be two-way possibly applicable to some human, and contingently inapplicable to some human. Thus, with contingent major, negative necessity minor, we have the premise pair

Animal *pp i/o* White
White *N o* Human

And with the same arrangement, except with an affirmative minor, we have

Animal *pp i/o* White
White *N i* Human

But this would be a very odd sort of example for Aristotle to have used. Not that he can't or doesn't in fact use examples that are, even from his own point of view, counterfactual; but within the *Prior Analytics* itself, White is a stock example of an *accident* of humans – of a two-way-possibly-applying characteristic. My suspicion is that Aristotle failed to mention this unusual implication of his counterexamples simply because he overlooked it. One might, of course, retain the example and ask the reader to imagine (perhaps even *per impossibile*) that White is necessarily (in)applicable to some humans, and contingently applicable to others. More plausible, to my mind, would be to acknowledge an oversight on Aristotle's part and alter the example by substituting, on good Aristotelian precedent, Bird for Human, for Aristotle often cites the case of ravens as necessarily black, swans as necessarily white, with, presumably, the whiteness of some species of birds being a contingent matter. Then this highly efficient proof of mass invalidity will work in the way Aristotle plainly had in mind.

50. Robin Smith's explanation that Aristotle would not accept second-figure $A_n$, $A_{pp}/E_{pp}$ because "he never accepts the deduction of a negative conclusion from affirmative premises" (*Notes*, p. 138) seems to me unlikely because

$E_{pp}$ is, as Aristotle has pointed out, *positive*, and equivalent to $A_{pp}$. It is important to notice that although he does rule out negative assertoric and necessity conclusions on those grounds (38b13–17), he then adds that there will be no conclusion of two-way possibly not applying, and demonstrates that claim by giving a counterexample, rather than by pointing out that both premises are affirmative (38b17–21).

51. Yet another worry – this time a false alarm – about Aristotle's presentation of certain concrete counterinstances in chapters A.19 and A.20: Jonathan Barnes, Susan Bobzien, Keven Flannery, and Katerina Ierodiakonou maintain (Alexander of Aphrodisias, *On Aristotle's Prior Analytics I.1–7*, pp. 11–14) that Alexander seriously misunderstands how counterinterpretations work, insofar as he seems to think, e.g., 'Every medicine is a science' follows from 'No line is a science' and 'No medicine is a line'. That inference is obviously invalid. [What his example actually shows is that because the two premises in question are compatible with the universal affirmative 'Every medicine is a science' (in fact, all three propositions are true), they cannot entail any negative conclusion. Thus the example shows that premises of the form 'No *B* is *C*' and 'No *A* is *B*' do not entail either 'No *A* is *C*' or 'Some *A* is not *C*'.] Barnes *et al.* go on to say (p. 12, note 75) that Aristotle seems (at 38a29–31, b18–20, 39b3–6) to commit the same error. I find it hard to believe that either Alexander or Aristotle would believe that a universal affirmative assertoric conclusion would follow from two universal negative assertoric premises. However that may be, it can be shown that Aristotle does not commit that error in any of the three passages mentioned.

It is true that each of those three passages contains a clause that, taken out of context, could be read as saying that a certain conclusion follows from the premises given. But in context, Aristotle can hardly be understood that way, because (a) in all three passages he explicitly says that no conclusion follows from the premises, and (b) in all three cases the clauses at issue can easily be read so as to make the required point in a logically straightforward and unobjectionable manner.

Consider 39b2–6: "But when both [premises] are taken as indefinite or particular, *there will not be a syllogism.* For *A* necessarily applies to *B* and necessarily fails to apply to *B*. Terms for applying, Animal, Human, White; for not applying, Horse, Human, White, middle White" (emphasis added).

When he says there will be no syllogism, Aristotle means (here as elsewhere) that no *A–B* conclusion follows from the premises. So I do not think he can be saying in the next breath that '*A* necessarily applies to *B*' (and '*A* necessarily fails to apply to *B*' as well?) follows from the premises. What he does mean is simply that with the first set of terms, and with the sorts of premises given, it will be true that the premises hold and that '*A N a B*' also holds; with the second set of terms, the premises hold and '*A N e B*' also holds. These facts rule out the possibility of the premises entailing any negative *A–B* conclusion, then any positive *A–B* conclusion. Thus this counterinstance works in the same way as Aristotle's many uncontroversially correct ones do.

The second passage is 38b17–20: "But neither will there be a (conclusion that) *B* possibly fails to apply to each *C*. For given such premises [or terms] *B* will necessarily fail to apply to *C*, for example if *A* (middle) should be taken as White, that to which *B* applies, Swan, and *C*, Human."

The premise pair in question is second-figure '*A N a B*' and '*A pp a C*'. In the lines just quoted, Aristotle rules out a conclusion of the form '*B pp e C*'. Does he do this by (incorrectly) supposing that '*B N e C*' follows from the premises (as Barnes *et al.* suggest), or by (correctly) pointing out that with premises of the given sort, one can choose concrete terms for *A*, *B*, and *C* such that the premises are true and "*B* will necessarily fail to apply to *C*." The former reading seems to me to be ruled out by a look at the context of the lines in question. In the first place, Aristotle explicitly says at 38b14 that with these premises, "there will not be a syllogism"; he then says at b14–16 that there will be no negative conclusion, whether assertoric or necessary; and he concludes his discussion of this case with the remark that with premises of the sort in question, "there is no syllogism whatsoever" (*ouk ara ginetai sullogismos holōs*, 38b22–23).

The situation with regard to 38a29–31 is similar. Here Aristotle says, at a30, *sumbainei gar to B tōi C ex anangkēs mē huparchein.* Barnes *et al.* presumably take *sumbainei* to mean "it follows that" – as indeed it does in many other contexts. But the preceding line reads, "if the terms are like this, then there will not be a syllogism" (a29). In light of this, it seems to me that what Aristotle is saying at a30 must be that when the premises are of the sort given, then with an appropriate selection of concrete terms, "it results that" – i.e., it will hold that – *B* necessarily does not apply to any *C*. (I believe I owe this suggestion to Peter Geach.) Continuing now with a30ff., let *A* be White, that to which *B* applies be Human, and that to which *C* applies be Swan. These terms will make the point Aristotle needs. [Ross's solution (*Commentary*, p. 359), to read *sumbainei* as "it sometimes happens" (rather than as "it follows") seems to me less satisfactory.]

In sum, it seems to me that Aristotle does not commit the alleged error in any of these passages and that there is no such basis for suggesting that he has misunderstood his method of counterinstances.

52. Ross, *Commentary*, p. 363.
53. *Ibid.*
54. W. Wieland, "Möglichkeitsschlüsse," p. 151.
55. *Ibid.*
56. One might then wonder, especially in view of the fact that certain other moods with one necessity and one two-way possibility premise entailed an assertoric conclusion, whether by using this weakening procedure one forfeits the chance for, say, an assertoric conclusion to these syllogisms (necessity is clearly too much to hope for). Perhaps some other type of proof – *reductio*

or *ekthesis* – might yield an assertoric conclusion. But it was just proved that even with a strong *cop* minor, no assertoric (or necessity) conclusion could be derived.

## CHAPTER 7

1. The few exceptions are discussed in Chapter 8.
2. Again, I use "complete" and "perfect" indifferently as translations of *teleios.* John Corcoran and Timothy Smiley strongly prefer "complete" on grounds that it indicates something important about the "completion" (*epiteleisthai, teleiousthai, perainesthai*) of a syllogism, namely, that this consists in supplying additional steps so as to make a valid premise-conclusion argument (i.e., a set of premises and a conclusion that they imply) into a deduction (i.e., an extended discourse that makes it evident that a certain conclusion is implied by certain premises). See especially Corcoran, "Aristotle's Natural Deduction System," and T. Smiley, "What Is a Syllogism?" Cf. Robin Smith, *Notes*, p. 110.

   In fact, one could use the terms "perfect" and "perfecting" in this way, too, because "perfect" (as a translation of *teleios*) often means "not missing any parts." Thus, although I find the view of Corcoran, Smiley, and Smith attractive and plausible, I shall use both terms. (I shall not be discussing that view here, because it is a general one that does not concern modal syllogistic in particular; nor do Aristotle's modal chapters shed any distinctive light on the matter.)
3. G. Patzig, *Aristotle's Theory.*
4. *Ibid.*, p. 58, where Patzig also accounts for an exception to the rule.
5. *Ibid.*, pp. 51f.
6. Kneale and Kneale, *The Development of Logic*, p. 73.
7. Lynn Rose, *Aristotle's Syllogistic* (Springfield, IL: Thomas, 1968), p. 104. Because Rose does not cite Patzig, the relation "cites with approval" is manifestly not transitive.
8. G. Patzig, *Aristotle's Theory*, p. 52.
9. *Ibid.*
10. Kneale and Kneale, *The Development of Logic*, pp. 79f.
11. See Morris R. Cohen and E. Nagel, *An Introduction to Logic and Scientific Method* (New York: Harcourt, Brace, 1934), p. 87.
12. For example, Jonathan Lear, *Logical Theory*, pp. 6f.
13. L. Rose, *Aristotle's Syllogistic*, pp. 106f.
14. We have already met this economical method of refutation by counterexample. For assertoric premise pairs, Aristotle gives two trios of terms for a given premise pair, one of which makes both premises *and* '*A* all *C*' come out true, the other of which makes both premises *and* '*A* no *C*' come out

true. The former shows that the premises cannot entail any negative conclusion, and the latter that they cannot entail any positive conclusion.

15. David Ross, *Commentary*, sec, e.g., pp. 338f. Elsewhere he speaks simply of Aristotle's "generalizations" (p. 314, on 29a19–27).
16. Robin Smith asserts that 24b26–30 (the definition of 'applies to all' quoted earlier) already "contains what later became known as the *dictum de omni et nullo.*" Although I am in basic agreement with him, I have taken things more slowly so as to determine as securely as possible where and on what grounds one could see the *dictum* at work in *Pr. An.* A.4.
17. Patzig, *Aristotle's Theory*, p. 65.
18. *Ibid.*
19. *Ibid.*, p. 66.
20. *Ibid.*
21. *Ibid.*, p. 63.
22. *Ibid.*, p. 66.
23. Notice that this also removes any need to suppose with Patzig that the perfection of Barbara *N, N/N* depends on that of Barbara *N, A/N* – a view that is not supported by the text.
24. Thus we arrive at the sort of formula Łukasiewicz used in defending Barbara *NAN* (*Aristotle's Syllogistic*, p. 184).
25. Some qualification is necessary: The principle is general enough to cover the kinds of predicative relations recognized by Aristotle, but would not accommodate others that might be imagined. For example, as Michael Woods reminds me, the principle would not work if $R_1$, $R_2$, and $R_3$ *all* stood for 'probably applies to'. (It will work, however, if $R_1$ is 'probably applies to' and $R_2$ is 'applies' or 'necessarily applies' – i.e., if the minor premise simply "brings the *C*'s under the *B*'s.") Aristotle might well object that such probability assertions do not constitute, even in part, a predicative relation between a predicative and each of a number of individual subjects. (He might also say that probabilities are in fact *epistemic*, asserting what is likely to be the case given the state of our knowledge, rather than what is the case with regard to any specific individual subject.)

    However, the genuinely Aristotelian concept of a connection obtaining 'by nature and for the most part' can raise similar questions. For discussion of these questions, and the relation of this concept to Aristotelian science, see Chapter 6 herein.

## CHAPTER 8

1. Two smaller examples: He was not interested in identifying all possible *C–A* conclusions (where *A* is the major term, and *C* the minor), as opposed to *A–C* ones. Sometimes when it is clear that a *C–A*, but not an *A–C*, con-

clusion will follow, Aristotle actually says "There will be no syllogism." Where both sorts would follow, he seems to be looking only for (or at least concerned only with) the *A–C* conclusion. Second, he could have used ampliation to generate many more valid syllogisms by applying the operation to necessity propositions or even assertoric ones.

2. Becker, *Möglichkeitsschlüsse*, tables II (following p. 24) and III (following p. 88); I. M. Bochenski, *Ancient Formal Logic*, chart of "The Aristotelian Modal Syllogistic Laws" (following p. 62); Ross, *Commentary*, p. 286; Storrs McCall, *Aristotle's Modal Syllogisms*, esp. pp. 43, 45, 76, 83–6, 92; Robin Smith, *Notes*, Appendix I, pp. 229–35.
3. Paul Thom points this out in criticism of Johnson's formal model of Aristotle's logic in "The Two Barbaras," p. 136.
4. See Ulrich Nortmann, "Ueber die Stärke der aristotelischen Modallogik."
5. The main valid conversions in the system are plain *A, E*, and *I*, strong *cop* necessity *A, E*, and *I*, and the direct term-term (or type I) versions of two-way possibility propositions. (We saw also that certain ampliated propositions not investigated by Aristotle would convert.) This means, of course, that some valid moods (e.g., Cesare $N_w, A/N_w$) must be validated by means other than those Aristotle invokes, because the modal conversions on which his proofs depend are invalid.
6. It would also be possible to construct simple formal models for various Aristotelian modal logics. One could, for example, extend Robin Smith's categorical model for plain syllogistic (Aristotle, *Prior Analytics.* Indianapolis: Hackett, 1989, pp. ix–xxiv) in obvious ways, depending on which sorts of premise pairs one wanted to include. But at this point such a model would not in itself add to our understanding of Aristotle's aims or methods.
7. One might speculate that despite having some interest in the question, he did not pursue the matter (in part) because the pattern followed in the assertoric case would be blocked in most modal cases. The reduction of Darii and Ferio to second-figure Camestres and Cesare *via reductio* would not work (within Aristotle's system) in any case in which modal Darii or Ferio contained a necessity or two-way possibility conclusion, because the reducing syllogism would contain either a one-way possibility premise or a disjunctive necessity premise.

# Select bibliography


Alexander of Aphrodisias. *In Aristotelis Analyticorum priorum librum I commentarium*, ed. M. Wallies. CAG II, 1 (Berlin 1883).

*On Aristotle's Prior Analytics I.1–7*, trans. J. Barnes, S. Bobzien, K. K. Flannery, S. J., and K. Ierodiakonou (London 1991).

Ammonius. *Commentary on Aristotle's De Interpretatione*, ed. A. Busse, CAG IV, 5.

Angelelli, Ignacio. "The Aristotelian Modal Syllogistic in Modern Modal Logic," in *Konstruktionen versus Positionen. Beiträge zur Diskussion um die Konstruktive Wissenschaftstheorie*, ed. K. Lorenz (Berlin: Walter de Gruyter, 1979), pp. 176–215.

Bäck, Allan. "Philoponus on the Fallacy of Accident." *Ancient Philosophy* 7(1987):131–46.

Balme, David. "Aristotle's Use of Division and Differentiae," in Gotthelf and Lennox (1987), pp. 69–89.

Barnes, Jonathan. *Aristotle's Posterior Analytics* (Oxford University, Press, 1975).

"Aristotle's Theory of Demonstration." *Phronesis* 14(1969):123–52 [reprinted in *Articles on Aristotle I*, ed. J. Barnes, M. Schofield, and R. Sorabji (London 1975)].

"Proof and the Syllogism," in Berti (1981), pp. 17–59.

"Sheep Have Four Legs," in *Proceedings of the World Congress on Aristotle* (Athens: Ministry of Culture and Science, 1981–3).

Becker, Albrecht. *Die aristotelische Theorie der Möglichkeitsschlüsse: Eine logisch-philologische Untersuchung der Kapitel 13–22 von Aristotle's Analytica Priora I* (Berlin: Junker & Dünnhaupt, 1933).

Berti, E., ed. *Aristotle on Science: The Posterior Analytics.* Proceedings of the 8th Symposium Aristotelicum (Padova 1981).

Bochenski, Innocentius M. *A History of Formal Logic*, trans. Ivo Thomas, 2nd ed. (New York 1970).

*Ancient Formal Logic* (Amsterdam: North Holland, 1963).

"Notes historiques sur les propositiones modales." *Revue des Science Philosophiques et Theologiques* 26 (1937):673–92.

Broadie, Sarah Waterlow. *Passage and Possibility: A Study of Aristotle's Modal Concepts* (Oxford University Press, 1986).

Brunschwig, Jacques. *Aristotle Topiques I. Livres I–IV* (Paris: Budé, 1967).

Cohen, Morris R., and Nagel, E. *An Introduction to Logic and Scientific Method* (New York: Harcourt, Brace, 1934).

Cooper, John. "Hypothetical Necessity," in *Aristotle on Nature and Living Things. Philosophical and Historical Studies Presented to David Balme on His Seventieth Birthday*, ed. A. Gotthelf (Pittsburgh: Mathesis, 1985).

Corcoran, John, ed. *Ancient Logic and Its Modern Interpretations* (Dordrecht 1974).

"Aristotle's Natural Deduction System," in Corcoran (1974), pp. 85–131.

Demos, Raphael. "The Structure of Substance According to Aristotle." *Philosophy and Phenomenological Research* 5(1944/45):255–68.

Ferejohn, Michael. "Aristotle on Necessary Truth and Logical Priority." *American Philosophical Quarterly* 18(1981):285–93.

"Essentialism in Aristotle's Organon." Ph.D. dissertation, University of California at Irvine (1976).

*The Origins of Aristotelian Science* (New Haven: Yale University Press, 1991).

Forbes, G. *The Metaphysics of Modality* (Oxford University Press, 1985).

Frede, Michael. "Categories in Aristotle," in his *Essays in Ancient Philosophy* (Minneapolis: University of Minnesota Press, 1987), pp. 29–48.

"Stoic vs. Aristotelian Syllogistic." *Archiv für Geschichte der Philosophie* 56(1974):1–32 [reprinted in *Essays in Ancient Philosophy* (Minneapolis 1987)].

Geach, Peter. Unpublished (and incomplete) commentary on *Prior Analytics* A.

"History of the Corruptions of Logic," in *Logic Matters* (Berkeley 1972), 44–61.

Gotthelf, Allan, and Lennox, James G. *Philosophical Issues in Aristotle's Biology* (Cambridge 1987).

Hamelin, O. *Le Systeme D'Aristote* (Paris 1920).

Hamlyn, D. W. "Aristotle on Predication." *Phronesis* 5(1961):110–26.

Hintikka, Jaakko. *Time and Necessity: Studies in Aristotle's Theory of Modality* (Oxford: Clarendon Press, 1973).

"On Aristotle's Modal Syllogistic," in Hintikka (1973), pp. 135–46.

"An Aristotelian Dilemma." *Ajatus* 22(1959):87–92.

"Aristotle's Different Possibilities." *Inquiry* 3(1960):18–28.

Johnson, Fred. "Models for Modal Syllogisms." *Notre Dame Journal of Formal Logic* 30(1989):271–83.

Kneale, Martha, and Kneale, William. *The Development of Logic* (Oxford University Press, 1962).

Kneale, William. "Modality *de re* and *de dicto*," in *Logic, Methodology, and Philosophy of Science*, ed. E. Nagel, P. Suppes, and A. Tarski (Stanford University Press, 1962), pp. 622–33.

Kosman, L. A. "Aristotle on Incontrovertible Modal Propositions." *Mind* NS 79(1970):254–8.

"Aristotle's First Predicament." *Review of Metaphysics* 20(1967):483–506.

"Necessity and Explanation in Aristotle's *Analytics*," in *Biologie, Logique, et Metaphysique chez Aristote*, Seminaire CNRS-N.S.F. 1987 (Paris 1990), pp. 349–64.

Kripke, Saul. "Naming and Necessity," in *Semantics of Natural Languages*, ed. G. Harman and D. Davidson (Dordrecht: Reidel, 1972), pp. 253–355, 763–9.

"Semantical Considerations on Modal Logic." *Acta Philosophica Fennica* 16(1963):83–94 [reprinted in *Reference and Modality*, ed. L. Linsky (Oxford University Press, 1971)].

Lear, Jonathan. *Aristotle and Logical Theory* (Cambridge University Press, 1980).

Łukasiewicz, Jan. *Aristotle's Syllogistic from the Standpoint of Modern Formal Logic*, 2nd ed. (Oxford University Press, 1957).

McCall, Storrs. *Aristotle's Modal Syllogisms* (Amsterdam: North Holland, 1963).

Maier, H. *Die Syllogistic des Aristoteles*, 3 vol. (Tübingen 1896–1900).

Mignucci, Mario. *On a Controversial Demonstration of Aristotle's Modal Syllogistic: An Enquiry on Prior Analytics A.15* (Padua 1972).

Moody, Ernest. "Medieval Logic," in *Encyclopedia of Philosophy*, vol. 4, ed. Paul Edwards (New York 1967), pp. 528–34.

Mueller, Ian. "Stoic and Peripatetic Logic." *Archiv für Geschichte der Philosophie* 51(1969):173–87.

Nortmann, Ulrich. "Ueber die Stärke der aristotelischen Modallogik." *Erkenntnis* 32(1990):61–82.

Owen, G. E. L. "Inherence." *Phronesis* 10(1965):97–105.

Patterson, Richard. "Aristotle's Perfect Syllogisms." *Synthese* 96(1993):359–77.

"Conversion Principles and the Basis of Aristotle's Modal Logic." *History and Philosophy of Logic* 11(1990):151–72.

"The Case of the Two Barbaras: Basic Approaches to Aristotle's Modal Logic." *Oxford Studies in Ancient Philosophy* 7(1989):1–40.

Patzig, Günther. *Aristotle's Theory of the Syllogism*, trans. J. Barnes (Dordrecht: Reidel, 1968).

Plantinga, Alvin. *The Nature of Necessity* (Oxford 1974).

Putnam, Hilary. "The Meaning of 'Meaning'," in *Minnesota Studies in the Philosophy of Science VII*, ed. K. Gunderson (Minneapolis: University of Minnesota Press, 1975) [reprinted in Putnam, *Mind, Language, and Reality. Philosophical Papers*, vol. 2 (Cambridge University Press, 1975), pp. 215–71].

Rescher, Nicholas. "Aristotle's Theory of Modal Syllogisms and Its Interpretation," in *The Critical Approach to Science and Philosophy: Essays in Honor of Karl Popper*, ed. M. Bunge (New York: Collier-Macmillan, 1964), pp. 152–77.

Rose, Lynn E. *Aristotle's Syllogistic* (Springfield, IL: Thomas, 1968).

Ross, David. *Aristotle, Prior and Posterior Analytics: A Revised Text with Introduction and Commentary* (Oxford University Press, 1949).

Smiley, Timothy. "What Is a Syllogism?" *Journal of Philosophical Logic* 2(1973):136–54.

Smith, Robin. *Aristotle, Prior Analytics* (Indianapolis: Hackett, 1989).

"What Is Aristotelian *Ecthesis?*" *History and Philosophy of Logic* 3(1982): 113–27.

Solmsen, Friedrich. *Die Entwicklung der aristotelischen Logik und Rhetorik* (Berlin 1929).

Sorabji, Richard. *Necessity, Cause, and Blame: Perspectives on Aristotle's Theory* (Ithaca, NY: Cornell University Press, 1980).

Striker, Gisela. "Notwendigkeit mit Lücken." *Neue Hefte für Philosophie* 24/25(1985):146–64.

Thom, Paul. "The Two Barbaras." *History and Philosophy of Logic* 12(1991): 135–49.

van Rijen, Jeroen. *Aspects of Aristotle's Logic of Modalities* (Dordrecht: Kluwer, 1989).

Waitz, T. *Aristotelis Organon*, 2 vol. (Leipzig 1844–6).

White, Nicholas. "Origins of Aristotle's Essentialism." *Review of Metaphysics* 26(1972–3):57–85.

Wieland, Wolfgang. "Die aristotelische Theorie der Notwendigkeitsschlüsse." *Phronesis* 11(1966):35–60.

"Die aristotelische Theorie der Möglichkeitsschlüsse." *Phronesis* 17(1972): 124–52.

"Die aristotelische Theorie der Syllogismen mit modal gemischten Prämissen." *Phronesis* 20(1975):77–92.

"Die aristotelische Theorie der Konversion von Modalaussagen." *Phronesis* 25(1980):109–16.

Wiggins, David. "The *de re* 'Must': A Note on the Logical Form of Essentialist Claims," in *Truth and Meaning*, ed. G. Evans and J. McDowell (Oxford University Press, 1976), pp. 285–312.

# Index

# Index